Jan C. Willems (Ed.)

From Data to Model

With 35 Figures

Springer-Verlag Berlin Heidelberg New York
London Paris Tokyo Hong Kong

Professor Jan C. Willems
Department of Mathematics
University of Groningen
P.O. Box 800
9700 AV Groningen
The Netherlands

ISBN-13:978-3-642-75009-0 e-ISBN-13:978-3-642-75007-6
DOI: 10.1007/978-3-642-75007-6

Softcover reprint of the hardcover 1st edition 1989

2142/7130-543210

PREFACE

The problem of obtaining dynamical models directly from an observed time-series occurs in many fields of application. There are a number of possible approaches to this problem. In this volume a number of such points of view are exposed: the statistical time series approach, a theory of guaranted performance, and finally a deterministic approximation approach.

This volume is an out-growth of a number of get-togethers sponsered by the Systems and Decision Sciences group of the International Institute of Applied Systems Analysis (IIASA) in Laxenburg, Austria. The hospitality and support of this organization is gratefully acknowledged.

Jan Willems
Groningen, the Netherlands
May 1989

TABLE OF CONTENTS

LINEAR SYSTEM IDENTIFICATION - A SURVEY

M. DEISTLER

Abstract

In this paper we give an introductory survey on the theory of identification of (in general MIMO) linear systems from (discrete) time series data. The main parts are: Structure theory for linear systems, asymptotic properties of maximum likelihood type estimators, estimation of the dynamic specification by methods based on information criteria and finally, extensions and alternative approaches such as identification of unstable systems and errors-in-variables.

Keywords

Linear systems, parametrization, maximum likelihood estimation, information criteria, errors-in-variables.

1. INTRODUCTION

The problem of deducing a good model from data is a central issue in many branches of science. As such problems are often far from being trivial and on the other hand often have a lot of common structure, systematic formal approaches for their solution have been developed. A large part of statistics, parts of system theory (namely system identification) and of approximation theory are concerned with this topic.

Here a special, but important case is considered, namely identification of *linear systems* from (equally spaced discrete) *time series data.* Both with respect to the existing body of theories and with respect to applications, linear system identification is quite an extensive subject now. The most important applications are signal processing (e.g. speech processing, sonar and radar applications), control engineering, econometrics, time series analysis of geophysical and metereological data, and the analysis of medical and biological time series (e.g. EEG analysis). In different areas emphasis has been put on different problems (and there still seems to be lack of communication between scientists working in those areas). For instance in modern system and control theory, a lot of emphasis has been put on the structure theory for linear multi-input multi-output (MIMO) systems, in signal processing on on-line algorithms for real time calculation and in statistical time series analysis on asymptotic properties of (mainly off-line) estimation procedures.

Linear system identification has many different aspects and facets depending among others on the goals one wants to achieve, on the amount of a priori information available, on the nature of data and on the way that noise is modelled. Nevertheless in the last twenty years something like a "mainstream" theory has been developed.

In system identification one has to specify:

(i) The *model class* i.e. the class of all a priori feasible systems which are candidates to be fitted to the data.

(ii) The *class of observations* $(y(t))$.

(iii) The *identification procedure* which is a rule (in the automatic case a function) attaching to every finite part of the data of the form $(y(t)|t=1\ldots T)$ a system from the model class.

The actual problem of linear system identification, however, has much additional structure. We now describe the basic assumptions and ingredients of the mainstream approach. At the end of our contribution we indicate some deviations from this approach.

(i) The systems contained in the model class are (in general MIMO) causal, stable, finite dimensional and time-invariant linear dynamic systems. Here in addition we restrict ourselves to the discrete-time case, where the range of time points are the integers $\mathbb{Z}$. The two most important system representations in this case are the state-space and the ARMA(X) representation. For simplicity and since the differences are minor (see e.g. Hannan and Deistler, 1988, Chapter 2 for a discussion) we only discuss the second case here, i.e. the case where

$$a(z)y(t) = b(z)\varepsilon(t) \tag{1.1}$$

where $y(t)$ is the s-dimensional output, $\varepsilon(t)$ is the m-dimensional input, z is used for a complex variable as well as for the delay operator (i.e. $z(y(t)|t\in\mathbb{Z}) = (y(t-1|t\in\mathbb{Z}))$ and finally where

$$a(z) = \sum_{j=0}^{p} A(j)z^j, A(j)\in\mathbb{R}^{s\times s}, b(z) = \sum_{j=0}^{q} B(j)z^j, B(j)\in\mathbb{R}^{s\times m} \tag{1.2}$$

With the exception of the last section unless the contrary is stated explicitely we will assume

$$\det a(z) \neq 0 \qquad |z| \leq 1 \tag{1.3}$$

and we will only consider the steady state solution

$$y(t) = \sum_{j=0}^{\infty} K(j)\varepsilon(t-j) \tag{1.4}$$

of (1.1), where

$$\sum_{j=0}^{\infty} K(j)z^j = k(z) = a^{-1}(z)b(z) \tag{1.5}$$

Thus we restrict ourselves to the stable steady state case.

(ii) Every reasonable identification procedure has to separate the "essential" part from the "noisy" part of the data. For instance, for an ARMAX system, where in general the data will not exactly fit to the deterministic part of such a system, a decision has to be made what is attributed to the deterministic part and what is attributed to noise. A basic decision that has to be made is whether we should (explicitely) model noise or not. In statistics this is an old question and the answer to it constitutes dividing line between descriptive and inferential statistics.

Here we give a stochastic model for the noise part, and thus, from this point of view, our problem becomes part of inferential statistics. In this case, additional a priori assumptions on the stochastic noise process, such as stationarity and ergodicity have to be imposed, in order to make inference a sensible task. The advantage of such a way of noise modelling is that the quality of identification procedures can be evaluated in a formal-mathematical way, for instance by deriving asymptotic properties of estimators. On the other hand, such a priori assumptions on the noise are not innocent and in actual applications the question has to be posed whether such a priori assumptions can be justified, or at least whether such a stochastic noise process provides a meaningful "test case" for the evaluation of identification procedures. These questions in particular have be posed in applications such as in econometrics or control engineering where there is rarely any stochastic theory or even vague a priori reasoning about the nature of noise.

(iii) The next question is, how the deterministic system should be embedded in its stochastic "environment". In mainstream analysis all of the noise is added to the equations or (which is the same in most respects) to the outputs, whereas the inputs are assumed to be observed without noise. This can be modelled by distinguishing between observed inputs and unobserved noise inputs in the vector $\varepsilon(t)$. In addition in this approach, the noise process is assumed to be uncorrelated with the observed inputs. If the contrary is not stated explicitely, here, for simplicity we will assume $m=s$ and that $\varepsilon(t)$ will consist of unobserved white noise errors only, i.e.

$$E\,\varepsilon(t)=0, \qquad E\,\varepsilon(s)\varepsilon'(t)=\delta_{st}.\Sigma \tag{1.6}$$

In this case (1.1) is called an ARMA system and its solution (1.4) is called an ARMA process. As is well known such a process is stationary with spectral density given by

$$f_y(\lambda)=(2\pi)^{-1}.k(e^{-i\lambda}).\Sigma.k(e^{-i\lambda})^*$$

(where * denotes the conjugate transpose). In addition we assume

$$k(0)=I, \qquad \Sigma>0 \tag{1.7 a,b}$$

and the miniphase condition

$$\det b(z) \neq 0 \qquad |z| < 1 \tag{1.8}$$

As is well known, for given f_y assumptions (1.7 a) and (1.8) are costless. As is also well known, under (1.7), (1.8), k and Σ are uniquely determined from f_y. For the additional complications arising in the ARMAX case, the reader is referred to Hannan and Deistler (1988).

(iv) For many cases discussed in this paper, the decision, which system has to be chosen, given the data, is based on optimizing a function which, in general, describes a certain trade off between goodness of fit of a system to the data and the complexity of the system. Thus we have to introduce a measure for goodness of fit, a measure for the complexity of a system and we have to formulate the trade off between the contradictory goals to maximize goodness of fit and to minimize the complexity of the system used. Clearly these choices are very much related to measures for the quality of inference procedures.

In the mainstream approach the (Gaussian) likelihood or a function of the (one step ahead) prediction error variance-covariance matrix are used as measures for goodness of fit and the quality of (parameter) estimators is described in terms of consistency and relative asymptotic efficiency.

In case of "small" model classes only goodness of fit is optimized. Measures of complexity are used in addition, in particular if the original model class is so large that it has to be broken up into subclasses and the subclass has to be determined from the data too. Since in a "large" model class measures of goodness of fit alone, such as the likelihood would tend to overfit the sample, such a measure of fit has to be "penalized" by a term measuring complexity of a system usually, in terms of the dimension of the parameter space. This is explained in detail in Section 4.

Let us consider the case, where the (original) model class is T_A, the set of all ARMA systems (a,b) (satisfying our assumptions) for given s (but for arbitrary p,q) i.e. where we have no a priori assumptions besides the general ones mentioned above. By U_A we denote the set of all transfer functions $a^{-1}.b$ corresponding to T_A and by $\pi: T_A \rightarrow U_A$ we denote the mapping defined by $\pi(a,b) = a^{-1}.b$. Two ARMA systems are called observationally equivalent if they have the same transfer function k. A set $T \subset T_A$ is called *identifiable* if π restricted to T is injective; in this case the mapping $\psi: \pi(T) \rightarrow T: \psi(\pi(a,b)) = (a,b)$ is called an *(ARMA-)parametrization* of $V = \pi(T)$. For

$(a,b)\in T$, in general not all entries of the parameter matrices $A(j)$, $B(j)$ may be needed for a unique description of (a,b) due to constraints. A vector τ of entries of the $A(j)$, $B(j)$, such that $(a,b)\in T$ is uniquely determined from τ, and such that τ (whose dimension is kept constant over T) has a minimal number of entries, is called a vector of *free parameters* for T (or for $\pi(T)$). We will identify (a,b) with τ.

Every parametrization of U_A has the disadvantage that the corresponding parameter space T is infinite dimensional and clearly finite dimensional parameter spaces are more convenient for inference. What is even more cumbersome is the fact that there exists no continuous parametrization of U_A. For these reasons, U_A (and T_A) is broken into parts U_α, $\alpha\in I$, in a way that every such part can be parametrized separately, by $\psi_\alpha:U_\alpha\rightarrow T_\alpha$ say, in a convenient way.

For the sake of mathematical convenience, we may decompose an identification procedure into three steps. The first step is to determine the subclass U_α, or the index α, characterizing this subclass, from the data. Here α is a multi-index of integers, in the scalar case $(s=1)$ the usual choices are $\alpha=(p,q)$ or $\alpha=(n);n=\max(p,q)$. The determination of α sometimes is called dynamic specification. Here we will almost exclusively deal with automatic procedures for dynamic specification which are in particular inference procedures based on optimization of a function describing a certain trade off between goodness of fit and complexity as has been mentioned above. However it should be emphasized that (besides the case where suitable a priori information about α is available from "physical" theories and where therefore the first step is omitted), in particular for the scalar case, dynamic specification may also be performed by non-automatic procedures (where subjective judgement based on certain patterns is involved), the most prominent of which is the Box-Jenkins procedure (Box and Jenkins 1970). Once α has been determined, for mathematical convenience, estimation of the free parameters τ and $\sigma(\Sigma)$ (where $\sigma(\Sigma)$ is the vector of on and above diagonal elements of Σ) may be decomposed into two further steps namely estimation of the transfer function k (by $\hat{k}$ say) and of $\sigma(\Sigma)$ (by $\sigma(\hat{\Sigma})$) and, finally the realization of the estimated transfer function to obtain the parameter estimator $\hat{\tau}=\psi_\alpha(\hat{k})$. Whereas the second step is concerned with statistics in the strict sense [namely with extraction of information from data], the third is concerned with (deterministic) realization and only properties of the parametrization are relevant. In order to estimate k and Σ in the second step usually a criterion for goodness of fit, such as the likelihood is optimized. The decomposition of the problem into these two further steps is based on the observation that most of these criteria only depend on τ via k.

The structure of the problem of identification of linear systems (when the original model class is T_A[or U_A]) can be schematically represented by the following figure:

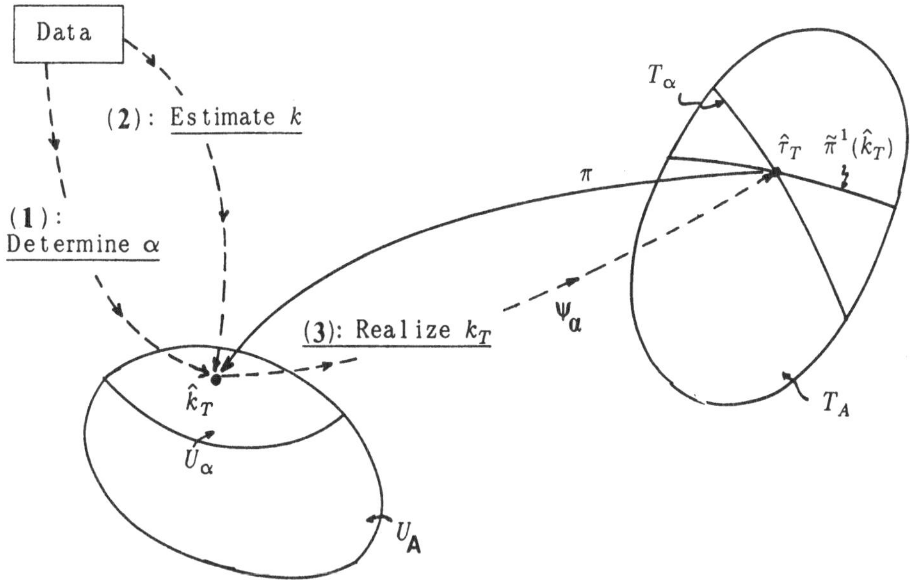

Fig. 1: The Structure of Linear System Identification

2. REALIZATION AND PARAMETRIZATION

As has been pointed out already U_A has to be broken into parts U_α, $\alpha \in I$, in order to allow for a convenient parametrization $\psi_\alpha : U_\alpha \rightarrow T_\alpha$. Clearly, there are many different ways to define such parts. From the point of view of identification some desirable properties of such parameter spaces and parametrizations are:

(i) T_α is identifiable; i.e. the mapping $\psi_\alpha : U_\alpha \rightarrow T_\alpha : \psi_\alpha(\pi(a,b)) = (a,b)$; $(a,b) \in T_\alpha$ exists.

(ii) T_α can be embedded into an Euclidian space $\mathbb{R}^{d_\alpha}$, i.e.the parameter space is finite dimensional; in addition T_α should contain an open set of $\mathbb{R}^{d_\alpha}$.

(iii) An important property of ψ_α is its continuity in the sense that T_α is endowed with the relative Euclidean topology and U_A is endowed with the relative topology of $(\mathbb{R}^{s \times s})^{\mathbb{N}}$, where the transfer functions are identified with their power series coefficients $(K(j) \mid j \in \mathbb{N})$. The latter topology is called the *pointwise topology* T_{pt} and is quite natural in our context, since the maximum likelihood estimators of the transfer functions k can be shown to be consistent in this sense. As is clear immediately, continuity of the mapping ψ_α relating the external characteristics k to the internal characteristics $\tau \leftrightarrow (a,b)$ makes the identification problem well posed and implies consistency for the estimators of τ for every estimation method (as the maximum likelihood method) which gives consistent estimators of k. As will be discussed later, also openness of U_α in $\bar{U}_\alpha$ is desirable. Note that in our analysis we do not need to show that the mapping relating *second moments* of $(y(t))$ to parameters τ is continuous, since the starting point of the analysis is consistency of ***transfer functions***. For asymptotic normality of the estimators of τ, some differentiability properties are required.

(iv) A reasonable requirement is that the set of all U_α, $\alpha \in I$ is a cover for U_A, i.e. $\bigcup_{\alpha \in I} U_\alpha = U_A$.

(v) There is a certain trade off between the size of the cover U_α, $\alpha \in I$ and the dimension of the corresponding parameter spaces for the U_α. Vaguely speaking a coarser cover would tentatively make the determination of α simpler but would give a larger dimension of the parameter space T_α actually used and thus more components the parameter vector τ have to be estimated, which would cause a certain "efficiency loss". Another H in certain sense, reasonable requirement seems to be that the cover is minimal in the sense that no element of the cover can be removed without loosing the covering property.

In particular for the multi-output $(s>1)$ case, there is a number of different parametrizations which are used, the most important of which are Echelon canonical forms, the overlapping parametrization of the manifold of all systems of order n and monic (in the sense that $a(0)=I$ holds) ARMA systems with prescribed column degrees. [there is a large number of references to this, see Hannan and Deistler 1988 and the references therein]. We will only describe *Echelon-forms* here. We begin with a transfer function of the form

$$\tilde{k}(z) = k(z^{-1}) = \sum_{j=0}^{\infty} K(j) z^{-j} \tag{2.1}$$

rather than with $k(z)$, for mathematical convenience. Causality of k means that $\tilde{k}$ is proper [i.e. $\lim_{|z|\to\infty} \tilde{k}(z)$ is finite]. An ARMA system $(\tilde{a},\tilde{b})$ corresponding to $\tilde{k}$ (i.e. $\tilde{a}^{-1}.\tilde{b}=\tilde{k}$) then [in an obvious notation] is of the form

$$\sum_{j=0}^{\tilde{p}} \tilde{A}(j)y(t+j) = \sum_{j=0}^{\tilde{q}} \tilde{B}(j)\varepsilon(t+j) \tag{2.2}$$

Let

$$H = \begin{pmatrix} K(1),K(2),\ldots \\ K(2),K(3),\ldots \\ \ldots\ldots\ldots\ldots \end{pmatrix}$$

denote the (block) Hankel matrix of k. Then from a comparison of coefficients corresponding to negative powers of z in $\tilde{a}(z).k(z^{-1})$ we obtain

$$(\tilde{A}(o),\tilde{A}(1),\ldots).H = 0 \tag{2.3}$$

As is well known, since k is rational, the rank of H is finite and furthermore this rank is equal to the *order* i.e. the degree of $\det\tilde{a}$ for any (left) coprime$(\tilde{a},\tilde{b})$ [i.e. $\tilde{a},\tilde{b}$ have no nonunimodular common matrix polynomial (left) divisor; a polynomial matrix u is called unimodular if $\det u = \text{const} \neq 0$] corresponding to $\tilde{k}$. Let $M(n)$ denote the set of all $k \in U_A$ such that H has rank n. Further, let $h(i,j)$ denote the jth row in the ith block of rows of H. Due to the block Hankel structure of H, the first rows (in natural ordering) of H which form a basis for the row space of H are of the form

$$h(1,1),\ldots,h(n_1,1),h(1,2),\ldots,h(n_2,2),\ldots,h(1,s),\ldots,h(n_s,s)$$

for a suitable chosen multi-index $\alpha = (n_1,\ldots,n_s)$; these $n_1\ldots n_s$ are called the *Kronecker indices*. Clearly $n = n_1+\ldots+n_s$. Expressing the respective first linear dependent rows in terms of the preceeding elements from this basis, we obtain

$$-h(n_i+1,i) = \sum_{j=1}^{s} \sum_{\mu=1}^{n_{ij}} \tilde{a}_{ij}(\mu-1)h(\mu,j) \quad , \qquad i = 1\ldots s \tag{2.4}$$

where

$$n_{ij} = \begin{cases} \min\ (n_i+1,n_j) & \text{for} \quad j<i \\ \min\ (n_i,n_j) & \text{for} \quad j \geq i \end{cases}$$

Equations (2.4) define unique coefficients $\tilde{a}_{ij}(\mu)$ and they can be considered as special relations of the form (2.3) where $\tilde{a}_{ij}(\mu)$ is the (i,j)

element of $\tilde{A}(\mu)$, $\tilde{a}_{ii}(n_i)=1$, $i=1\ldots s$ and all other elements are equal to zero.

By this procedure, for every $\tilde{k}\leftrightarrow k\in U_A$ we have defined (unique) Kronecker indices $\alpha=(n_1\ldots n_s)$ and a corresponding unique ARMA realization $(\tilde{a},\tilde{b})$, with

$$\tilde{b}=\tilde{a}.\tilde{k} \tag{2.5}$$

where

$$(\tilde{a},\tilde{b}) \text{ is (left) coprime} \tag{2.6}$$

and (with $\delta(p)$ denoting the degree of polynomials)

$$\begin{array}{lll} \delta(\tilde{a}_{ij})\leq\sigma(\tilde{a}_{ii})=n_i & ; & j\leq i \\ \delta(\tilde{a}_{ij})<\delta(\tilde{a}_{ii}) & & j>i \\ \delta(\tilde{a}_{ji})<\delta(\tilde{a}_{ii}) & & j\neq i \\ \delta(\tilde{b}_{ij})\leq\delta(\tilde{a}_{ii}) & & \end{array} \tag{2.7}$$

the row–end matrices in $\tilde{a}$ and $\tilde{b}$ are the same.

Such a unique realization is called the Echelon form. As can be shown, conversely every ARMA system satisfying (2.6) and (2.7) is in Echelon form.

An ARMA realization for k then is obtained from

$$(a(z),b(z))=\mathrm{diag}\{z^{n_i}\}(\tilde{a}(z^{-1}),\tilde{b}(z^{-1})) \tag{2.8}$$

and this is called the the *reversed Echelon form.* For reversed Echelon form we have:

$$(a,b) \text{ is (left) coprime} \tag{2.9}$$

and

$A(0)[=B(0)]$ is lower triangular and all its diagonal elements are equal to one;
the degree of the ith row is n_i;
$z^{n_i-n_{ij}}$ divides $\tilde{a}_{ij}$. (2.10)

Let U_α denote the set of all $k\in U_\alpha$ with Kronecker indices $\alpha=(n_1,\ldots,n_s)$, and T_α denote the set of all $(a,b)\in T_A$ satisfying (2.6), (2.7) and (2.8). For
$(a,b)\in T_\alpha$ a vector $\tau\in\mathbb{R}^{d_\alpha}$ of free parameters consisting of all elements of

$(\tilde{a},\tilde{b})$ which are not explicitely restricted by (2.7) is defined where

$$d_\alpha = \Big(\sum_{i=1}^{s} n_i\Big)(s+1) + \sum_{i,\,j\,:j<i} (\min(n_i,n_j)+\min(n_j,n_i+1)) \tag{2.11}$$

Then by the procedure described above in introducing (reversed) Echelon form we have defined a parametrization $\psi_\alpha : U_\alpha \to T_\alpha$. By $\bar{A}$ we denote the closure of the set A, and by $\beta = (m_1 \ldots m_s) \le \alpha = (n_1 \ldots n_s)$ we mean $m_i \le n_i, i = 1 \ldots s$. $\beta < \alpha$ is to indicate that $m_i < n_i$ for at least one i holds. For the next theorem we do not impose assumptions (1.3) and (1.8). We have:

Theorem 2.1:

(i) T_α is open and dense in $\mathbb{R}^{d_\alpha}$

(ii) $\psi_\alpha : U_\alpha \to T_\alpha$ is a $(T_{pt}-)$ homeomorphism

(iii) $\{U_\alpha | \sum_{i=1} n_i = n\}$ is a disjoint partition of $M(n)$ containing $\binom{n+s-1}{s-1}$ elements

(iv) $\pi(\bar{T}_\alpha) = \bigcup_{\beta \le \alpha} U_\beta$

(v) For every $k \in U_\beta$, $\beta \le \alpha$, the class of all observationally equivalent ARMA systems in $\bar{T}_\alpha$ is an affine subspace of dimension

$$\sum_{i=1}^{s} \sum_{j=1}^{s} (n_{ij} - n'_{ij})$$

where

$$n'_{ij} = \begin{cases} \min(n_i+1, m_j) & \text{for} \quad j<i \\ \min(n_i, m_j) & \text{for} \quad j \ge i \end{cases}$$

(vi) U_α is $(T_{pt}-)$ open in $\bar{U}_\alpha$

(vii) $\pi(\bar{T}_\alpha) \subset \bar{U}_\alpha$ and equality holds for $s=1$

A similar result can be shown for the overlapping parametrization of $M(n)$ or for monic ARMA systems with prescribed column degrees (see e.g. Deistler 1983, Hannan and Deistler 1988, Deistler and Wang 1988). The implications of such results for estimation will be discussed in the next section.

3. ESTIMATION FOR GIVEN DYNAMIC SPECIFICATION

In most cases the estimators - at least asymptotically - only exploit information from the data $y(t), t=1\ldots T$, via their second moments

$$\hat{K}(s) = \begin{cases} (T)^{-1} \cdot \sum_{t=1}^{T-s} y(t+s)y'(t) & , \quad 0 \le s < T \\ \hat{K}'(-s) & , \quad 0 > s > -T \\ 0 & , \quad |s| \ge T \end{cases} \tag{3.1}$$

Clearly, these second moments can be "realized" by a moving average system of order $T-\Lambda$. [Note that typically, e.g. for the Gaussian case no data $y(t), t=1\ldots T$ in a deterministic sense could ever be incompatible with any system; by "realize" here we meant that we can find a system whose population second moments are given by (3.1)]. Such a system estimator however has two disadvantages. Typically it would "overfit" the data [i.e. it would use too many parameters for description] and second $\hat{K}(s)=0$ for $|s| \ge T$, in general, is not a "good" extrapolation. So we have to "smooth" the $\hat{K}(s)$, $|s| < T$ by using (in general) less parameters for their (approximate) description and at the same time we have to extrapolate these values for $|s| \ge T$. This can also be understood as a smoothing of the periodogram

$$I(\omega) = (2\pi)^{-1} \sum_{s=-T}^{T} \hat{K}(s) e^{i\omega s} \tag{3.2}$$

by rational approximation. In addition, in general, the empirical second moments are not contained in the class of (population) second moments corresponding to the class $T_\alpha \times \underline{\Sigma}$ under consideration, so that estimation can be understood as approximating the empirical second moments of the data by an element corresponding to $T_\alpha \times \underline{\Sigma}$. Here $\underline{\Sigma} = \{\Sigma \in \mathbb{R}^{s\times s} | \Sigma > 0, \Sigma' = \Sigma\}$.

In mainstream theory the Gaussian maximum likelihood estimator (MLE) is the prototype estimator. Under Gaussian assumptions $-2T^{-1}$ times the logarithm of the likelihood of $y(1),\ldots,y(T)$ is given up to a constant by

$$\hat{L}_T(\tau,\Sigma) = T^{-1} \log \det \Gamma_T(\tau,\Sigma) + T^{-1} y_T' \Gamma_T^{-1}(\tau,\Sigma) y_T \tag{3.3}$$

Here $y_T = (y'(1),\ldots,y'(T))'$ denotes the stacked vector of the data and

$$\Gamma_T(\tau,\Sigma) = \left[\int e^{-i\lambda(r-t)} . f(\lambda;\tau,\Sigma) \; d\lambda \right]_{r,t=1\ldots T} \tag{3.4}$$

denotes the matrix of second moments of a vector $(y'(1)\ldots y'(T))'$ made from an ARMA process with parameters τ,Σ [correspondingly $f(\lambda;\tau,\Sigma)$ denotes the spectral density of such a process]. Since no confusion can arise, $\hat{L}_T$ is also called the likelihood function. Evidently $\hat{L}_T$ depends on the parameters τ only via k and thus we can define a likelihood by.

$$L_T(\pi(\tau),\Sigma) = \hat{L}_T(\tau,\Sigma) \tag{3.5}$$

This "coordinate-free" likelihood will prove to be mathematically convenient since certain statistical properties of MLE's can be analysed in terms of transfer functions.

If $U \subset U_A$ is the set of transfer functions considered, the MLE's $\hat{k}_T, \hat{\Sigma}_T$ [over $U \times \underline{\Sigma}$] are defined as

$$(\hat{k}_T,\hat{\Sigma}_T) = \underset{(k,\Sigma)\in U\times\underline{\Sigma}}{\arg\min} \; L_T(k,\Sigma) \tag{3.6}$$

In general it is not clear whether L_T has a minimum over $U \times \underline{\Sigma}$ (see e.g. Deistler and Pötscher 1984). What is much more important and cumbersome is that in general no explicit expression for the MLE will exist. Clearly in such a situation finite sample properties of the estimators would be hard to obtain. However the asymptotic analysis of the MLE's in this case has reached a certain stage of completeness now, see e.g. Hannan 1973, Dunsmuir and Hannan 1976, Hannan and Deistler 1988.

As far as consistency is concerned the main complications arise due to the noncompactness of the "natural" parameter spaces. For given $U \subset U_A$ under consideration let $\bar{U}$ denote its ($T_{pt}-$) closure, $\hat{U}$ the set of all $k \in \bar{U}$ which have no pole for $|z| = 1$ and U^* the set of all $k \in \hat{U}$ which have no zero for $|z| = 1$. We have (see Dunsmuir and Hannan 1976 Hannan and Deistler 1988).

Theorem 3.1. Let the true system satisfy

$$k_0 \in U^* \tag{3.7}$$

let

$$\lim_{T\to\infty} T^{-1} \sum_{t=1}^{T} \varepsilon(t+s)\varepsilon(t) = \delta_{s,0}\cdot\Sigma_0 \qquad \text{a.s.} \tag{3.8}$$

and let $\bar{U} \subset \bar{M}(n)$ for a suitable n. Then the MLE's over $\hat{U} \times \underline{\Sigma}$ are strictly consistent, i.e.

$$\lim_{T\to\infty} (\hat{k}_T,\hat{\Sigma}_T) = (k_0,\Sigma_0) \qquad \text{a.s.} \tag{3.9}$$

Thus consistency of the MLE's holds under fairly general conditions. For a consistency proof in the ARMAX case see Hannan and Deistler 1988.

If the data are not generated by a system contained in the model class U^* but by a general linear regular stationary process in Wold representation

$$y(t) = k_0(z)\varepsilon(t) \tag{3.10}$$

with

$$k_0(z) = \sum_{j=0}^{\infty} K(j)z^j \quad ; \qquad \sum_{j=0}^{\infty} \|K(j)\|^2 < \infty$$

then still a generalized consistency result (see e.g. Ljung 1978, Pötscher 1987, Hannan and Deistler 1988) in the following sense holds: Let D denote the subset of $\bar{U}\times\underline{\Sigma}$ where the "asymptotic form" of the likelihood

$$L(k,\Sigma) = \log\det\Sigma + (2\pi)^{-1}\int_{-\pi}^{\pi} tr\{(k\Sigma k^*)^{-1}(k_0\Sigma_0 k_0^*)d\lambda \tag{3.11}$$

attains its minimum over $\bar{U}\times\underline{\Sigma}$. As can be shown, $L(k,\Sigma)$ is the (a.s.) limit of $L_T(k,\Sigma)$ (for $T\to\infty$) and L is a measure of goodness of fit of a system to the complete (infinite) observations. D then is the set of all (k,Σ) which are the best approximations within $\bar{U}\times\underline{\Sigma}$ to the true system (k_0,Σ_0). Now the MLE's $\hat{k}_T,\hat{\Sigma}_T$ can be shown to be (a.s.) convergent to the set D. This is an important generalization of the consistency result of Theorem 3.1 since in many cases the true system may be of higher order or even not rational and this result indicates that in such cases the MLE's still give good approximations to the true system. In a certain sense this idea is related to robustness. As has been pointed out first by Kabaila (1983), D may consist of more than one point. However (Ploberger 1982) for the usual parameter spaces (e.g. for $\bar{U}_\alpha$ corresponding to Echelon forms), there is at least a neighborhood of $\bar{U}_\alpha\times\underline{\Sigma}$ [corresponding to the weak topology of spectral measures] such that if (k_0,Σ_0) is in this neighborhood, the best approximation within $\bar{U}_\alpha\times\underline{\Sigma}$ is unique (see Fig. 2)

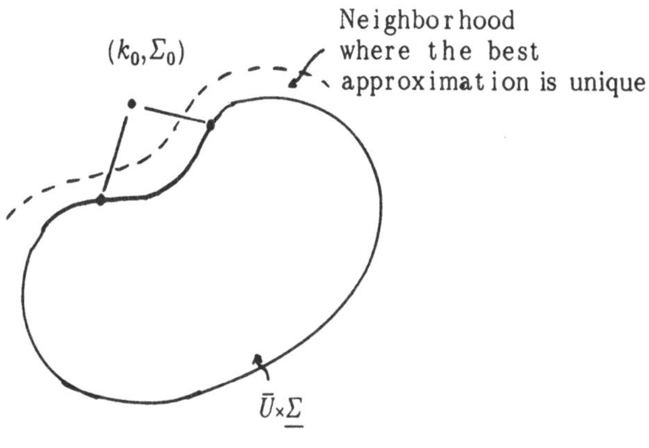

Fig. 2: Some aspects of approximation of (k_0,Σ_0) within $\bar{U}\times\underline{\Sigma}$

Let us stress again the general nature of the approach described above. In particular besides the boundedness of the degrees of the ARMA systems considered (i.e. $\bar{U}\subset M(n)$, for some n) no assumption has been imposed on the "parameter space" U (which here is a set of transfer functions). By the coordinate-free nature of the results, we had not to care about questions of existence and continuity of parametrizations. In particular, we were able to analyse the cases where k_0 is contained in the boundary U^*-U and also [since certain boundary points in the process of the optimization of the likelihood cannot be excluded a priori] the optimization of the likelihood is performed over $\hat{U}\times\underline{\Sigma}$ rather than over $U\times\underline{\Sigma}$.

However, actual calculation of the MLE's has to be performed in coordinates and in addition in many cases the parameters τ are of direct interest. Therefore we now consider estimation of the true parameter τ_0. Let $U=U_\alpha$ i.e. the set of all transfer functions $k\in U_A$ with Kronecker indices $\alpha=(n_1\ldots n_s)$ [as discussed in Section 2] and let $\psi_\alpha:U_\alpha\to T_\alpha$ be the corresponding parametrization [alternatively other standard parametrizations such as the overlapping parametrization of the manifold $M(n)$ or monic ARMA systems with prescribed column degrees may be chosen]. Then if $\hat{k}_T$ is the MLE [or any other consistent estimator] for k and if $\hat{k}_T\in\pi(\bar{T}_\alpha)$, we define a (nonnecessarely unique) MLE [or correspondingly another estimator] $\hat{\tau}_T$ of τ as any $\hat{\tau}_T\in\mathbb{R}^{d_\alpha}$ which satisfies $\pi(\hat{\tau}_T)=\hat{k}_T$. Clearly if $\hat{k}_T\in U_\alpha$, then $\hat{\tau}_T$ is uniquely given by $\psi_\alpha(\hat{k}_T)$. Investigating the behaviour of these parameter estimators we have to distinguish the following three cases:

(i) If the dynamic specification is correct in the sense that $k_0 \in U_\alpha$ holds, then $\hat{k}_T \to k_0$ and the openness of U_α in $\bar{U}_\alpha$ (Theorem 2.1) imply $\hat{k}_T \in U_\alpha$ from a certain T_0 onwards, and thus, at least for $T > T_0$, $\hat{\tau}_T = \psi_\alpha(\hat{k}_T)$ exists [Note that T_0 in general depends on the point ω in the sample space]. The continuity of ψ_α (Theorem 2.1) then implies

$$\lim_{T\to\infty} \hat{\tau}_T = \tau_0 = \psi_0(k_0) \tag{3.12}$$

and thus (under the conditions of Theorem 3.1), the MLE's $\hat{\tau}_T$ are strongly consistent for the parameter τ.

(ii) Next, we consider the case where $k_0 \in \pi(\bar{T}_\alpha) - U_\alpha$ holds, i.e. where there is a $\beta < \alpha$ such that $k_0 \in U_\beta$ (see Theorem 2.1). In this case k_0 corresponds to an equivalence class [containing more than one element] on the boundary of T_α, and the likelihood function $\hat{L}_T$ [when defined on $\bar{T}_\alpha \times \underline{\Sigma}$] is constant along this equivalence class [for any Σ]; moreover its asymptotic form L [which again here is considered as being defined on $\bar{T}_\alpha \times \underline{\Sigma}$], attains its minimum over this equivalence class [for Σ_0]. It might be the case that for $\hat{k}_T \to k$, the corresponding $\hat{\tau}_T$ will converge to infinity, without converging to the 'true' equivalence class. However, if we impose suitable prior bounds on the norm of the elements of $\bar{T}_\alpha$, then the [not necessarily unique] $\hat{\tau}_T$ will converge to the true equivalence class, but not necessarily to a fixed point within this class. Thus an identification algorithm may search along this class.

(iii) Finally we consider the case $k_0 \in \bar{U}_\alpha - \pi(\bar{T}_\alpha)$, which can only occur for $s > 1$. In this case k_0 corresponds to the point of infinity in the one point compactification of $\bar{T}_\alpha$; even if $\hat{k}_T \in U_\alpha$, $T \in \mathbb{N}$ holds, then $\hat{k}_T \to k_0$ implies $\|\psi_\alpha(\hat{k}_T)\| \to \infty$.

In order to discriminate between different consistent estimators and in order to obtain an approximate distribution for the parameter estimators, in the asymptotic analysis central limit theorem are provided (see e.g. Dunsmuir and Hannan 1976, Hannan and Deistler 1988).

For a central limit theorem we have to consider a parameter space $T \times \underline{\Sigma}$ (and not $U \times \underline{\Sigma}$) and we have to impose additional assumptions: First the parameter space $T \subset \mathbb{R}^d$ has to be open [this is not an essential assumption; for boundary points the limiting distribution would not be Gaussian]. For standard parameter spaces, such as T_α, we have to strengthen (1.8) to $\det b(z) \neq 0$, $|z| \leq 1$, in order to ensure openness. Also, in addition to the assumptions of Theorem 3.1, the process generating the data is assumed to satisfy the following conditions: $\varepsilon(t)$ is strictly stationary and

$$E\{\varepsilon(t)|\mathcal{F}_{t-1}\} = 0 \qquad (3.13)$$

where $\mathcal{F}_t$ is the σ-algebra generated by $\varepsilon(s)$, $s \leq t$.

Condition (3.13) seems to be quite natural in our context, since it is equivalent to the condition that the best (in least squares sense) predictor $E(y(t)|\mathcal{F}_{t-1})$ of $y(t)$ from its past $y(t-1), y(t-2), \dots$ is equal to the best linear predictor of $y(t)$ given its past, and since in cases where the difference between these two predictors is substantial, nonlinear, rather than linear systems should be used.

Theorem 3.2. Let the true system satisfy $\tau_0 \in T_\alpha$; then under the assumptions of Theorem 3.1, the assumptions above, and under the assumption

$$E\{\varepsilon(t)\varepsilon'(t)|\mathcal{F}_{t-1}\} = \Sigma_0 \qquad (3.14)$$

the vector $T^{1/2}(\hat{\tau}_T - \tau_0)$ has a Gaussian limiting distribution (with mean zero and with covariance matrix given by (the inverse of the Fisher information as):

$$\left\{ \left[\frac{1}{2} \frac{\partial}{\partial \tau_i \partial \tau_j} \left((2\pi)^{-1} \int_{-\pi}^{\pi} tr\{(k\Sigma k)^{-1}(k_0 \Sigma_0 k_0)\} d\lambda \right) \right]_{i,j=1..d} \right\}^{-1} \qquad (3.15)$$

Here τ_i is the i-the entry of τ. If in addition

$$E\{\varepsilon_j(t)^4\} < \infty, \qquad j = 1..s \qquad (3.16)$$

[where $\varepsilon_j(t)$ is the j-th entry of $\varepsilon t)$] and

$$E\{\varepsilon_i(t)\varepsilon_j(t)\varepsilon_k(t)|\mathcal{F}_{t-1}\} = E\varepsilon_i(t)\varepsilon_j(t)\varepsilon_k(t) \qquad 1 \leq i,j,k \leq s \qquad (3.17)$$

hold then also the on - and above diagonal elements of $T^{1/2}(\hat{\Sigma}_T - \Sigma_0)$ have a Gaussian limiting distribution.

From Theorems 3.1 and 3.2 we see that asymptotic properties of MLE's obtained from a Gaussian likelihood are also valid for a class of non-Gaussian data. For instance if the data are generated by what is sometimes called a linear process, i.e. a process of the form

$$y(t) = \sum_{j=0}^{\infty} K_0(j)\varepsilon(t-j)$$

where $(\varepsilon(t))$ is a sequence of independent (not only uncorrelated) identically distributed random variables then (3.14) is fulfilled and $T^{1/2}(\hat{\tau}_T - \tau_0)$ will have a normal limiting distribution given by (3.15) independent of the actual distribution of the $\varepsilon(t)$. Clearly, if the actual distribution of $\varepsilon(t)$ were known, for the non Gaussian case, the actual (non Gaussian) likelihood would give estimators that have a smaller limiting variance covariance matrix than (3.15). As is well known, for Gaussian processes, the Gaussian MLE's are asymptotically efficient. By the last theorem we see that the Gaussian case is the worst case among all processes satisfying (3.14) and thus Gaussian likelihood estimation can be interpreted as minimization of the worst asymptotic variance covariance matrix.

4. DYNAMIC SPECIFICATION

In most applications the dynamic specification is not known a priori and has to be determined from the data. The development and evaluation of data-based procedures for dynamic specification constitutes one of the most important contributions to the subject during the last twenty years.
Theses procedures may be classified into non-automatic and automatic ones. In the non-automatic case subjective decisions have to be made at a certain stage. A particulary successful procedure of this kind was developed by Box and Jenkins (1970) for the SO case. The advantage of automatic procedures is that they do not require a large amount of experience.

First, as the perhaps most important case we consider the problem of estimating the order n. The classical procedure for choosing a model is the maximum likelihood method. However, since $\bar{M}(n_1) \subset \bar{M}(n_2)$ for $n_1 < n_2$ holds and $M(n_1)$ has smaller dimension than $M(n_2)$, the likelihood method will usually choose the largest allowed order [the same, more generally is true for every criterion which only contains a goodness of fit term]. The common procedures for order estimation are based on minimizing a criterion of the form

$$A(n) = \log \det \hat{\Sigma}_T(n) + (2ns)\frac{c(T)}{T} \quad ; \quad 0 \le n \le N \tag{4.1}$$

where $\hat{\Sigma}_T(n)$ is the MLE of Σ_0 over $\hat{M}(n) \times \underline{\Sigma}$ with sample size T, and N is a prescribed upper bound for the order and $c(T)$ is a prescribed function. Criteria of the form (4.1) have been mentioned already in the introduction. The first term of the righthand side of (4.1), namely $\log \det \hat{\Sigma}_T(n)$ is a measure for goodness of fit of a system to the data. For given T, $\log \det \hat{\Sigma}_T(n)$ will be decreasing for increasing n. The idea is, that this increase will be not so "significant" beyond the true order n_0 (if there is any), compared with the case when we are below the true order and that this "nonsignificant" decrease can be compensated by the "penalty term"

$$(2ns)\frac{c(T)}{T}$$

which contains the dimension $2ns$ of $M(n)$ as a measure of complexity. However, criteria of the form (4.1) are also meaningful for the case where the true system is infinite dimensional. N in (4.1) may depend on sample size T too (Hannan and Deistler 1988). Another interpretation of $A(n)$ is that it provides a tradeoff between bias (due to "underfitting") and efficiency loss by using too many parameters.

Clearly, $c(T)$ describes the tradeoff between goodness of fit and complexity in (4.1). The most common choices for $c(T)$ are $c(T) = 2$, in which case $A(n)$ is called the *AIC criterion* $AIC(n)$and $c(T) = c.\log T$, $c \ge 1$ and then $A(n)$ is called the *BIC* criterion $BIC(n)$.

The actual choice of $c(T)$ can be motivated by a number of partially different ideas. Akaike (1969) (1977) described *AIC* from an entropy maximization principle or from ideas of optimal out of sample forecasting (see also Bhansali 1986, Findley 1985). Rissanen (1983) (1986) derived *BIC* from coding theory.

The asymptotic properties of order estimators based on (4.1) have been derived in Hannan (1980) (1981) for the case where a (finite) true order n_0 exists:

Theorem 4.1. Let $k_0 \in M(n_0)$; then under all assumptions in theorem 3.1. and under the additional assumptions (3.13), (3.14), (3.16)

$$\det k(z) \neq 0 \quad , \quad |z| < 1+\delta \quad \text{for some } \delta > 0$$

and in some coordinate system the norm of every τ is bounded a priori, the following results hold:

(i) If $c(T)/T \to 0$ (for $T \to \infty$) and $\liminf_{T\to\infty}[c(T)/\log T] > 0$

then

$$\hat{n}_T \to n_0$$

(ii) If $c(T)/T \to 0$ and $c(T) \uparrow \infty$, then

$$\hat{n}_T \to n_0 \text{ in probability}$$

(iii) If $\limsup_{T\to\infty} c(\mathrm{T}) < \infty$ then

$$\lim_{\delta\to 0} \lim_{T\to\infty} P\{\hat{n}_T > n_0\} = 1$$

Thus in particular *AIC* gives no consistent estimator $\hat{n}_T$ for n_0. However, as has been shown by Shibata (1980), *AIC* has an optimality property if the true system is infinite dimensional.

The Kronecker indices α can also be estimated by a criterion of the form (4.1), in particular $A(n)$ gives consistent estimators of the Kronecker indices under analogous conditions as in the theorem above see Hannan and Kavalieris (1984).

Alternative inference procedures for dynamic specification are based on the investigation of the linear independence relations of an estimate of the block Hankel matrix H. Such an approach is appropriate in particular if for given n, the local coordinates in the overlapping parametrization of $M(n)$ have to be estimates, since in the case a criterion of the form (4.1) fails.

5. ALTERNATIVE APPROACHES AND EXTENSIONS

Here we give a short summary of some extensions and alternatives to the mainstream approach.

5.1. Identification of Unstable Systems

In many applications, the data show apparent non-stationarities which can be removed applying transformations such as detrending by trendregressions or (iterated) differencing before the actual identification procedure is applied. Clearly differencing removes a particular kind of instability [associated with unit roots of $\det a(z)$] however, a more general approach seems to be preferable.

For the case of unstable systems, i.e., if $\det a(z)$ has roots on or within the unit circle [and when causal solutions are considered], a complete theory is still not available.

For the *scalar* ($s=1$) *autoregressive* case

$$y(t)=\alpha_1 y(t-1)+\ldots+\alpha_p y(t-p)+\varepsilon(t) \tag{5.1}$$

the following properties of the least squares estimator for $\tau=(\alpha_1,\ldots,\alpha_p)$, namely

$$\hat{\tau}_T=\left(\sum_{t=1}^{T} y_{t-1}y_{t-1}'\right)^{-1}\left(\sum_{t=1}^{T} y_{t-1}y_t\right) \tag{5.2}$$

where $y_t=(y(t),\ldots y(t-p+1))$, have been derived (under some additional assumptions):

(i) $\hat{\tau}_T$ is strictly consistent (Lai and Wei 1983).

(ii) For the special case $p=1$ and $\alpha_1=1$, i.e.

$$y(t)=y(t-1)+\varepsilon(t) \tag{5.2}$$

the limiting distribution of $\hat{\tau}_T(=\hat{\alpha}_{1,T})$ obeys the relation

$$T(\hat{\tau}_T-1)\xrightarrow{L}\tfrac{1}{2}(W^2(1)-1)/\int_0^1 W^2(t)dt \tag{5.3}$$

where $W(t)$ is a standard Brownian motion and where $\xrightarrow{L}$ indicates weak convergence of the distributions. This in particular shows that the convergence rate [for consistency] is T [rather than $T^{\frac{1}{2}}$ which is true for the stable case] and that the limiting distribution is no longer normal in general. The faster rate of convergence is quite plausible, since the regressor $y(t-1)$ becomes large in relation to the stationary error $\varepsilon(t)$. The result (5.3) is due to White (1958); this case was treated in a number of further papers, e.g. in Dickey and Fuller (1979).

(iii) The most general results seem to be those of Chan and Wei (1986). They deal with the case where all roots of $a(z)$ are on or inside the unit circle and they derive the limiting distribution of $\hat{\tau}_T$ and characterize them as a functional of stochastic integrals.

Another case of special unstable systems, namely the case of *cointegration* has attracted considerable attention in econometrics recently, see e.g. Engle and Granger (1987): Consider a nonstationary vector process $y(t)$, whose first differences $(1-z)y(t)$ are stationary [and linearly regular]. Such a process $y(t)$ is called cointegrated, if there exists a nonzero vector $a \in \mathbb{R}^s$ such that $a'y(t)$ is stationary. The interpretation is that a represents the (static) equilibrium solution of the system [where $a'y(t)$ is a stationary error which is smaller than the components of the variables]. This kind of models seems to be suited for a number of econometric applications, where in most cases the observed variables show trends in mean and variances but where there is some economic long-term "mechanism" "stabilizing" a certain linear combination of the components [such that it becomes relatively small]. An example for this would be if $y(t)$ contained consumption and income and the linear combinations correspond to a (static) comsumption function, or if $y(t)$ contained supply-side and demand-side variables for a market tending to equilibrium.
If we write

$$(1-z)y(t) = c(z)\varepsilon(t)$$

where $c(z)\varepsilon(t)$ is stationary and in *Wold* representation, and $c(z) = c(1) + (1-z)c^*(z)$, then we obtain

$$y(t) = (1-z)^{-1}c(1)\varepsilon(t) + c^*(z)\varepsilon(t) \tag{5.4}$$

From (5.4) we see that $y(t)$ is cointegrated iff $c(1)$ is singular. $\hat{y}(t) = (1-z)^{-1}c(1)\varepsilon(t)$ may be considered as unobserved "true" variables [since they satisfy the exact relation $a'\hat{y}(t) = 0$] and clearly they are generated by a vector autoregression, where all roots of $\det a(z)$ are equal to one; the second part on the r.h.s. of (5.4) are the stationary errors.
Estimators for a and tests for cointegration are considered e.g. in Engle and Granger (1987) and Phillips and Ouliaris (1986). Typically, here again the rate of consistency is T and the limiting distributions are obtained (via functional central limit theorems) from stochastic integrals.

5.2. Alternative Measures of Goodness of Fit

In particular in control engineering in many cases uniform approximation of transfer functions, in the sense that approximation in the norm $\sup_{\omega\in[\pi,\pi]} \|k(e^{-i\omega})\|$ is considered, is appropriate. However for such an approximation actual calculation would be difficult to perform. Balanced realizations and Hankel norm approximations are relatively easy to calculate and it is still possible to derive error bounds in the uniform norm for them (Glover 1984). However, most of the work done in this area commences from a known true transfer-function, rather than from data, and there are only a few results available on the statistical properties of procedures commencing from data, e.g. via a first estimate of the second moments.

5.3. Errors-in-Variables

Consider an ARMAX system, i.e.

$$a(z)y(t) = d(z)x(t) + b(z)\varepsilon(t) \tag{5.6}$$

where $d(z) = \Sigma D(j)x(t-j)$, $D(j) \in \mathbb{R}^{s\times m}$ and $x(t)$ are observed inputs where $E\varepsilon(s)x'(t) = 0$ for all s and t. ARMAX modelling, or more general errors-in-equations modelling is the "conventional" approach to embed a deterministic (input-output) system into a stochastic environment. However, there is a certain amount of unsymmetry in this way of modelling, since first we have to know a-priori the classification into inputs and outputs and second, and even more important, all of the noise is added to the equations or (for our analysis) equivalently to the outputs. *Linear errors-in-variables* (EV) modelling provides a more general way of modelling of the form:

$$w(z)\hat{z}(t) = 0 \quad ; \quad w(z) = \sum_{j=-\infty}^{\infty} W(j)z^j \; ; \; W(j) \in \mathbb{R}^{s\times(s+m)} \tag{5.7}$$

$$z(t) = \hat{z}(t) + w(t) \tag{5.8}$$

where $z(t)$ is the stacked vector of all observations at time t, i.e. $z(t) = (x(t)', y(t)')'$; $\hat{z}(t)$ is the corresponding vector of, in general unobserved, true, variables (which are related by the deterministic system (5.7) and $w(t)$ is a noise vector, where noise is added, in general, to each component. The main cases, when this more general EV setting is appropriate are:

(i) If we are interested in the "true" system generating the data, rather than in encoding the data by system parameters, and is we cannot be sure a priori that the inputs are not corrupted by noise.

(ii) If we have no a priori classification of the observed variables into inputs and outputs or if even the number of outputs (i.e. the number of equations) is not known a priori and thus has to be determined from the data. Clearly $z(t)$ could also be modelled by a (vector) ARMA system, however in general, this leads to parameter spaces with dimension being considerably higher compared to the corresponding EV system.

(iii) Under certain additional assumptions on the noise structure EV-models are equivalent to dynamic principal component models or to dynamic factor analysis models. If we assume that the noise components are mutually uncorrelated then the model provides a decoupling of common and individual effects between the variables, where all common effects are attributed to the system.

One of the main problems in this context is identifiability of transfer functions (see e.g. Kalman 1982, Deistler and Anderson 1988, Picci and Pinzoni 1986). The statistical analysis is far from being complete.

REFERENCES

Akaike, H. (1969). Fitting autoregressive models for prediction. *Ann. Inst. Statist. Math.* **21**, 243–247.

Akaike, H. (1977). On entropy maximisation principle. In *Applications of Statistics* (ed. P.R. Krishnaiah), 27–41. Amsterdam, North-Holland.

Bhansali, R.J. (1986). The criterion autoregressive transfer function of Parzen. *J. Time Series Anal.* **7**, 79–104.

Box, G.E.P., and Jenkins. G.M. (1970). *Timc Series Analysis, Forecasting and Control*, San Fransisco, Holden Day.

Chan, N.A., and Wei, C.Z. (1986). Limiting distributions of least squares estimates of unstable autoregressive processes. To appear.

Deistler, M. (1983). The properties of the parametrization of ARMAX systems and their relevance for structural estimation. *Econometrica* **51**, 1187–1207.

Deistler, M, and Pötscher, B.M. (1984). The behavior of the likelihood function for ARMA models. *Adv. Appl. Probab.* **16**, 843–865.

Deistler, M., and Anderson, B.D. (1988). Linear dynamic errors invariables models: some structure theory. To appear in *J. Econometrics*.

Deistler, M., and Wang, Liqun (1987). The common structure of parametrizations for linear systems. To appear.

Dickey, D.A., and Fuller, W.A. (1979). Distribution of the estimators for autoregressive time series with a unit root. *J. Amer. Statist. Assoc.* **74**,

427–431.
Dunsmuir, W., and Hannan, E.J. (1976). Vector linear time series models. *Adv. Appl. Probab.* **8**. 339–364.
Engle, R., and Granger, C.W.J. (1987). Co-integration and error-correction: Representation, estimation and testing. *Econometrica* **55**, 251–276.
Findley, D.F. (1985). On the unbiasedness property of *AIC* for exact or approximating linear stochastic time series models. *J. Time Series. Anal.* **6**, 229–252.
Glover, K. (1984). All optimal Hankel-norm approximations of linear multivariable systems and their L^{∞} error bounds. *Internat. J. Control* **39**. 1115–1193.
Hannan, E.J. (1973). The asymptotic theory of linear time series models. *J. Appl. Probab.* **10**, 130–145.
Hannan, E.J. (1980). The estimation of the order of an ARMA process. *Ann. Statist.* **8**, 1071–1081.
Hannan, E.J. (1981). Estimating the dimension of a linear system. *J. Multivariate Anal.* **11**, 459–473.
Hannan, E.J., and Deistler, M. (1988). The Statistical Theory of Linear Systems, New York, John Wiley.
Hannan, E.J., and Kavalieris, L. (1984). Multivariate linear time series models. *Adv. Appl. Prob.* **16**, 492–561.
Hannan, E.J., and Rissanen, J. (1982). Recursive estimation of ARMA order. *Biometrika* **69**, 81–94.
Kabaila, P. (1983). Parameter values of ARMA models minimising the one-step-ahead prediction error when the true system is not in the model set. *J. Applied Prob.* **20**, 405–408.
Kalman, R.E. (1982). System identification from noisy data, in (A. Bednarak and L. Cesari, eds) *Dynamical Systems II, a University of Florida International Symposium*, Academic Press, New York.
Lai, T.L., and Wei, C.Z. (1983). Asymptotic properties of general autoregressive models and strong consistency of least-squares estimates of their parameters. *J. Multivariate Anal.* **13**, 1–23.
Ljung, L. (1978). Convergence analysis of parametric identification methods. *IEEE Trans. Autom. Control* **AC-23**, 770–783.
Phillips, P.C.B., and Ouliaris, S. (1986). Testing for cointegration. *Cowles Foundation Discussion Paper* **809**.
Picci, G., and Pinzoni, S. (1986). Dynamic factor-analysis models for stationary processes. *IMA Math. Control and Information* **3**, 185–210.
Ploberger, W. (1982). Slight misspecifications of linear systems. In *Operations Research in Progress* (eds. G. Feichtinger and P. Kall), 413–424. Dordrecht, The Netherlands, D. Reidel.
Pötscher, B.M. (1987). Convergence results for maximum likelihood type estimators in multivariable ARMA models. *J. Multivariate Anal.* **21**. 29–52.
Rissanen, J. (1983). Universal prior for parameters and estimation by minimum description length. *Ann. Statist.* **11**, 416–431.
Rissanen, J. (1986). Stochastic complexity and modeling. *Ann. Statist.* **14**, 1080–1100.
Shibata, R. (1980). Asymptotically efficient selection of the order of the model for estimating parameters of a linear process. *Ann. Statist.* **8**, 147–164.
Solo, V. (1986). topics in advanced time series analysis. In *Lectures in Probability and Statistics* (eds. G. del Pino and R. Rebodedo). Berlin, Springer-Verlag.
White, J.S. (1958). The limiting distribution of the serial correlation coefficient in the explosive case. *Ann. Math. Statist.* **23**, 1188–1237.
Willems, J.C. (1986). From time series to linear system, Part I: Finite dimensional linear time invariant systems. *Automatica*, **22**, 561–580.

A TUTORIAL ON HANKEL-NORM APPROXIMATION

KEITH GLOVER

Abstract

A self-contained derivation is presented of the characterization of all optimal Hankel-norm approximations to a given matrix-valued transfer function. The approach involves a state-space characterization of all-pass systems as in the author's previous work, but has been greatly simplified. A section of preliminary results is included giving general results on linear fractional transformations, Hankel operators and all-pass systems. These results then can be applied to give the characterization of all optimal Hankel-norm approximations of a given stable transfer function. Frequency response bounds for these approximations are then derived from finite rank perturbation results.

Keywords

Hankel norm, Hardy spaces, $\mathcal{H}_\infty$, model order reduction, rational approximation.

1 INTRODUCTION

An important question when modelling dynamic systems is whether a model can be simplified without undue loss of accuracy. A measure of the complexity of a linear state-space model,

$$\dot{x}(t) = Ax(t) + Bu(t) \tag{1.1}$$

$$y(t) = Cx(t) + Du(t) \tag{1.2}$$

is the dimension, n, of its state vector $x(t)$. In (1.1)-(1.2), $u(t) \in \mathbf{C}^m$, $x(t) \in \mathbf{C}^n$, $y(t) \in \mathbf{C}^p$ for all t, and A, B, C are complex matrices of compatible dimensions. Low order models will give more efficient simulations and, for example, control system design calculations.

Approximating (1.1)-(1.2) by a reduced order system,

$$\dot{\hat{x}}(t) = \hat{A}\hat{x}(t) + \hat{B}u(t) \tag{1.3}$$

$$\hat{y}(t) = \hat{C}\hat{x}(t) + \hat{D}u(t) \tag{1.4}$$

where $\hat{x}(t) \in \mathbf{C}^k$, $k < n$, is termed a *model reduction problem.* Substantial progress has been made on problems of this type in recent years by the use of truncated balanced realizations as introduced by Moore (1981) and optimal Hankel-norm approximations as given in Adamjan, Arov and Krein (1971). The first method truncates states from a particular realization but has not been shown to be optimal in any sense; whereas the second method minimizes a specific norm of the error between (1.1)-(1.2) and (1.3)-(1.4). Both methods have been shown to give excellent results in many application areas. If we define the corresponding transfer functions as

$$G(s) = D + C(sI - A)^{-1}B$$

$$\hat{G}(s) = \hat{D} + \hat{C}(sI - \hat{A})^{-1}\hat{B}$$

then we might consider minimizing a variety of norms on the error system, $G(s) - \hat{G}(s)$. The induced norm corresponding to $\mathcal{L}_2$-norms on the signals is the H_∞-norm of $(G(s) - \hat{G}(s))$, denoted $\|G - \hat{G}\|_\infty$. One of the reasons for the success of the above two methods is that both have been shown to be close to optimal with respect to the H_∞-norm [see Enns (1984) and Glover (1984)].

Glover (1984) gave a characterizations of all optimal Hankel-norm approximations of a given $G(s)$ together with an upper bound on $\|G - \hat{G}\|_\infty$. The approach taken involved some lengthy calculations and it is the primary purpose of the present paper to re-derive many of these results in a self-contained but more efficient manner, hence giving greater insight into the technique and its derivation.

Background to the problem can be found in Glover (1984) and reference to more recent works, especially that of Ball and Ran (1986), can be found in Francis (1987) together with its application to H_∞-control problems. The approach to be described here was also partly presented in Glover(1987) and Glover, Curtain, and Partington (1988).

In section 2 a number of background results will be stated and for completeness most will be derived. Section 3 then considers a sub-optimal Hankel-norm approximation problem, whereas section 4 considers the optimal case. Section 5 then derives the H_∞-norm upper bounds.

The following notation will be used. For $A \in \mathbf{C}^{n\times m}$, A' denotes its complex conjugate transpose and $A^\dagger$ denotes its pseudo inverse. $\mathbf{C}_+$ and $\mathbf{C}_-$ denote the open right and left half planes respectively. $\mathcal{RH}^{p\times m}_{\infty,+}$ denotes the space of proper rational $p\times m$-matrix-valued functions of $s \in \mathbf{C}$, analytic in $\mathbf{C}_+$ (i.e. poles in $\mathbf{C}_-$). Similarly $\mathcal{RH}^{p\times m}_{\infty,-}$ functions are analytic in $\mathbf{C}_-$, and $Q \in \mathcal{RH}^{p\times m}_{\infty,-(k)}$ implies that $Q = G + F$ for $G \in \mathcal{RH}^{p\times m}_{\infty,+}$, $F \in \mathcal{RH}^{p\times m}_{\infty,-}$ with G of McMillan degree $\leq k$. For $Q \in \mathcal{RH}^{p\times m}_{\infty,-(k)}$

$$\|Q\|_\infty := \sup_\omega \bar{\sigma}(Q(j\omega))$$

where $\bar{\sigma}$ denotes the maximum singular value. State-space realizations are denoted

$$G(s) = D + C(sI - A)^{-1}B = \left[\begin{array}{c|c} A & B \\ \hline C & D \end{array}\right]$$

and

$$G(s)^\sim = G(-\bar{s})' = \left[\begin{array}{c|c} -A' & C' \\ \hline -B' & D' \end{array}\right].$$

[Note that in contrast to Francis (1987), $\mathcal{RH}_\infty$ will include rational functions with complex coefficients.]

2 PRELIMINARIES

A number of results on the manipulation of matrices and systems will be required. Since some results are not easily accessible, proofs or outlines of proofs will also be included for tutorial reasons when appropriate.

2.1 Unitary dilation of matrices

A standard approach to representing a contraction is to imbed it in a unitary operator (a dilation of the contraction). Constant matrices will be considered first.

Lemma 2.1 *Let $D_{11} \in \mathbf{C}^{p\times m}$ satisfy $D_{11}'D_{11} \leq I$ with the nullity of $I - D_{11}'D_{11} = r$. Then there exist $D_{12} \in \mathbf{C}^{p\times(p-r)}$, $D_{21} \in \mathbf{C}^{(m-r)\times m}$, $D_{22} \in \mathbf{C}^{(m-r)\times(p-r)}$ such that*

$$\begin{aligned} D_{12}D_{12}' &= I - D_{11}D_{11}', \\ D_{21}'D_{21} &= I - D_{11}'D_{11}, \\ D_{22} &= -D_{21}D_{11}'(D_{12}^L)' = -(D_{21}^R)'D_{11}'D_{12}, \quad \text{where } D_{12}^L D_{12} = I = D_{21}D_{21}^R, \end{aligned}$$

and $D := \begin{bmatrix} D_{11} & D_{12} \\ D_{21} & D_{22} \end{bmatrix}$ *is unitary.*

Proof. With $D_{22} := -D_{21}D_{11}'(D_{12}^L)'$ note that

$$\begin{aligned} D_{21}'D_{22} &= -D_{21}'D_{21}D_{11}'(D_{12}^L)' \\ &= (D_{11}'D_{11} - I)D_{11}'(D_{12}^L)' \\ &= -D_{11}'D_{12}D_{12}'(D_{12}^L)' = -D_{11}'D_{12} \\ D_{22}'D_{22} &= D_{12}^L D_{11}D_{21}'D_{21}D_{11}'(D_{12}^L)' \\ &= D_{12}^L D_{11}(I - D_{11}'D_{11})D_{11}'(D_{12}^L)' \\ &= D_{12}^L D_{12}D_{12}'D_{11}D_{11}'(D_{12}^L)' \\ &= I - D_{12}'D_{12} \end{aligned}$$

which, together with the definition of D_{21}, verifies that $D'D = I$. $DD' = I$ then gives the other expression for D_{22}. □

Lemma 2.2 *Let $B \in \mathbf{C}^{n\times m}$, $C \in \mathbf{C}^{p\times n}$ have rank r and satisfy $C'C = BB'$. Then there exists a unitary $D \in \mathbf{C}^{(m+p-r)\times(m+p-r)}$ where $D = \begin{bmatrix} D_{11} & D_{12} \\ D_{21} & 0 \end{bmatrix}$, $D_{11} = -C'^{\dagger}B = -CB'^{\dagger}$ such that $\begin{bmatrix} C' & 0 \end{bmatrix} D + \begin{bmatrix} B & 0 \end{bmatrix} = 0$.*

Proof. Let $C = U_1\Sigma_1V_1'$ with $U_1'U_1 = I$, $V_1'V_1 = I$, $\det\Sigma_1 \neq 0$ be the SVD of C. Define $W_1 = B'V_1\Sigma_1^{-1}$. Then

$$\begin{aligned}
W_1'W_1 &= \Sigma_1^{-1}V_1'BB'V_1\Sigma_1^{-1} = I, \\
D_{11} &= -C'^{\dagger}B = -U_1\Sigma_1^{-1}V_1'B = -U_1W_1', \\
D_{11}B' &= -U_1\Sigma_1^{-1}V_1'C'C = -U_1\Sigma_1^{-1}V_1'V_1\Sigma_1^2V_1' = -C, \\
C'D_{11}D_{11}'C &= V_1\Sigma_1^2V_1' = C'C \\
&\Rightarrow (C'D_{11}+B)(D_{11}'C+B') = 0 \Rightarrow C'D_{11}+B = 0, \\
B &= V_1\Sigma_1W_1'; \;\; B'^{\dagger} = V_1\Sigma_1^{-1}W_1', \\
D_{11} &= -CB'^{\dagger}
\end{aligned}$$

The construction is completed by choosing D_{12}, D_{21} such that $\begin{bmatrix} U_1, & D_{12} \end{bmatrix}$ and $\begin{bmatrix} W_1, & D_{21}' \end{bmatrix}$ are unitary. □

2.2 Linear fractional maps

Consider the feedback system of Fig. 1.

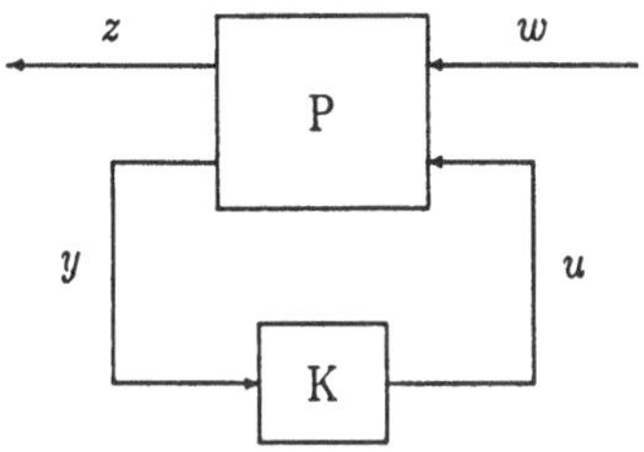

Figure 1: The linear fractional map

We will refer to the transfer function from w to z as the *linear fractional map* of K with coefficient matrix P, denoted

$$\mathcal{F}_l(P,K) = P_{11} + P_{12}K(I - P_{22}K)^{-1}P_{21}$$

where

$$P = \begin{bmatrix} P_{11} & P_{12} \\ P_{21} & P_{22} \end{bmatrix}, \quad P_{11} : p_1 \times m_1, \; P_{22} : p_2 \times m_2$$

and dimensions are compatible. Similarly feedback around the upper loop is denoted $\mathcal{F}_u(P,J) = P_{22} + P_{21}J(I - P_{11}J)^{-1}P_{12}$. Redheffer (1960) proves many results on such transformations, some of which are now given. Notice that for the feedback loop to be well-posed, the condition $\det(I - P_{22}(\infty)K(\infty)) \neq 0$ is required.

Theorem 2.3 *Let* $\det(I - P_{22}K)(\infty) \neq 0$. *Then*

(a) *If* $\|P\|_\infty \leq 1$, $\|K\|_\infty \leq 1$, *then*

$$\|\mathcal{F}_l(P,K)\|_\infty \leq 1.$$

(b) *If* $P^\sim P = I$ *and* $K^\sim K = I$ *then*

$$[\mathcal{F}_l(P,K)]^{\sim}\mathcal{F}_l(P,K) = I.$$

(c) If $PP^{\sim} = I$ *and* $KK^{\sim} = I$ *then*

$$\mathcal{F}_l(P,K)[\mathcal{F}_l(P,K)]^{\sim} = I.$$

(d) If P_{21} *has generically full row rank with* $P^{\sim}P = I$ *and* $\|K\|_\infty > 1$ *then*

$$\|\mathcal{F}_l(P,K)\|_\infty > 1.$$

(e) If P_{12} *has generically full column rank with* $PP^{\sim} = I$ *and* $\|K\|_\infty > 1$ *then*

$$\|\mathcal{F}_l(P,K)\|_\infty > 1.$$

(f) If rank $P_{21}(j\omega) = p_2 \ \forall\, \omega \ \in \mathbf{R} \cup \infty$ *with* $P^{\sim}P = I$ *then* $\|\mathcal{F}_l(P,K)\|_\infty < 1$ *if and only if* $\|K\|_\infty < 1$.

(g) If rank $P_{12}(j\omega) = m_2 \ \forall\, \omega \ \in \mathbf{R} \cup \infty$ *with* $PP^{\sim} = I$ *then* $\|\mathcal{F}_l(P,K)\|_\infty < 1$ *if and only if* $\|K\|_\infty < 1$.

Proof. Consider the system of Fig. 1, which will be well-posed by assumption, and consider the signals w, u, z, y at frequency ω.

(a) $|w|^2 + |u|^2 \geq |z|^2 + |y|^2$ since $\|P\|_\infty \leq 1$. Further, $\|K\|_\infty \leq 1 \ \Rightarrow \ |u|^2 \leq |y|^2$ and hence $|w|^2 + |u|^2 \geq |z|^2 + |u|^2$ for all ω and the result follows.

(b) $u = Ky \Rightarrow |u|^2 = y'K'Ky = |y|^2$, also $|z|^2 + |y|^2 = |w|^2 + |u|^2 = |w|^2 + |y|^2 \Rightarrow |z|^2 = |w|^2 \ \forall \ \omega, y$ and the result follows.

(c) is the dual of *(b)*

(d) Consider a frequency ω such that $\bar{\sigma}(K(j\omega)) > 1$ and $P_{21}(j\omega)$ has full row rank. Then there exists $\hat{y}$ such that $\hat{u} = K(j\omega)\hat{y}$, $|\hat{u}| > |\hat{y}|$. Now let $w = P_{21}(j\omega)^{\dagger}(\hat{y} - P_{22}(j\omega)\hat{u})$; then $y = P_{21}w + P_{22}u = P_{21}P_{21}^{\dagger}(\hat{y} - P_{22}K\hat{y}) + P_{22}Ky \Rightarrow y = \hat{y}, u = \hat{u}$, since $P_{21}P_{21}^{\dagger} = I$. Hence $|z|^2 + |y|^2 = |w|^2 + |u|^2 > |w|^2 + |y|^2 \Rightarrow |z| > |w|$ for this w and ω and the result follows.

(e) is the dual of *(d)*

(f) and (g) Follow in the same way as (d).

□

It is therefore seen that if P is an all-pass system then the feedback system will have norm strictly less than unity if and only if the feedback term satisfies $\|K\|_\infty < 1$. The location of the closed-loop poles can also be deduced as follows:

Lemma 2.4 *Let* $P \in \mathcal{RH}_{\infty,-(k)}$ *and* $K \in \mathcal{RH}_{\infty,-(\ell)}$ *satisfy* $\|P_{22}K\|_\infty < 1$. *Then*

$$\mathcal{F}_l(P,K) \in \mathcal{RH}_{\infty,-(k+\ell)}.$$

Proof. A proof is given in Glover *et al.* (1988) and just observes that the open-loop poles move continuously to the closed-loop poles as the feedback gain is increased, but cannot cross the imaginary axis due to the condition on $P_{22}K$. □

Furthermore the location of any cancellations in the feedback can be examined as follows [a similar result is in Limebeer and Hung (1987)].

Lemma 2.5 *Let P have the state-space realization*

$$P = \left[\begin{array}{c|cc} A & B_1 & B_2 \\ \hline C_1 & D_{11} & D_{12} \\ C_2 & D_{21} & D_{22} \end{array}\right],$$

where rank $D_{12} = m_2$, rank $D_{21} = p_2$, $B_2 = B_{20}D_{12}$, $C_2 = D_{21}C_{20}$, and let K have a minimal realization. Then

(a) All unobservable modes of the natural realization of $\mathcal{F}_l(P,K)$ are contained in $\lambda_i(A - B_{20}C_1)$.

(b) All uncontrollable modes of the natural realization of $\mathcal{F}_l(P,K)$ are contained in $\lambda_i(A - B_1C_{20})$.

The natural realization of $\mathcal{F}_l(P,K)$ refers to the feedback connection of the realizations for P and K.

Proof. Let K have the minimal realization $K = \left[\begin{array}{c|c} \hat{A} & \hat{B} \\ \hline \hat{C} & \hat{D} \end{array}\right]$. Then the state-space equations for the closed loop are:

$$\begin{aligned} \mathcal{F}_l(P,K) &= \left[\begin{array}{cc|c} A + B_2\hat{D}L_1C_2 & B_2L_2\hat{C} & B_1 + B_2\hat{D}L_1D_{21} \\ \hat{B}L_1C_2 & \hat{A} + \hat{B}L_1D_{22}\hat{C} & \hat{B}L_1D_{21} \\ \hline C_1 + D_{12}L_2\hat{D}C_2 & D_{12}L_2\hat{C} & D_{11} + D_{12}\hat{D}L_1D_{21} \end{array}\right] \\ &=: \left[\begin{array}{c|c} A_c & B_c \\ \hline C_c & D_c \end{array}\right], \end{aligned}$$

where $L_1 := (I - D_{22}\hat{D})^{-1}$, $L_2 := (I - \hat{D}D_{22})^{-1}$.

Suppose $\mathcal{F}_l(P,K)$ has unobservable state $(x', y')'$ and mode λ; then the $P - B - H$ test [Kailath (1980)] gives

$$\begin{aligned} & \begin{bmatrix} A_c - \lambda I \\ C_c \end{bmatrix} \begin{bmatrix} x \\ y \end{bmatrix} = 0 \\ \Rightarrow\ & \begin{bmatrix} A - \lambda I & B_2 \\ C_1 & D_{12} \end{bmatrix} \begin{bmatrix} x \\ \hat{D}L_1C_2x + L_2\hat{C}y \end{bmatrix} = 0 \\ \Rightarrow\ & \begin{bmatrix} A - \lambda I & B_{20} \\ C_1 & I \end{bmatrix} \begin{bmatrix} x \\ D_{12}\hat{D}L_1C_2x + D_{12}L_2\hat{C}y \end{bmatrix} = 0 \\ \Rightarrow\ & (A - B_{20}C_1 - \lambda I)x = 0 \end{aligned}$$

If $x = 0$ then $\hat{C}y = 0$ and $\hat{A}y = 0$ which contradicts $(\hat{A}, \hat{C})$ being completely observable. Hence $x \neq 0$ and $\lambda \in \lambda_i(A - B_{20}C_1)$ and part *(a)* is proven. Part *(b)* is a dual result. □

The following corollary is now an immediate consequence of Lemmas 2.4 and 2.5.

Corollary 2.6 *Let* $P \in \mathcal{RH}_{\infty,-(k)}$, $P \notin \mathcal{RH}_{\infty,-(k-1)}$, *have a state-space realization as in Lemma 2.5 with*

$$\begin{aligned} \operatorname{Re}\lambda_i(A - B_{20}C_1) &\geq 0 \\ \operatorname{Re}\lambda_i(A - C_{20}B_1) &\geq 0 \end{aligned}$$

and let $K \in \mathcal{RH}_{\infty,-(\ell)}$, $K \notin \mathcal{RH}_{\infty,-(\ell-1)}$, *and* $\|KP_{22}\|_\infty < 1$. *Then*

$$\begin{aligned} \mathcal{F}_l(P,K) &\in \mathcal{RH}_{\infty,-(k+\ell)} \\ \mathcal{F}_l(P,K) &\notin \mathcal{RH}_{\infty,-(k+\ell-1)}. \end{aligned}$$

The following lemma concerns the inversion of a linear fractional map.

Lemma 2.7 *Let* P *and* K *be rational transfer function matrices, and let* $G = \mathcal{F}_l(P,K)$. *Then*

(a) If P *and* K *are proper with* $\det(I - P_{22}K)(\infty) \neq 0$ *then* G *is proper.*

(b) If P_{12} *and* P_{21} *have generically full column and row rank respectively, then* $\mathcal{F}_l(P,K) = \mathcal{F}_l(P,K_2)$ *implies that* $K_1 = K_2$.

(c) If P *and* G *are proper,* $\det P(\infty) \neq 0$, $\det\left(P + \begin{bmatrix} G & 0 \\ 0 & 0 \end{bmatrix}\right)(\infty) \neq 0$ *and* P_{12} *and* P_{21} *are square and invertible for almost all* s, *then* K *is proper and*

$$K = \mathcal{F}_u(P^{-1}, G)$$

Proof.

(a) is immediate from the definition of $\mathcal{F}_l(P,K)$.

(b) follows from the identity

$$\mathcal{F}_l(P,K_1) - \mathcal{F}_l(P,K_2) = P_{12}(I - K_2P_{22})^{-1}(K_1 - K_2)(I - P_{22}K_1)^{-1}P_{21}$$

(c) Let $Q = P^{-1}$, which will be proper since $\det P(\infty) \neq 0$, and define

$$\begin{aligned} K &= \mathcal{F}_u(Q,G) = Q_{22} + Q_{21}G(I - Q_{11}G)^{-1}Q_{12} \\ &= [Q_{22}Q_{12}^{-1}(I - Q_{11}G) + Q_{21}G](I - Q_{11}G)^{-1}Q_{12} \\ &= P_{12}^{-1}(G - P_{11})(I - Q_{11}G)^{-1}Q_{12} \end{aligned}$$

This expression is well-posed and proper since at $s = \infty$

$$\begin{aligned} \det(I - Q_{11}G) &= \det\left(I - \begin{bmatrix} I & 0 \end{bmatrix} P^{-1} \begin{bmatrix} I & 0 \end{bmatrix}' G\right) \\ &= \det\left[P^{-1}\left(P - \begin{bmatrix} I & 0 \end{bmatrix}' G \begin{bmatrix} I & 0 \end{bmatrix}\right)\right] \\ &\neq 0. \end{aligned}$$

We also need to ensure that $\mathcal{F}_l(P,K)$ is well-posed:

$$\begin{aligned} I - P_{22}K &= (I - P_{22}Q_{22}) - P_{22}Q_{21}G(I - Q_{11}G)^{-1}Q_{12} \\ &= P_{21}Q_{12} + P_{21}Q_{11}G(I - Q_{11}G)^{-1}Q_{12} \\ &= P_{21}(I - Q_{11}G)^{-1}Q_{12} \end{aligned}$$

and $\det(I - P_{22}K) \neq 0$ since P_{21}^{-1} exists and $Q_{12}^{-1} = P_{12} - P_{11}P_{21}^{-1}P_{22}$. Hence the LFT are both well-posed and we immediately obtain that $\mathcal{F}_l(P,K) = G$ as required on substituting for K and $(I - P_{22}K)$ as above. □

Remark 2.1 The proof of part *(c)* was primarily to show that the feedback systems were well-posed. A simple interpretation of the result is given by considering the signals in the feedback systems, assuming they are well-posed, as follows:

$$\begin{aligned} \begin{bmatrix} z \\ y \end{bmatrix} &= P \begin{bmatrix} w \\ u \end{bmatrix}, \quad u = Ky \\ z &= \mathcal{F}_l(P, K)w = Gw \end{aligned}$$

hence

$$\begin{aligned} \begin{bmatrix} w \\ u \end{bmatrix} &= P^{-1} \begin{bmatrix} z \\ y \end{bmatrix}, \quad z = Gw \\ u &= \mathcal{F}_u(P^{-1}, G)w \\ \Rightarrow K &= \mathcal{F}_u(P^{-1}, G) \end{aligned}$$

2.3 Hankel Operators

It is now well-known that Hankel operators play an important role in model reduction and H_∞ design [see Francis (1987), Glover (1984), and the references therein]. General results on Hankel operators, particularly for the infinite rank case, can be found in the books by Power (1982) and Partington (1988).

Let the Hankel operator, Γ, corresponding to the stable system $G(s) = C(sI - A)^{-1}B$ be defined as

$$\Gamma_G : L^2(0,\infty) \to L^2(0,\infty) : \quad u \to \int_0^\infty h(t+\tau)u(\tau)\,d\tau$$

where $h(t) = Ce^{At}B$.

The rank of Γ_G is the McMillan degree of G and it will have a singular value decomposition

$$\Gamma_G(u) = \sum_{i=1}^{n} \sigma_i \langle u, v_i \rangle w_i$$

where the σ_i are the ordered (Hankel) singular values, also denoted $\sigma_i(G)$, and (v_i, w_i) the corresponding Schmidt pairs. Let the controllability and observability Gramians X, Y be given by the unique solutions to the Lyapunov equations,

$$\begin{aligned} AX + XA' + BB' &= 0 \\ A'Y + YA + C'C &= 0. \end{aligned}$$

Further, let $XYx_i = \sigma_i^2 x_i$, $x_i'Yx_i = 1$. Then it is easily verified [see Glover (1984)] that

$$\begin{aligned} v_i(t) &= B' \exp(A't) Y x_i \sigma_i^{-1} \\ w_i(t) &= C \exp(At) x_i. \end{aligned}$$

Now let us consider the Hankel-norm approximation, that is, approximating Γ_G by $\Gamma_{\hat{G}}$ of rank $k < n$. The main result of Adamjan, Arov and Krein (1971) is that $\inf \|\Gamma_G - \Gamma_{\hat{G}}\| = \sigma_{k+1}(G)$, and the derivation in Glover (1984) to derive all solutions to this problem is quite involved. Sections 3 and 4 will give a much more economical derivation based on the results in this section. The present derivation will, however, still be based on the central all-pass construction in Glover (1984).

First, a general result on approximating operators (not necessarily Hankel operators) is given.

Lemma 2.8 *Let $\Gamma : X_1 \to X_2$ be an operator on the Hilbert spaces X_1 and X_2 with Schmidt vectors (v_i, w_i):*

$$\begin{aligned} \Gamma v_i &= \sigma_i w_i, \quad i = 1,2,... \\ \Gamma^* w_i &= \sigma_i v_i, \quad i = 1,2,... \end{aligned}$$

where

$$\sigma_i \geq \sigma_{i+1}, |w_i| = |v_i| = 1 \quad \text{for all } i$$

If $\hat{\Gamma} : X_1 \to X_2$ is of rank k then $\|\Gamma - \hat{\Gamma}\| \geq \sigma_{k+1}$. Further if $\sigma_k > \sigma_{k+1}$ and $\|\Gamma - \hat{\Gamma}\| = \sigma_{k+1}$ then $(\Gamma - \hat{\Gamma})v_{k+1} = \sigma_{k+1} w_{k+1}$.

Proof. The proof is taken from Partington (1988, Theorem 6.14). Let P be the projection from X_2 onto span$(w_1, w_2, ..w_{k+1})$; then

$$\|P(\Gamma - \hat{\Gamma})\| \leq \|\Gamma - \hat{\Gamma}\|.$$

Consider the following restriction of $P\hat{\Gamma}$:

$$P\hat{\Gamma} : \text{lin span } (v_1, .., v_{k+1}) \to \text{lin span } (w_1, ..w_{k+1})$$

which has rank $\leq k$ and hence there exists $x \in \ker(P\hat{\Gamma})$, $\|x\| = 1$ say $x = \sum_{i=1}^{k+1} a_i v_i$ with $\sum_{i=1}^{k+1} a_i^2 = 1$.

$$\begin{aligned} P\Gamma(x) &= \sum_{i=1}^{k+1} a_i \sigma_i w_i \\ \|\Gamma - \hat{\Gamma}\|^2 &\geq \|P\Gamma(x) - P\hat{\Gamma}(x)\|^2 = \|P\Gamma(x)\|^2 = \sum_{i=1}^{k+1} \sigma_i^2 a_i^2 \geq \sigma_{k+1}^2 \end{aligned}$$

Further if $\|\Gamma - \hat{\Gamma}\| = \sigma_{k+1}$ and $\sigma_k > \sigma_{k+1}$, then

$$a_1 = a_2 = \ldots = a_k = 0, \quad |a_{k+1}| = 1 \Rightarrow x = a_{k+1} v_{k+1}$$

Also since $\|\Gamma x - \hat{\Gamma} x\| \leq \sigma_{k+1}$ and $\Gamma x = \Gamma v_{k+1} = \sigma_{k+1} w_{k+1}$, $\langle w_{k+1}, \hat{\Gamma} v_{k+1} \rangle = 0$. Then $\hat{\Gamma} v_{k+1} = 0$ and the result follows. □

Specialising this result to Hankel operators and interpreting it in the frequency domain gives the following result [see Francis (1987, page 71) for the $k = 0$ case].

Lemma 2.9 *Let the Hankel operator Γ_G have Schmidt pairs as above with $\sigma_k > \sigma_{k+1}$. Let $Q \in \mathcal{RH}_{\infty,-(k)}$ be such that*

$$\|G + Q\|_\infty = \sigma_{k+1} \;;$$

then

$$\begin{aligned} (G+Q)V(-s) &= \sigma_{k+1} W(s) \\ W^\sim(G+Q) &= \sigma_{k+1} V^T(s) \end{aligned}$$

where

$$\begin{aligned} V(s) &= \text{Laplace transform of } v_{k+1}(t) \in H_2(\text{rhp}) \\ W(s) &= \text{Laplace transform of } w_{k+1}(t) \in H_2(\text{rhp}). \end{aligned}$$

Note that for $XYx_i = \sigma_i^2 x_i$,

$$\begin{aligned} V(-s) &= B'(-sI - A')^{-1} Y x_{k+1} \sigma_{k+1}^{-1} \\ W(s) &= C(sI - A)^{-1} x_{k+1} \end{aligned}$$

Proof. Let $Q = -\hat{G} + F$ with $\hat{G}$ rational of McMillan degree k and $F \in \mathcal{RH}_{\infty,-}$. $\|G - \hat{G} + F\|_\infty = \sigma_{k+1}$ implies that $\|\Gamma_G - \Gamma_{\hat{G}}\| \le \sigma_{k+1}$ and hence by Lemma 2.8,

$$(\Gamma_G - \Gamma_{\hat{G}})v_{k+1} = \sigma_{k+1}w_{k+1},$$

and recalling that the Hankel operator, Γ_G, is equivalent to a Toeplitz operator with symbol, G, followed by a projection [Francis (1987)] we have in the frequency domain that

$$(G(s) - \hat{G}(s))V(-s) = \sigma_{k+1}W(s) + U(-s)$$

where

$$U(s), V(s), W(s) \in H_2, \quad \|V\|_2 = \|W\|_2 = 1.$$

Hence

$$(G + Q)V(-s) = \sigma_{k+1}W(s) + U(-s) + F(s)V(-s)$$

and $\|G + Q\|_\infty = \sigma_{k+1}$ implies

$$\begin{aligned} \|(G+Q)V(-s)\|_2 &\le \sigma_{k+1} \\ U(-s) + F(s)V(-s) &\in H_2^\perp \text{ implies that} \\ \|(G+Q)V(-s)\|_2^2 &= \sigma_{k+1}^2\|W\|_2^2 + \|U(-s) + F(s)V(-s)\|_2^2 \\ &= \sigma_{k+1}^2 + \|U(-s) + F(s)V(-s)\|_2^2 \\ &\le \sigma_{k+1}^2 \end{aligned}$$

Therefore, $U(-s) + F(s)V(-s) = 0$ and the result follows. Similarly for the dual result. □

Note that in the case when G is scalar that Lemma 2.9 implies that

$$G + Q = \sigma_{k+1}W(s)/V(-s)$$

and the difficulty is to demonstrate that $Q \in H^\infty_{-(k)}$ [Adamjan *et al.* (1971)].

2.4 All-pass systems

The approach taken in Glover (1984) to optimal Hankel norm approximation is to construct an augmented all-pass error system, and then to connect a contraction around the augmented system to generate all solutions. A characterization of all-pass systems is given in Glover (1984, Theorem 5.1) and is now re-stated.

Lemma 2.10

(a) Let $G(s) = D + C(sI - A)^{-1}B$ be a minimal realization. Then $GG^\sim = G^\sim G = I$ if and only if $\exists\ X = X'$, $Y = Y'$ such that

(i) $XY = I$

(ii) $DD' = I$

(iii) $AX + XA' + BB' = 0$

(iv) $DB' + CX = 0$

(v) $D'D = I$

(vi) $A'Y + YA + C'C = 0$

(vii) $D'C + B'Y = 0$

(b) Conditions (ii) - (iv) above imply $GG^{\sim} = I$

(c) Conditions (v) - (vii) above imply $G^{\sim}G = I$

Note that stability is not assumed and parts *(b)* and *(c)* do not need minimality.

An all-pass dilation of transfer functions can be obtained as follows and entirely analogously to Lemma 2.1.

Lemma 2.11 *Let* $\|G_{11}\|_\infty \leq 1$ *then defining*

$$\begin{array}{ll} G_{12}: & G_{12}G_{12}^{\sim} = I - G_{11}G_{11}^{\sim} \\ G_{21}: & G_{21}^{\sim}G_{21} = I - G_{11}^{\sim}G_{11} \end{array}$$

where G_{12} *and* $G_{21}^{\sim}$ *are of generically full column rank. Then*

$$G_{22} = -G_{21}G_{11}^{\sim}(G_{12}^{L})^{\sim} = -(G_{21}^{R})^{\sim}G_{11}^{\sim}G_{12}$$

makes

$$G := \begin{bmatrix} G_{11} & G_{12} \\ G_{21} & G_{22} \end{bmatrix} \text{all-pass}$$

Proof. The proof is identical to Lemma 2.1 except that we take a generic point on $s = j\omega$. This then gives $G^{\sim}G = I$ for almost all $s = j\omega$ and hence for all s. □

2.5 Alternative Linear Fractional Transformations

An alternative approach to many of the results stated in this section is via coprime factorizations over $\mathcal{RH}_{\infty,-}$ (see Vidyasagar(1985)), although it is usual to consider factorizations over $\mathcal{RH}_{\infty,+}$ in control problems. A *right coprime factorization* of G over $\mathcal{RH}_{\infty,-}$ is given by $G = NM^{-1}$ where $N, M \in \mathcal{RH}_{\infty,-}$ and there exist $X, Y \in \mathcal{RH}_{\infty,-}$ such that the following right Bezout identity or right Diophantine identity is satisfied:

$$XN + YM = I$$

If $G \in \mathcal{RH}_{\infty,-(k)}$, $G \notin \mathcal{RH}_{\infty,-(k-1)}$ with $G = NM^{-1}$ as above, then $\det M(s)$ will have precisely k zeros (including multiplicities) in $\mathbf{C}_-$, or equivalently, since M has no poles in $\mathbf{C}_-$ the principle of the argument gives that the winding number of $\det M(s)$ about the origin, as s traverses the Nyquist D contour, is equal to k, (see Vidyasagar(1985) for more details). Hence the McMillan degree of the stable part of G can be determined.

When P_{21} is invertible for almost all s then the following alternative form of the linear fractional transformation can be used:

$$\begin{aligned} \mathcal{F}_l(P, K) &= T_\Theta(K) \\ &:= (\theta_{11}K + \theta_{12})(\theta_{21}K + \theta_{22})^{-1} \\ &= (\theta_{11}U + \theta_{12}V)(\theta_{21}U + \theta_{22}V)^{-1} \end{aligned}$$

where $K = UV^{-1}$ is a right coprime factorization over $\mathcal{RH}_{\infty,-}$. It is straightforward to verify that P and Θ are related as follows:

$$\begin{aligned}\Theta &= \begin{bmatrix}\theta_{11} & \theta_{12}\\ \theta_{21} & \theta_{22}\end{bmatrix}\\ &= \begin{bmatrix}P_{12}-P_{11}P_{21}^{-1}P_{22} & P_{11}P_{21}^{-1}\\ -P_{21}^{-1}P_{22} & P_{21}^{-1}\end{bmatrix}\\ &= \left[\begin{array}{c|cc} A-B_1D_{21}^{-1}C_2 & B_2-B_1D_{21}^{-1}D_{22} & B_1D_{21}^{-1}\\ \hline C_1-D_{11}D_{21}^{-1}C_2 & D_{12}-D_{11}D_{21}^{-1}D_{22} & D_{11}D_{21}^{-1}\\ -D_{21}^{-1}C_2 & -D_{21}^{-1}D_{22} & D_{21}^{-1}\end{array}\right].\end{aligned}$$

This representation is used extensively in the literature and the monographs of Dym(1989) and Helton(1987) contain a wealth of results in this area.

Now let us consider Corollary 2.6 in this framework. The assumptions that $B_2 = B_{20}D_{12}$ and $C_2 = D_{21}C_{20}$ imply that,

$$G = \mathcal{F}_l(P, K) = T_\Theta(D_{12}KD_{21})$$

where,

$$\Theta = \left[\begin{array}{c|cc} A-B_1C_{20} & B_{20}-B_1D_{220} & B_1\\ \hline C_1-D_{11}C_{20} & I-D_{11}D_{220} & D_{11}\\ -C_{20} & -D_{220} & I\end{array}\right]$$

and

$$D_{220} := D_{21}^{\dagger}D_{22}D_{12}^{\dagger} \Rightarrow D_{22} = D_{21}D_{220}D_{12}$$

and it is easy to verify that,

$$\Theta^{-1} = \left[\begin{array}{c|cc} A-B_{20}C_1 & B_{20} & B_1-B_{20}D_{11}\\ \hline -C_1 & I & -D_{11}\\ C_{20}+D_{220}C_1 & D_{220} & I-D_{220}D_{11}\end{array}\right].$$

The assumptions that Re $\lambda_i(A - B_1C_{20}) > 0$ and Re $\lambda_i(A - B_{20}C_1) > 0$ imply that Θ, $\Theta^{-1} \in \mathcal{RH}_{\infty,-}$, which are the fundamental assumptions being made. Now let $D_{12}KD_{21}$ have right coprime factorization UV^{-1} with $XU+YV = I$ and $U, V, X, Y \in \mathcal{RH}_{\infty,-}$, then

$$G = (\theta_{11}U + \theta_{12}V)(\theta_{21}U + \theta_{22}V)^{-1}$$

is a right coprime factorization of G since

$$\begin{bmatrix} X & Y \end{bmatrix}\Theta^{-1}\begin{bmatrix}\theta_{11}U+\theta_{12}V\\ \theta_{21}U+\theta_{22}V\end{bmatrix} = I.$$

The above winding number result, together with the identities,

$$\begin{aligned}\det(\theta_{21}U+\theta_{22}V) &= \det(\theta_{22})\det(V)\det(I+\theta_{22}^{-1}\theta_{21}UV^{-1})\\ &= \det(\theta_{22})\det(V)\det(I-P_{22}K)\\ \det(\theta_{22}) &= \frac{\det(sI-A)}{\det(sI-A+B_1C_{20})}\end{aligned}$$

and $\|P_{22}K\|_\infty < 1$ gives that the number of poles in $\mathbf{C}_-$ for G is precisely $k + \ell$.

Results analagous to Theorem 2.3 involve so so-called J-inner functions ($J := \begin{bmatrix} I & 0\\ 0 & -I\end{bmatrix}$). Θ is J-inner if $\Theta^{\sim}J\Theta = J$ for $s = j\omega$, with $\Theta \in \mathcal{RH}_{\infty,-}$, (again $\mathcal{RH}_{\infty,+}$ is more commonly used). For such Θ then T_Θ maps the unit ball onto the unit ball. This representation is more natural in, for example, the work of Ball and Ran (1986) on Hankel-norm approximation and Green *et al.* (1988) on a generalization to the Nehari problem.

3 SUB-OPTIMAL HANKEL-NORM APPROXIMATIONS

In this section the problem of approximating a Hankel operator Γ_G of rank n by $\Gamma_{\hat{G}}$ of rank $k < n$ will be considered. Indeed, all solutions $\Gamma_{\hat{G}}$ to the problem

$$\|G - \hat{G}\|_H = \|\Gamma_G - \Gamma_{\hat{G}}\| < \sigma \tag{3.5}$$

for some σ will be solved. It will be shown that this problem is equivalent to finding all $Q \in \mathcal{RH}^{p\times m}_{\infty,-(k)}$ such that

$$\|G + Q\|_\infty < \sigma. \tag{3.6}$$

This equivalence is a consequence of a theorem due to Nehari (1957) for which we will give an independent derivation.

3.1 All-pass dilations

Firstly note that for $G \in \mathcal{RH}^{p\times m}_{\infty,+}$ of degree n and $Q \in \mathcal{RH}^{p\times m}_{\infty,-(k)}$, $Q = -F - \hat{G}$ for $F \in \mathcal{RH}^{p\times m}_{\infty,-}$, $\hat{G} \in \mathcal{RH}^{p\times m}_{\infty,+}$ we have

$$\|G + Q\|_\infty \geq \|G - \hat{G}\|_H = \|\Gamma_G - \Gamma_{\hat{G}}\| \geq \sigma_{k+1}(G). \tag{3.7}$$

The first inequality is standard since the Hankel operator is a restriction of the convolution operator [see for example Francis (1987) or Glover (1984, Lemma 6.2)]. The final inequality follows from Lemma 2.8 since $\text{rank}(\Gamma_{\hat{G}}) \leq k$. Hence in order for (3.5) to have a solution, $\sigma > \sigma_{k+1}$ is required. Further it will be assumed that $\sigma < \sigma_k$ and without loss of generality that $\sigma = 1$ (a scaling of G can achieve this). That is,

$$\sigma_k > 1 > \sigma_{k+1}.$$

We will now construct $J \in H^{(p+m)\times(m+p)}_{\infty,-(k)}$ such that $G_a + J$ is all-pass, where

$$G_a = \begin{bmatrix} G & 0 \\ 0 & 0 \end{bmatrix} = \left[\begin{array}{c|cc} A & B & 0 \\ \hline C & 0 & 0 \\ 0 & 0 & 0 \end{array}\right] = \left[\begin{array}{c|c} A & B_a \\ \hline C_a & 0 \end{array}\right] \tag{3.8}$$

$$J = \left[\begin{array}{c|c} \hat{A} & \hat{B} \\ \hline \hat{C} & D_e \end{array}\right] = \left[\begin{array}{c|cc} \hat{A} & \hat{B}_1 & \hat{B}_2 \\ \hline \hat{C}_1 & D_{11} & D_{12} \\ \hat{C}_2 & D_{21} & D_{22} \end{array}\right] \tag{3.9}$$

$$E := G_a + J = \left[\begin{array}{c|c} A_e & B_e \\ \hline C_e & D_e \end{array}\right] = \left[\begin{array}{cc|c} A & 0 & B_a \\ 0 & \hat{A} & \hat{B} \\ \hline C_a & \hat{C} & D_e \end{array}\right] \tag{3.10}$$

Now from Lemma 2.10, E will be all-pass if there exists $X_e = X_e'$ such that

$$A_e X_e + X_e A_e' + B_e B_e' = 0 \tag{3.11}$$
$$D_e D_e' = I \tag{3.12}$$
$$D_e B_e' + C_e X_e = 0 \tag{3.13}$$

Now let X and Y be the controllability and observability Gramians of G satisfying

$$AX + XA' + BB' = 0 \tag{3.14}$$
$$A'Y + YA + C'C = 0 \tag{3.15}$$

so that $\sigma_i^2(G) = \lambda_i(XY)$. The (1,1) block of (3.11), bearing in mind the form of A_e in (3.10), gives that $\begin{bmatrix} I & O \end{bmatrix} X_e \begin{bmatrix} I & O \end{bmatrix}' = X$. Further $X_e^{-1}(3.11)X_e^{-1}$ and (3.13) give

$$X_e^{-1}A_e + A_e'X_e^{-1} + C_e'C_e = 0 \tag{3.16}$$

and hence $\begin{bmatrix} I & O \end{bmatrix} X_e^{-1} \begin{bmatrix} I & O \end{bmatrix}' = Y$. Let us now postulate a form for X_e, given by

$$X_e = \begin{bmatrix} X & I \\ I & YZ^{-1} \end{bmatrix}; \quad X_e^{-1} = \begin{bmatrix} Y & -Z' \\ -Z & ZX \end{bmatrix} \tag{3.17}$$

where $Z := XY - I$.

Although this form for X_e is apparently taken 'out of the air', its form is fixed once the dimension of $\hat{A}$ is chosen to be that of A and the (1,2) block of X_e is assumed to be nonsingular (which is then transformed to the identity by a similarity transformation on the realization of J). Lemma 8.2 in Glover (1984) in fact generates all possible X_e but the present approach does not require this. All that is required is the particular candidate solution in (3.17). Now let us solve for $\hat{A}$, $\hat{B}$, and $\hat{C}$ given some unitary D_e.

$\hat{C}$ is obtained from the (1,1) block of (3.13); $\hat{B}$ from the (1,1) block of (3.13) $\times X_e^{-1}$, $\hat{A}$ from the (2,1) block of (3.11).

$$\begin{aligned} \hat{C} &= -C_aX - D_eB_a' && (3.18)\\ \hat{B} &= Z'^{-1}(YB_a + C_a'D_e) && (3.19)\\ \hat{A} &= -A' - \hat{B}B_a' && (3.20)\\ &= -Z'^{-1}A'Z' + Z'^{-1}C_a'\hat{C} && (3.21) \end{aligned}$$

(3.21) is obtained from the (1,2) block of (3.16) and will be valid once (3.11) and (3.13) are verified.

(3.18) and (3.19) give that

$$(3.13) \times \begin{bmatrix} I & Y \\ O & -Z \end{bmatrix} = 0 \Rightarrow (3.13) = 0.$$

(3.14) and (3.19) give (3.11) $\times \begin{bmatrix} I \\ O \end{bmatrix} = 0$, and (3.15) gives that

$$\begin{bmatrix} I & O \end{bmatrix} X_e^{-1}(3.11)X_e^{-1} \begin{bmatrix} I & O \end{bmatrix}' = 0$$

and hence

$$\begin{bmatrix} I & O \\ Y & -Z' \end{bmatrix} (3.11) \begin{bmatrix} I & Y \\ O & -Z \end{bmatrix} = 0$$

which implies that (3.11) is satisfied. Hence the required all-pass equations are satisfied, and given X_e there are precisely the correct number of equations to generate $\hat{A}$, $\hat{B}$, and $\hat{C}$. Furthermore $\hat{A}$ will have $\leq k$ eigenvalues in the open left half plane since YZ'^{-1} has k positive eigenvalues,

$$\hat{A}YZ^{-1} + Z'^{-1}Y\hat{A}' + \hat{B}\hat{B}' = 0 \tag{3.22}$$

and by Theorem 3.3(2) in Glover (1984). A final property of J that will be required in Theorem 3.2 to characterize all solutions is that, for D_{12} and D_{21} invertible,

$$\hat{A} - \hat{B}_1D_{21}^{-1}\hat{C}_2 = -A' - \hat{B}_1B' - \hat{B}_1D_{21}^{-1}(-D_{21}B') = -A' \tag{3.23}$$

from (3.20) and (3.18). Similarly (3.21) and (3.19) give

$$\hat{A} - \hat{B}_2D_{12}^{-1}\hat{C}_1 = -Z'^{-1}A'Z' + Z'^{-1}C'\hat{C}_1 + Z'^{-1}C'D_{12}D_{12}^{-1}\hat{C}_1 = -Z'^{-1}A'Z' \tag{3.24}$$

The following theorem can now be stated:

Theorem 3.1 *Given $G \in \mathcal{RH}_{\infty,+}^{p\times m}$ defined by (3.8) then:*

(a) There exists $Q \in \mathcal{RH}_{\infty,-(k)}^{p\times m}$ such that $\|G+Q\|_\infty < 1$ iff $\sigma_{k+1}(G) = \lambda_{k+1}^{1/2}(XY) < 1$, where X and Y are given by (3.14) and (3.15).

(b) If $\sigma_k(G) > 1 > \sigma_{k+1}(G)$ then J defined by (3.9), (3.14)-(3.20) satisfies $J \in \mathcal{RH}_{\infty,-(k)}^{p\times m}$.

Proof. If $\|G+Q\|_\infty < 1$ then (3.7) implies that $\sigma_{k+1}(G) < 1$. Conversely, if $\sigma_{k+1}(G) < 1 < \sigma_k(G)$ then the construction of J has been shown to yield $J \in \mathcal{RH}_{\infty,-(k)}^{p\times m}$, with (G_a+J) all-pass. Furthermore $J_{12}(j\omega)$ is full rank for all ω (including ∞) since J_{12}^{-1} has 'A-matrix' $(\hat{A} - \hat{B}_2 D_{12}^{-1}\hat{C}_1) = -A'$ by (3.23) and hence J_{12} has no zeros on the imaginary axis since A is stable. Hence $\|G - J_{11}\|_\infty < 1$. If $\sigma_i > 1 > \sigma_{i+1} = \sigma_k = \sigma_{k+1}$ for some $i < k$ then the same construction can be used with k replaced by i, again giving J_{11} as a suitable Q. □

3.2 Characterization of all solutions

Once the all-pass dilation of Theorem 3.1 has been constructed, the results of section 2 can be applied to show that all solutions are characterized as follows.

Theorem 3.2 *Given $G \in \mathcal{RH}_{\infty,+}^{p\times m}$ defined by (3.8) with $\sigma_k(G) > 1 > \sigma_{k+1}(G)$, then all $Q \in \mathcal{RH}_{\infty,-(k)}^{p\times m}$ such that*

$$\|G+Q\|_\infty < 1 \tag{3.25}$$

are given by

$$Q = \mathcal{F}_l(J,\Phi), \; \Phi \in \mathcal{RH}_{\infty,-}^{p\times m}, \|\Phi\|_\infty < 1. \tag{3.26}$$

where J is defined in (3.9), (3.14)-(3.20) with D_{12} and D_{21} invertible.

Proof. Let $Q \in \mathcal{RH}_{\infty,-(k)}^{p\times m}$ be such that (3.25) holds. Then (3.26) has a solution for some rational proper Φ by Lemma 2.7 on noting that

$$\det D_e' \det\left(D_e + \begin{bmatrix} G(\infty) & 0 \\ 0 & 0 \end{bmatrix}\right) = \det \begin{bmatrix} I + D_{11}'G(\infty) & 0 \\ D_{12}'G(\infty) & I \end{bmatrix} \neq 0$$

since $\bar{\sigma}(D_{11}'G(\infty)) < 1$. Furthermore, (3.25) and (3.26) imply that

$$G + Q = \mathcal{F}_l(J + G_a, \Phi)$$

with $\|G+Q\|_\infty < 1$ and $J + G_a$ all-pass. Hence Theorem 2.3 implies that $\|\Phi\|_\infty < 1$. Finally Corollary 2.6 can be applied to $Q = \mathcal{F}_l(J,\Phi)$ to give that $\Phi \in \mathcal{RH}_{\infty,-(0)}$ since $Q \in \mathcal{RH}_{\infty,-(k)}$, $J \in \mathcal{RH}_{\infty,-(k)}$ and $J \notin H_{\infty,-(k-1)}$ (since $\|G + J_{11}\|_\infty < 1 < \sigma_{k-1}$, and the realization of J satisfies (3.23) and (3.24)). □

4 OPTIMAL HANKEL-NORM APPROXIMATIONS

In the limit as $\sigma_{k+1}(G) \to 1$ the characterization of all solutions in Theorem 3.2 becomes degererate because the term $Z = (XY - I)$ becomes singular. It is possible to rewrite the equations for J in descriptor form as in Safonov *et al.* (1987), and this will show that the optimal solutions are no longer strictly proper. The characterization of all-pass systems can also be done for descriptor systems and this approach is taken in Glover *et al* (1989) for an $\mathcal{H}_\infty$ control problem. To characterize all optimal solutions we will exploit the constraint given by Lemma 2.8 on all $(G+Q)$ such that $\|G+Q\|_\infty = \sigma_{k+1}$, where $Q \in \mathcal{RH}_{\infty,-(k)}$, and involving the Schmidt vectors of Γ_G. Suppose that σ_{k+1} has multiplicity r and that $\sigma_{k+1} = 1$.

Let the corresponding controllability and observability Gramians be

$$X = \begin{bmatrix} I_r & 0 \\ 0 & X_2 \end{bmatrix}, \quad Y = \begin{bmatrix} I_r & 0 \\ 0 & Y_2 \end{bmatrix} \tag{4.1}$$

after a suitable change of state coordinates, with

$$\{\lambda_i(X_2Y_2)\} = \left\{\sigma_1^2, \sigma_2^2, \ldots, \sigma_{k-1}^2, \sigma_{k+r+1}^2, \ldots, \sigma_n^2\right\}$$

The Laplace transforms of the Schmidt vectors of Γ_e corresponding to σ_{k+1} are then

$$\begin{aligned} W_i(s) &= C(sI-A)^{-1}e_i, \quad i = 1, 2, \ldots, r \\ V_i(-s) &= B'(-sI-A')^{-1}e_i \quad i = 1, 2, \ldots r \end{aligned}$$

where e_i are the standard basis vectors. Hence from Lemma 2.9 if $\|G+Q_i\| = \sigma_{k+1}$ for $Q_i \in \mathcal{RH}_{\infty,-(k)}$ and $i = 1, 2$, then for $W := [W_1, W_2, \ldots, W_r]$, $V := [V_1, \ldots, V_r]$,

$$W^\sim(Q_1 - Q_2) = 0 \tag{4.2}$$

$$(Q_1 - Q_2)V(-s) = 0 \tag{4.3}$$

In order to characterize all optimal solutions, suppose that we can construct $J^o \in \mathcal{RH}_{\infty,-(k)}^{(p+m-\ell)\times(p+m-l)}$, where ℓ is assumed to be the generic ranks of both W and V, with $J_{22}^o(\infty) = 0$, such that $G_a^o + J^o$ is all-pass, where $G_a^o = \begin{bmatrix} G & 0 \\ 0 & 0 \end{bmatrix} \in \mathcal{RH}_{\infty,+}^{(p+m-\ell)\times(p+m-\ell)}$. A set of solutions would then be given by

$$Q = \mathcal{F}_l(J^o, \Phi), \quad \Phi \in \mathcal{RH}_{\infty,-}, \|\Phi\|_\infty \le 1,$$

since $G + Q = \mathcal{F}_l(G_a^o + J^o, \Phi)$ so that $\|G+Q\|_\infty \le 1$ by Theorem 2.3 and $Q \in \mathcal{RH}_{\infty,-(k)}$ by Lemma 2.4. Now suppose that $Q \in \mathcal{RH}_{\infty,-(k)}$ and $\|G+Q\|_\infty \le 1$; then (4.2) and (4.3) together with $\|G_a^o + J^o\|_\infty = 1$ imply that

$$\begin{aligned} &W^\sim(Q - J_{11}^o) = 0 \\ &W^\sim J_{12}^o = 0 \\ &(Q - J_{11}^o)V(-s) = 0 \\ &J_{21}^o V(-s) = 0 \end{aligned}$$

Furthermore J_{12}^o and J_{21}^o have generically full column and row ranks respectively, so that for a generic point s, $Q - J_{11}^o \in \{\text{null space of } W^\sim\} \supset \{\text{range space } J_{12}^o\}$, but these two spaces will both have dimension $p - \ell$ and are hence equal; similarly for J_{21}^o. Hence the equation

$$Q - J_{11}^o = J_{12}^o \Psi J_{21}^o$$

has a rational solution Ψ, which will be proper. $\Phi(I-J_{22}^{o}\Phi)^{-1}=\Psi$ is achieved by setting $\Psi=(I+\Psi J_{22}^{o})^{-1}\Psi$, which is well-posed since $J_{22}^{o}(\infty)=0$ and this satisfies $Q=\mathcal{F}(J^{o},\Phi)$. Theorem 2.3 and Corollary 2.6 can then be applied to prove that $\Phi\in\mathcal{RH}_{\infty,-}$, $\|\Phi\|_{\infty}\leq 1$.

It only remains to construct J^{o} and verify its properties and this is a minor variation of the all-pass construction of section 3 and gives the following results.

Let the realization of G be partitioned conformally with X and Y as

$$G=\left[\begin{array}{c|c}A&B\\\hline C&0\end{array}\right]=\left[\begin{array}{cc|c}A_{11}&A_{12}&B_{1}\\A_{21}&A_{22}&B_{2}\\\hline C_{1}&C_{2}&0\end{array}\right]$$

The Lyapunov equations for X and Y then give

$$-A_{11}-A_{11}'=B_{1}B_{1}'=C_{1}'C_{1}$$

and hence by Lemma 2.2 there exists a unitary $D_{e}^{o}=\begin{bmatrix}D_{11}&D_{12}\\D_{21}&0\end{bmatrix}\in\mathbf{C}^{(p+m-\ell)\times(p+m-\ell)}$ where $\ell=\operatorname{rank}C_{1}=\operatorname{rank}B_{1}$, such that

$$\begin{bmatrix}C_{1}'&0\end{bmatrix}D_{e}^{o}+\begin{bmatrix}B_{1}&0\end{bmatrix}=0$$

A suitable value for X_{e}^{o}, the solution to the all-pass equations, is given by

$$X_{e}^{o}=\begin{bmatrix}I&0&0\\0&X_{2}&I\\0&I&Y_{2}Z_{2}^{-1}\end{bmatrix},\quad (X_{e}^{o})^{-1}=\begin{bmatrix}I&0&0\\0&Y_{2}&-Z_{2}'\\0&-Z_{2}&Z_{2}X_{2}\end{bmatrix}\tag{4.4}$$

It is then a straightforward exercise to verify that the all-pass equations are satisfied by the following realization of J^{o}:

$$\begin{aligned}J^{o}&=\left[\begin{array}{c|cc}-A_{22}'-\hat{B}_{1}^{o}B_{2}'&\hat{B}_{1}^{o}&Z_{2}'^{-1}C_{2}'D_{12}\\\hline -C_{2}X_{2}-D_{11}B_{2}'&D_{11}&D_{12}\\-D_{21}B_{2}'&D_{21}&0\end{array}\right]\\\hat{B}_{1}^{o}&=Z_{2}'^{-1}(Y_{2}B_{2}+C_{2}'D_{11})\end{aligned}$$

This realization of J^{o} clearly satisfies the required stability assumptions for Corollary 2.6. Furthermore, the generic rank of $W\geq\operatorname{rank}\lim_{s\to\infty}sW=\operatorname{rank}C_{1}=\ell$ and since $W^{\sim}J_{12}^{o}=0$, W has generic rank ℓ. Hence the characterization of all solutions is proven. This result is now stated without the $\sigma_{k+1}=1$ assumption which is removed by a simple scaling.

Theorem 4.1 *Let $G\in\mathcal{RH}_{\infty,+}^{p\times m}$ satisfy $\sigma_{k}(G)>\sigma_{k+1}(G)$. Then there exists a $Q\in\mathcal{RH}_{\infty,-(k)}^{p\times m}$ such that $\|G+Q\|_{\infty}\leq\sigma$ if and only if $\sigma\geq\sigma_{k+1}(G)$. Furthermore all solutions to*

$$\|G+Q\|_{\infty}\leq\sigma=\sigma_{k+1}(G)$$

are given by

$$Q=\mathcal{F}_{l}(J^{o},\Phi),\quad \Phi\in\mathcal{RH}_{\infty,-}^{(m-\ell)\times(p-\ell)},\|\Phi\|_{\infty}\leq\gamma$$

where J^{o} is constructed as follows. Let $G=\left[\begin{array}{cc|c}A_{11}&A_{12}&B_{1}\\A_{21}&A_{22}&B_{2}\\\hline C_{1}&C_{2}&0\end{array}\right]$ be a realization of G with controllability and observability Gramians given by $\begin{bmatrix}\sigma I&0\\0&X_{2}\end{bmatrix}$ and $\begin{bmatrix}\sigma I&0\\0&Y_{2}\end{bmatrix}$, respectively, and with $Z_{2}=X_{2}Y_{2}-\sigma^{2}I$ invertible. Define $D_{e}=\begin{bmatrix}D_{11}&D_{12}\\D_{21}&0\end{bmatrix}\in\mathbf{C}^{(p+m-\ell)\times(p+m-\ell)}$ according to Lemma 2.2 where $\ell=\operatorname{rank}C_{1}=\operatorname{rank}B_{1}$, and

$$\begin{bmatrix} C_1' & 0 \end{bmatrix} D_e + \begin{bmatrix} B_1 & 0 \end{bmatrix} = 0.$$

Then J^o is given by

$$J^o = \left[\begin{array}{c|cc} -A_{22}' - \hat{B}_1^o B_2' & \hat{B}_1^o & Z_2'^{-1} C_2' D_{12} \\ \hline -C_2 X_2 - \sigma D_{11} B_2' & \sigma D_{11} & D_{12} \\ -D_{21} B_2' & D_{21} & 0 \end{array}\right]$$

$$\hat{B}_1^o = Z_2'^{-1}(Y_2 B_2 + \sigma C_2' D_{11})$$

The set of all $\hat{G} \in \mathcal{RH}_{\infty,+}$ of McMillan degree k such that $\|G - \hat{G}\|_H = \sigma_{k+1}(G)$ is given by $\hat{G} = -Q + F$ for $F \in \mathcal{RH}_{\infty,-}$, with Q as above.

5 FREQUENCY RESPONSE BOUNDS

Section 4 was concerned with finding $Q \in \mathcal{RH}_{\infty,-(k)}^{p\times m}$ such that $\|G + Q\|_\infty \leq \sigma_{k+1}$, the optimal achievable norm, and by (3.7) this implies that for $Q = -\hat{G} - F$ with $\hat{G} \in \mathcal{RH}_{\infty,+}$, $F \in \mathcal{RH}_{\infty,-}$ we have

$$\sigma_{k+1} \leq \|G - \hat{G}\|_H \leq \|G - \hat{G} - F\|_\infty = \sigma_{k+1}$$

and hence the characterization of all optimal Hankel-norm approximations is given by the causal part of $-Q$. The question now arises as to whether $\hat{G}$ is a good approximation to G in the H_∞-norm. The results of this section will now re-derive some of those of Glover (1984) but in a more efficient manner. The basic approach is to exploit the optimality of $\hat{G} + F$ and to show that $\|F\|_\infty$ can be bounded.

In order to bound $\|F\|_\infty$ we will first re-state Corollary 9.3 from Glover (1984).

Lemma 5.1 *Let $G(s) \in \mathcal{RH}_{\infty,+}^{p\times m}$ have Hankel singular values $\sigma_1 > \sigma_2 \cdots > \sigma_N$, where each σ_i has multiplicity r_i, and let $G(\infty) = 0$. Then*

(a) $\|G\|_\infty \leq 2(\sigma_1 + \sigma_2 + \ldots + \sigma_N)$

(b) there exists a constant D such that $\|G - D\|_\infty \leq (\sigma_1 + \sigma_2 + \ldots + \sigma_N)$

Proof. The proof of this lemma just involves computing J^o in Theorem 4.1 for $k = n - r_N$. The form of X_e and X_e^{-1} then give that $J^o \in \mathcal{RH}_{\infty,+}$ and that $\sigma_i^2(J^o) = \lambda_i(Y_2 X_2)$, $\|G_a^o + J^o\|_\infty = \sigma_N$. Now J^o can be approximated in the same way and this repeated until just a constant remains. □

A lemma on all-pass systems is now stated.

Lemma 5.2 *Let $E = \left[\begin{array}{c|c} A & B \\ \hline C & D \end{array}\right]$ satisfy the all-pass equations of Lemma 2.10 and let A have dimension $n_1 + n_2$ with n_1 eigen-values strictly in the left half plane and $n_2 < n_1$ eigen-values strictly in the right half plane. If $E = G + F$ with $G \in \mathcal{RH}_{\infty,+}^{p\times m}$ and $F \in \mathcal{RH}_{\infty,-}^{p\times m}$ then,*

$$\sigma_i(G) = \begin{cases} 1 & i = 1, 2, \ldots, n_1 - n_2 \\ \sigma_{i-n_1+n_2}(F^\sim) & i = n_1 - n_2 + 1, \ldots, n_1 \end{cases}$$

In particular this result holds if $E = G + F$ is all-pass with $G \in \mathcal{RH}_{\infty,+}^{p\times m}$ of degree n_1, and $F \in \mathcal{RH}_{\infty,-}^{p\times m}$ of degree $n_2 < n_1$.

Proof. Firstly let the realization be transformed to,

$$E = \left[\begin{array}{cc|c} A_1 & 0 & B_1 \\ 0 & A_2 & B_2 \\ \hline C_1 & C_2 & D \end{array}\right] = \left[\begin{array}{c|c} A & B \\ \hline C & D \end{array}\right], \quad \operatorname{Re}\lambda_i(A_1) < 0, \quad \operatorname{Re}\lambda_i(A_2) > 0,$$

in which case $G = \left[\begin{array}{c|c} A_1 & B_1 \\ \hline C_1 & D \end{array}\right]$, $F = \left[\begin{array}{c|c} A_2 & B_2 \\ \hline C_2 & 0 \end{array}\right]$. The all-pass equations of Lemma 2.10 *(i)-(vii)* are then satisfied by a transformed X and Y, partitioned as,

$$X = \begin{bmatrix} X_1 & X_2 \\ X_2' & X_3 \end{bmatrix}, \quad Y = \begin{bmatrix} Y_1 & Y_2' \\ Y_2 & Y_3 \end{bmatrix}$$

$XY = I$ implies that,

$$\begin{aligned} \det(\lambda I - X_1Y_1) &= \det(\lambda I - (I - X_2Y_2)) \\ &= \det((\lambda - 1)I + X_2Y_2) \\ &= (\lambda - 1)^{n_1 - n_2} \det((\lambda - 1)I + Y_2X_2) \\ &= (\lambda - 1)^{n_1 - n_2} \det(\lambda I - Y_3X_3) \end{aligned}$$

The result now follows on observing that $\sigma_i(G) = \lambda_i(X_1Y_1)$ and $\sigma_i^2(F^\sim) = \lambda_i(X_3Y_3)$. The final statement then follows from Lemma 2.10 which gives the existence of suitable X and Y when the realization is minimal. □

Corollary 5.3 *Let G_a^o and J^o be as defined in Theorem 4.1 and write $J^o = \hat{G}_a^o + F_a^o$ with $\hat{G}_a^o \in \mathcal{RH}_{\infty,+}^{p\times m}$ and $F_a^o \in \mathcal{RH}_{\infty,-}^{p\times m}$. Then for $i = 1, 2, ..., 2k + r$,*

$$\sigma_i(G_a^o - \hat{G}_a^o) = \sigma_{k+1}(G),$$

and for $i = 1, 2, ..., n - k - r$,

$$\sigma_{i+3k+r}(G) \le \sigma_i(F_a^{o\sim}) = \sigma_{i+2k+r}(G_a^o - \hat{G}_a^o) \le \sigma_{i+k+r}(G)$$

Proof. The construction of J^o ensures that the all-pass equations are satisfied and an inertia argument easily establishes that the A-matrix has precisely $n + k$ eigen-values in the open lhp and $n - k - r$ in the open rhp. Hence Lemma 5.2 can be applied to give the equalities. The inequalities are standard results on the singular vaues of finite rank perturbations and follow from the mini-max definition of singular values, see for example Theorem 1.4 in Partington(1988). □

The following result can now be derived and is similar to Theorem 9.7 and Corollary 9.9 in Glover (1984).

Theorem 5.4 *Let $Q = \mathcal{F}_l(J^o, \Phi)$ be given by Theorem 4.1 for Φ a* constant *contraction, and let $Q = -\hat{G} - F$ for $\hat{G} \in \mathcal{RH}_{\infty,+}$, $F \in \mathcal{RH}_{\infty,-}$. Then*

(a) $\sigma_i(G - \hat{G}) \le \begin{cases} \sigma_{k+1}(G), & i = 1, 2, \ldots, 2k + r \\ \sigma_{i-k}(G) & i = 2k + r + 1, \ldots, n + k \end{cases}$

(b) $\sigma_i(G - \hat{G}) \ge \sigma_{i+k}(G) \quad i = 1, 2, \ldots, n - k$

(c) $\sigma_i(F^\sim) \le \sigma_{i+k+r}(G), \quad i = 1, 2, \ldots, n - k - r$

(d) there exists a D_0 such that

(i) $\delta := \|F - D_0\|_\infty \le \sum_{i=1}^{n-k-r} \sigma_i(F^\sim)$

(ii) $\|G - \hat{G} - D_0\|_\infty \le \sigma_{k+1}(G) + \delta \le \sigma_{k+1}(G) + \sum_{i=1}^{n-k-r} \sigma_{i+k+r}(G)$.

Proof.

(a) $\|G-\hat{G}\|_H = \sigma_{k+1}(G) \geq \sigma_i(G-\hat{G})$ for all i. Further, as in Corollary 5.3 for $i > 2k+r$,

$$\begin{aligned}\sigma_i(G-\hat{G}) &= \inf_{\deg(K_1)\leq i-1} \|G-\hat{G}-K_1\|_H \\ &\leq \inf_{\deg(K_2)\leq i-k-1} \|G-K_2\|_H \\ &= \sigma_{i-k}(G)\end{aligned}$$

(b) Standard finite rank perturbation result as in (a).

(c) By Lemma 2.1 we will dilate Φ to a unitary matrix, $\Phi_a = \begin{bmatrix} \Phi & \Phi_{12} \\ \Phi_{21} & \Phi_{22} \end{bmatrix}$, and observe that by Theorem 2.3

$$\begin{bmatrix} G+Q & Q_{12} \\ Q_{21} & Q_{22} \end{bmatrix} := \mathcal{F}_l\left(\begin{bmatrix} G+J_{11}^o & 0 & J_{12}^o & 0 \\ 0 & 0 & 0 & I \\ J_{21}^o & 0 & J_{22}^o & 0 \\ 0 & I & 0 & 0 \end{bmatrix}, \Phi_a\right)$$

is all-pass and satisfies an all-pass equation with X_e^o as in (4.4), with $Q_a = \begin{bmatrix} Q & Q_{12} \\ Q_{21} & Q_{22} \end{bmatrix}$ having the same state dimension as J^o (i.e. $n-k-r$) . Lemma 5.2 can now be applied to $(G_a+Q_a) = (G_a-\hat{G}_a-F_a)$ together with part (a) applied to $\sigma_i(G_a-\hat{G}_a)$, to give the result.

(d) This follows immediately from Lemma 5.1 and part *(c)*. □

Remark 5.1 If a non-constant Φ is used to generate Q, then weaker frequency response bounds are obtained by first dilating Q to an all-pass Q_a of the same degree as Q [see Glover (1984, Theorem 5.2)]. Then the bounds on $\sigma_i(F^{\sim})$ can be derived but will be weaker.

Remark 5.2 Note that Theorem 5.4(a) and Lemma 5.1 could be used to derive a bound on $\|G-\hat{G}-D_0\|_\infty$. However, this would give an extra term of $(2k+r)\sigma_{k+1}(G)$ and would be much weaker but would not depend on Φ being a constant.

Remark 5.3 Trefethen and Gutknetch(1983) have proposed using this method in the scalar case, which they call the Carathéodory Fejér method, for real rational approximation on $[-\epsilon,\epsilon]$ and for complex uniform rational approximation on the disk of radius ϵ. They obtain asymptotic results as $\epsilon \to 0$, essentially giving that $\sigma_{k+1} \sim O(\epsilon^{2k+1})$ and that $\|F\|_\infty \sim O(\epsilon^{4k+3})$. These estimates show that in this asymptotic sense the term F becomes insignificant and hence $\hat{G}$ gives an essentially optimal approximant in the $\mathcal{H}_\infty$-norm. Note that this asymptotic norm bound is substantially smaller than could be deduced from Theorem 5.4.

Table 1: Hankel singular values for the example

i	$\sigma_i(G)$	$\sigma_{i-2}(F_1^\sim)$	$\sigma_{i-3}(F_2^\sim)$	$\sigma_{i-4}(F_3^\sim)$	$\sigma_{i-5}(F_4^\sim)$	$\sigma_{i-6}(F_5^\sim)$	$\sigma_{i-7}(F_6^\sim)$
1	1.2473						
2	0.9714						
3	0.6770	0.4428					
4	0.4428	0.4152	0.1821				
5	0.2812	0.1783	0.1580	0.0940			
6	0.1783	0.1505	0.1460	0.0551	0.0497		
7	0.1170	0.0850	0.0057	0.0071	0.0356	0.0017	
8	0.0850	0.0444	0.0049	0.0070	0.0297	0.0015	0.0118
$\sigma_{k+1}(G)$		0.9714	0.6770	0.4428	0.2812	0.1783	0.1170
$\Vert G-\hat{G}_k-D_0\Vert_\infty$		2.2875	1.1738	0.6058	0.3962	0.1815	0.1288
$\sum_{i>k}\sigma_i(G)$		2.7527	1.7813	1.1043	0.6615	0.3803	0.2020

Figure 2: Error curve for $(G(j\omega)-\hat{G}_2(j\omega)-D_0)$

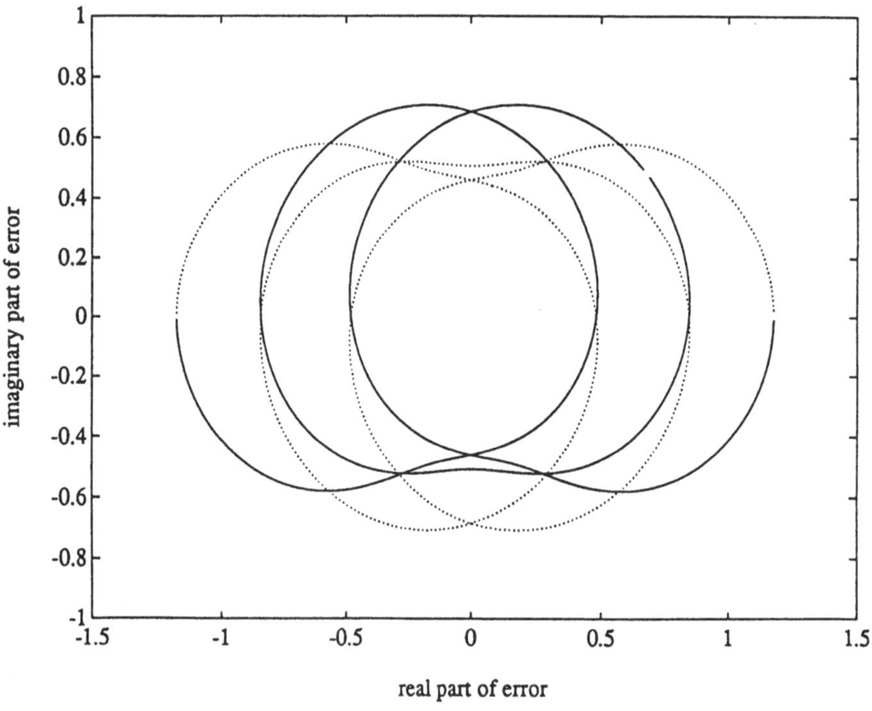

Example

We now give an example to illustrate the results of Theorem 5.4. Let

$$G(s) = \sum_{i=0}^{7} \frac{1}{1 + 10^{-i}s}$$

This is an example of a transfer function with a positive semidefinite symmetric Hankel operator, and hence its eigen-values equal its singular values. Also its poles and zeros are interlaced on the negative real axis and $\|G\|_\infty = 2\sum_{i\geq 1}\sigma_i(G)$. The singular values, $\sigma_i(G)$ are given in Table 1. The optimal Hankel norm approximants, $\hat{G}_k$, of degrees $k = 1, \ldots, 7$ were calculated together with the anti-causal terms F_k. Table 1 also gives the $\sigma_{i-k-1}(F_k^{\sim})$, verifying the inequalities of Corollary 5.3 which can in fact be shown to be always strict for systems with interlaced poles and zeros. The frequency response error is in fact given by

$$\|G - \hat{G} - D_0\|_\infty = \sigma_{k+1}(G) + \sum_{i>0}\sigma_i(F_k^{\sim})$$

with the bound of Theorem 5.4 *(d)(i)* and the first inequality of *(d)(ii)* both equalities. For small values of k the error curves, $(G(j\omega) - \hat{G}_k(j\omega) - D_0)$, are far from being circular, in contrast to Remark 5.3, and that for $k = 2$ is plotted in Figure 1.

This example has not been chosen to illustrate the utility of the method, since this is a very difficult system to approximate with its poles spanning 8 orders of magnitude. It has however been chosen to illustrate the theoretical bounds and the fact that they may be tight. The truncated balanced realization technique will give errors equal to $2\sum_{i>k}\sigma_i(G)$ on examples of this type.

REFERENCES

Adamjan, V.M., D.Z. Arov and M.G. Krein (1971). Analytic properties of Schmidt pairs for a Hankel operator and the generalized Schur-Takagi problem, *Math USSR Sbornik*, vol. 15, pp. 31–73.

Ball, J.A. and A.C.M. Ran (1986), Hankel norm approximation of a rational matrix function in terms of its realizations, in *Modelling, Identification and Robust Control* (C.I. Byrnes and A. Lindquist, eds.). North-Holland.

Dym, H. (1989), *J-Contractive Matrix Functions, Reproducing Kernel Hilbert Spaces and Interpolation*, under preparation.

Enns, D.F. (1984), Model reduction with balanced realisations: an error-bound and frequency-weighted generalization, *Proc. IEEE Conf. on Decision and Control*, Las Vegas NV, pp. 127–132.

Francis, B.A. (1987), *A Course in H_∞ Control Theory*. Springer-Verlag Lecture Notes in Control and Information Sciences, vol. 88.

Glover, K. (1984). All optimal Hankel-norm approximationsof linear multivariable systems and their $\mathcal{L}_\infty$-error bounds, *Int. J. Control*, vol. 39, pp. 1115-1193.

Glover, K. (1987), Model reduction: a tutorial on Hankel-norm methods and lower bounds on L_2 errors, *Proc. Xth Trienniel IFAC World Congress*, Pergamon Press, Munich, vol. X, pp. 288–293.

Glover, K., Curtain R.F. and J.R. Partington (1988), Realisation and approximation of linear infinite dimensional systems with error bounds, *SIAM J. Control and Optim.*, vol. 26, no. 4, pp. 863–898.

Glover, K., D.J.N. Limebeer, J.C. Doyle, E.M. Kasenally, and M.G. Safonov (1989). A characterization of all solutions to the four block general distance problem, under revision.

Green, M., K. Glover, D.J.N. Limebeer and J.C. Doyle (1988), "A J-spectral Factorization Approach to H_∞ Control", submitted.

Helton, J.W. (1987), *Operator Theory, Analytic Functions, Matrices, and Electrical Engineering*, American Mathematical Society CMBS Number 68.

Kailath, T. (1980), *Linear Systems.* Prentice-Hall.

Limebeer, D.J.N. and Y.S. Hung (1987). An analysis of pole-zero cancellations in H_∞-optimal control problems of the first kind, *SIAM J. Control Opt.*, vol.25, pp. 1457-1493.

Moore B.C. (1981), Principal component analysis in linear systems: controllability, observability and model reduction, *IEEE Trans. Auto. Cont.*, vol. AC–26, pp. 17–32.

Nehari, Z. (1957), On bounded bilinear forms, *Annals of Math.*, vol. 65, no. 1, pp. 155–162.

Partington, J.R. (1988), *An Introduction to Hankel Operators.* Cambridge University Press, London Mathematical Society Student Texts, vol. 13.

Power, S.C. (1982), *Hankel Operators on Hilbert Space.* Pitman.

Redheffer, R.M. (1960), On a certain linear fractional transformation, *J. Math. Phys.*, vol. 39, pp. 269–286.

Safonov, M.G., R.Y. Chiang and D.J.N. Limebeer (1987), Hankel model reduction without balancing: a descriptor approach, *Proc. 26th IEEE Conf. Dec. and Cont.*, Los Angeles.

Trefethen, L.N. and M. Gutknetch (1983), The Carathéodory-Fejér method for real rational approximation, *SIAM J. Numer. Anal.*, Vol. 20 No. 2, pp. 420-436.

A DETERMINISTIC APPROACH TO APPROXIMATE MODELLING

C. HEIJ AND J.C. WILLEMS

Abstract

In this paper we will describe a deterministic approach to time series analysis. The central problem consists of approximate modelling of an observed time series by means of a deterministic dynamical system. The quality of a model with respect to data will depend on the purpose of modelling. We will consider the purpose of description and that of prediction. We define the quality by means of complexity and misfit measures, expressed in terms of canonical parametrizations of dynamical systems. We give algorithms to determine optimal models for a given time series and investigate some consistency properties. Finally we present some simulations of these modelling procedures.

Keywords
Approximate modelling, time series analysis, dynamical systems, canonical forms, complexity, misfit, consistency.

1. INTRODUCTION

1.1. Modelling: specification and identification

The purpose of this paper is to describe a deterministic approach to time series analysis. This means that within the realm "from data to model", we will pay special attention to the case where the data consist of a sequence of observations over time and where the models consist of deterministic dynamical systems. Our approach to this particular modelling problem forms part of a more general modelling philosophy, which we will now describe.

Some of the essential factors which play a role in the problem of modelling data are depicted in figure 1. Two of the main aspects in approaching this problem are *specification* of the problem and, subsequently, *identification* of the model.

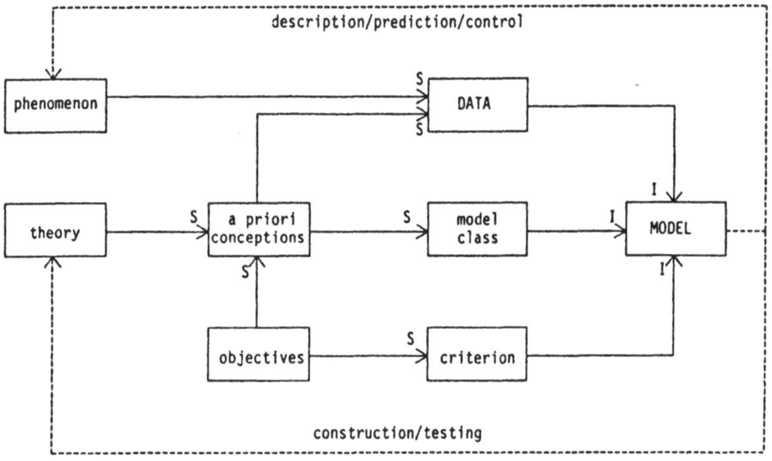

figure 1: modelling (S: specification; I: identification)

In general terms, the problem of modelling *data* consists of constructing a good *model* on the basis of these data. So the class of candidate models, i.e., the *model class*, has to be specified. Moreover, the quality of candidate models for modelling the data has to be assessed. This assessment, by means of a *criterion*, depends on the *objectives* underlying the modelling problem. An *identification procedure* describes the way a model is chosen (identified) from the model class, given the data. The aim is to construct the procedure in such a way that the identified models are of good quality with respect to the data, as measured by the criterion.

So in order to investigate the identification aspect of the data modelling problem it is necessary to specify the model class and the objectives. In modelling problems in general it is not known a priori which data will be included for identification of a model. This leads us to the specification aspect.

Often the primary objective of constructing a model is not only to model the data, but also to model a *phenomenon*. It then is supposed that the data somehow reflect the phenomenon. The phenomenon is then considered as a system which produces the data.

In the specification of the modelling problem one can incorporate prior knowledge concerning the phenomenon. This prior knowledge partly can be given by a *theory* concerning the phenomenon. Apart from this, one will impose restrictions partly based upon the objectives of modelling and partly for convenience. This leads to a collection of *a priori conceptions*, on the basis of which one decides which variables will be included in the model and what models will be considered. The identification problem is then specified.

Some of the main objectives of modelling are given in figure 1. On the one hand, an objective could be to model the phenomenon. One can think of description, prediction or control of the phenomenon. On the other hand, another objective could be to construct or test theories concerning the phenomenon.

It is beyond the scope of this paper to discuss fundamental problems of data, like the relationship between the phenomenon and the data and problems of data collection.

In the practice of modelling one often considers the specification aspect as part of the relevant scientific discipline and the identification aspect as a problem of constructing mathematical procedures. However, especially the choice of the model class also implies prior conceptions of a mathematical nature. The choice between deterministic and stochastic models forms a particular example.

We will illustrate the foregoing general description of the data modelling problem by means of five simple examples.

1.2. Example 1: a resistor

Suppose one wants to describe a resistor. On the basis of physical theory ("Ohm's law") one postulates a linear relationship between the voltage

(V) across and the electrical current (I) through the resistor, i.e., $V=I.R$ with $R\geq 0$ the resistance. A resistor is then described by a model R. So the model class is $\mathbb{R}_+$. To identify R, suppose one performs a number (n) of experiments with resulting voltage and current measurements $(\tilde{V}_i,\tilde{I}_i)$, $i=1,\ldots,n$. See figure 2.

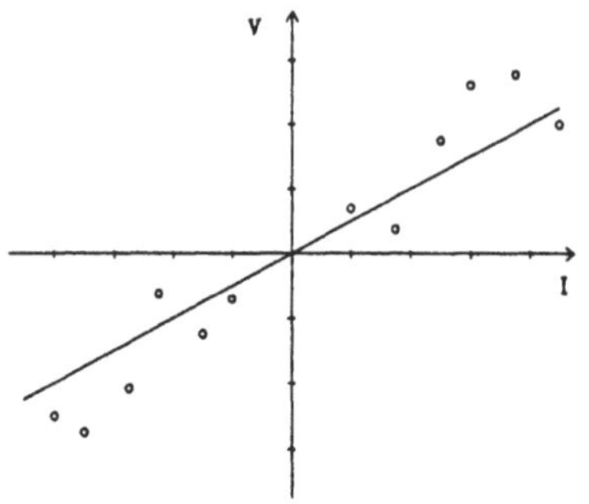

figure 2

The identification problem consists of choosing R on the basis of these data. In general there will exist no R such that $\tilde{V}_i=\tilde{I}_i.R$ for all $i=1,\ldots,n$. This can be due to inaccurate measurements and to the fact that the linear relationship is an idealization - though it may be an accurate one. A reasonable criterion could be total least squares.

So in this case, in order to describe the resistor, one uses physical theory to specify the model class and the data to be collected.

1.3. Example 2: eye colour

Suppose one wants to predict the colour of the eyes of a person. On the basis of biological theory (genetics) one postulates a specific probabilistic relationship between this colour and the colour of the eyes of the ancestors. Assume that the colour is either brown (1) or blue (0). As model class one could take $[0,1]$, where a particular model $p\in[0,1]$ means that p is the probability that the person has brown eyes. To identify p one collects data on the colour of the eyes of the parents, grandparents and so on. One then identifies p by means of elementary probabilistic calculations. See figure 3.

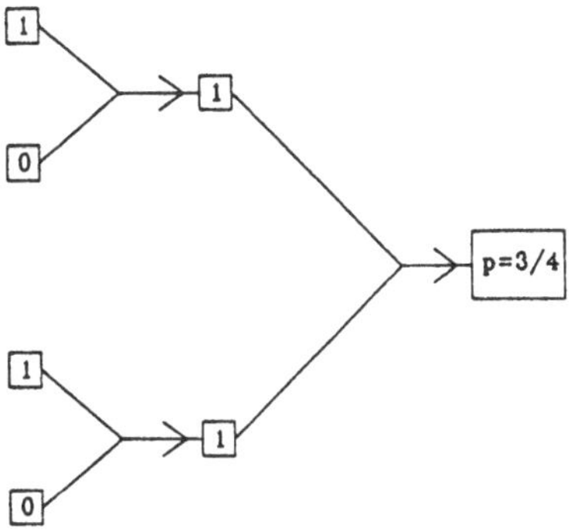

figure 3

One could now make a prediction for example by maximum likelihood, i.e., predict the colour to be brown if and only if $p > \frac{1}{2}$.

So in this case, in order to predict the eye colour, one uses biological theory to specify the identification and prediction problem.

1.4. Example 3: consumption

Suppose one wants to predict the consumption C_{t_0+1} for the coming year. On the basis of an economic theory one postulates that the dominant factor determining C_{t_0+1} is the income Y_{t_0} in the current year. Suppose data for consumption and income, $(\tilde{C}_t, \tilde{Y}_t)$, $t = s, s+1, \ldots, t_0$, are available. For convenience one could postulate an affine relationship between consumption in a year and income in the preceding year. The model class for example

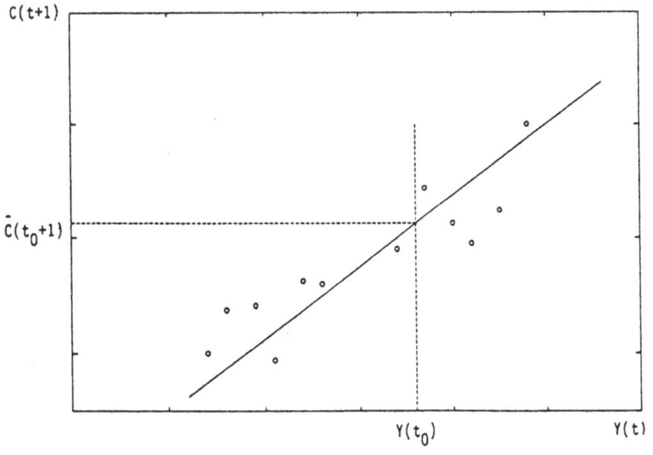

figure 4

could be $\mathbb{R}^2_+$, where the model (a,b) with $a,b \geq 0$ describes the postulated relationship $C_{t+1} = a + b.Y_t$. In order to identify a model one could use the data to estimate a and b for example by means of ordinary least squares. If the resulting estimates $\hat{a},\hat{b}$ indeed are nonnegative, one could predict C_{t_0+1} by means of $\hat{a}+\hat{b}.\tilde{Y}_{t_0}$. See figure 4.

So in this case, in order to predict consumption, one uses economic theory to specify the data. The choice of the model class is entirely a matter of convenience. If the estimated values $\hat{a},\hat{b}$ are not accepted as a reasonable description of consumptive behaviour one is ready to specify a different class of models, e.g., $C_{t+1} = \alpha + \beta.\log Y_t$.

1.5. Example 4: rainfall

Suppose one wants to control the water supply from a reservoir. The water of the reservoir is supplied to customers and replenished by rain. Suppose that one can construct a reasonable control strategy, once the rainfall is modelled.

If the climatological conditions are rather stable the rainfall could be viewed as a stationary stochastic process. As model class one could consider the class of Gaussian ARMA processes. Suppose that rainfall data $\{\tilde{r}(t);\ t_1 \leq t \leq t_2\}$ are available. To identify a model on the basis of these data one could consider the objective of simultaneous prediction of the rainfall for a number of periods in the future.

So in this case, in order to formulate the water supply problem in terms of only the rainfall, one has used prior knowledge of e.g. the demand pattern for water and of (stochastic) control theory. It is assumed that the rainfall can be modelled as a stationary stochastic process. This assumption is of a mathematical nature. It can be supported by arguing that the mechanism producing the rainfall is rather stable. This for example means that, although the rainfall is uncertain, some time averages of the rainfall are less uncertain.

1.6. Example 5: realization

Suppose one wants to interpolate n points $(x_i,y_i) \in \mathbb{R}^2$, $i = 1,\dots,n$, by means of a polynomial p of lowest possible degree. So the data consists of n points in $\mathbb{R}^2$ and the model class consists of polynomials. As a criterion to choose p one requires $y_i = p(x_i)$, $i = 1,\ldots,n$, and the degree of p has to be minimal.

So in this case the objective is to give an exact description of the data in a most simple way. This is an example of exact modelling or realization. The concepts of phenomenon or theory do not play a role in the specification of the modelling problem. The criterion is inspired by aesthetics or the desire to give a compact representation of the data.

1.7. Choice of model class

The foregoing examples especially are intended to illustrate the various considerations which can play a role in specifying the model class. In examples 1 and 2 well-established theories are used to choose the model class, one deterministic and the other probabilistic. In example 5 the choice is inspired by aesthetics. In examples 3 and 4 the choice of the model class reflects an aim of simplicity.

One of the crucial elements of the specification of modelling problems is the choice whether the model class should consist of *stochastic* or of *deterministic* models. In examples 1 and 2 the choice is based on a relevant scientific theory. In examples 3 and 4, like in the majority of modelling problems outside of the natural sciences, the choice is inspired by convenience. Moreover, the current practice seems to be to take the model to be stochastic. This implies that one introduces disturbances (noise) to explain the fact that in general the data do not satisfy simple, exact relationships. Moreover, it is nearly invariably assumed that the noise has a stable distribution over time, i.e., the disturbances form a stationary process.

This explanation of the discrepancy between the data and simple (deterministic) relations has two important implications. First, the model error is caused by disturbances of a stable nature, i.e., the relative frequency of the disturbance terms is assumed to be rather constant over time. Second, and based on this, the quality of proposed identification procedures is assessed on the basis of statistical criteria like unbiasedness, consistency and efficiency.

Clearly, this paradigm of stochastics often is a reasonable and convenient one. However, especially for complex phenomena, the fact that the data do not exactly satisfy simple deterministic relationships is often not due to disturbances or observation noise. Often the phenomenon simply is too complex to be modelled exactly within the model class. The models even deliberately are chosen to be simple. Both for human understanding and for practical implementation a simple, slightly inaccurate model of the phenomenon often is preferred above a complex, more accurate one. The central issue then is not noise or stochastics, but approximation.

1.8 Overview of the paper

To conclude the introduction we give an overview of the contents of the paper.

In section 2 we give a formal framework for approximate modelling, using the concepts of complexity and misfit. We illustrate this framework by some examples which play an important role in the sequel. In section 3 we describe the model class which we will consider in this paper, i.e., the class of deterministic dynamical systems. We will consider the objectives of description and prediction. Corresponding identification procedures are presented in section 6. These procedures solve an optimal approximate modelling problem, defined in terms of a utility of models. This utility depends on complexity and misfit measures, which are described in section 5. The complexity and misfit measures are expressed in terms of canonical representations of dynamical systems. These canonical forms reflect the objectives of description or prediction and are defined in section 4.

Section 7 describes the numerical algorithms corresponding to the modelling procedures of section 6. In section 8 we investigate some of the consistency properties of the procedures. The procedures have a clear optimality property as data modelling procedures. However, consistency analysis deals with the question whether the models identified by a procedure also are good models of the phenomenon. It is assumed that the phenomenon belongs to a certain class of systems, which does not need to coincide with the model class.

Section 9 contains some numerical simulations illustrating the deterministic approximate modelling procedures of section 6. Section 10 concludes the paper by summarizing the main results and indicating some topics of current research.

The main reference for the deterministic approach to approximate modelling as presented in this paper is Willems [15].

2. APPROXIMATE MODELLING

2.1 Complexity, misfit, utility

In the sequel of this paper we restrict attention to the identification aspect of the modelling problem. So we assume that one has specified the objectives of modelling, denoted by π, the model class, denoted by $\mathbb{M}$, and a set of conceivable data, denoted by D.

Definition 2-1 A data modelling *procedure* is a map $P:D\to 2^{\mathbb{M}}$.

In other words, a procedure associates with any data a set of models. Usually $P(d)$ will be a singleton, but it need not be.

The aim now is to construct procedures which are optimal in view of the objectives π. This means that for $d\in D$ the identified model(s) $P(d)$ should, within $\mathbb{M}$, reflect the data in a way which is optimal with respect to π.

A general objective is to construct models which are both simple and accurate. We will assume that the objectives π can be specified by a *complexity map* $c:\mathbb{M}\to C$ and a *misfit map* $\varepsilon:D\times\mathbb{M}\to E$. We assume the spaces C and E to be partially ordered. It is desirable to have models for which both the complexity and the misfit are small. However, these desires in general are competitive. We will therefore assume that π can be expressed by means of a *utility map* $u:C\times E\to U$, with U a partially ordered set. The aim then is to choose a model for which the complexity and misfit are such that the corresponding utility is maximal. For a partial ordering $\leq$ on U, $m\in U'\subset U$ is said to be a maximal element of U' if $\{u'\in U', m\leq u'\} \Rightarrow \{u'=m\}$.

Definition 2-2 The procedure P_u: $D\to 2^{\mathbb{M}}$ *corresponding* to the utility $u:C\times E\to U$ is defined by $P_u(d):=\operatorname{argmax}\{u(c(M),\varepsilon(d,M));\ M\in\mathbb{M}\}$ for $d\in D$.

So P_u assigns to data the set of models for which the utility is maximal. This clearly raises questions of existence and unicity of maximal elements.

In the remainder of this section we illustrate this approach by means of several examples. It will turn out that many classical identification procedures can be formalized in this context.

2.2. Exact modelling

In *exact modelling* one does not allow any misfit and wants to minimize the complexity. We consider three examples.

2.2.1. Synthesis problem

As a first example, consider a synthesis problem of electrical circuit theory. Suppose one wants to construct an electrical circuit with one external port with a prescribed current/voltage behaviour B. Here $B \subset (\mathbb{R}^2)^{\mathbb{R}}$ describes which current/voltage trajectories over time at the external port are compatible with the circuit. Moreover, suppose one wants to realize B by means of an RLC-network, i.e., only using resistors, inductors and capacitors. For an RLC-network with one external port, let $B(RLC)$ denote the current/voltage behaviour at the port and let $n(RLC)$ denote the total number of resistors, inductors and capacitors of the network.

The synthesis problem consists of finding an RLC-network with external behaviour B and such that $n(RLC)$ is as small as possible. So one allows no misfit and wants to minimize the complexity, measured by the number of constituent elements. This can be formulated in terms of a utility. Let $D = \mathbb{M}$ consist of the external current/voltage behaviours of RLC-networks with one external port. Define the complexity by $c(B(RLC)) := n(RLC)$ and the misfit by $\varepsilon(B,B') := +\infty$ if $B \neq B'$, $\varepsilon(B,B') := 0$ if $B = B'$. The synthesis problem then corresponds to the utility $u(n,\varepsilon) := -n-\varepsilon$.

2.2.2. Undominated unfalsified modelling

Let S be a set and let the set of conceivable data consist of finite tuples of observations in S, i.e., $D := \cup\{S^n; n \geq 1\}$. Let a model M consist of a subset $M \subset S$ and let $\mathbb{M} \subset 2^S$ denote a class of models.

A model M is called *unfalsified* by a measurement $d \in D$ if $d \subset M$. A model M is called *undominated* unfalsified in $\mathbb{M}$ for d if $d \subset M \in \mathbb{M}$ and $\{d \subset M' \in \mathbb{M}, M' \subset M\} \Rightarrow \{M' = M\}$. Define $P(d)$ as the collection of undominated unfalsified models in $\mathbb{M}$ for d. So P models d by models which are as small as possible in the sense of set inclusion. This could be expressed by means of the following utility. Let $\varepsilon(d,M) := 1$ if $d \not\subset M$, $\varepsilon(d,M) := 0$ if $d \subset M$ and define $c(M) := M$. Let $\underline{u} \notin \mathbb{M}$, $U := \mathbb{M} \cup \{\underline{u}\}$ and define the utility by $uu(M,1) := \underline{u}$ and $uu(M,0) := M$. Define a partial ordering $\leq$ on U as follows: $\underline{u} \leq M$ for all $M \in \mathbb{M}$ and for M_1, $M_2 \in \mathbb{M}$, $M_1 \leq M_2$ if and only if $M_1 \supset M_2$. Then P coincides with the procedure P_{uu} corresponding to the utility uu.

A special case of this arises if $S=(\mathbb{R}^q)^{\mathbb{Z}}$, so the data consists of a finite number of infinite time series in q real-valued variables. We will briefly return to this case in section 3.2. For a more thorough discussion we refer to Willems [16]. Here we only discuss a particular instance, known as the *minimal realization* problem.

In the minimal realization problem of linear systems theory the data set is $D=(\mathbb{R}^{p\times m})^{\mathbb{N}}$, where $\mathbb{N}:=\{1,2,3,\dots\}$. In this case the data $d\in D$ consists of an (impulse response) sequence $(G_k;\ k\in\mathbb{N})$ with $G_k\in\mathbb{R}^{p\times m}$, $k\in\mathbb{N}$. The model set consists of triples (A,B,C) with $A\in\mathbb{R}^{n\times n}$, $B\in\mathbb{R}^{n\times m}$, $C\in\mathbb{R}^{p\times n}$ for some $n\in\mathbb{N}$. The triple (A,B,C) is called a realization of $(G_k;\ k\in\mathbb{N})$ if $CA^{k-1}B=G_k$ for all $k\in\mathbb{N}$. It is called a minimal realization if n is as small as possible. For $d=(G_k;\ k\in\mathbb{N})$ and $M=(A,B,C)\in\mathbb{R}^{n\times n}\times\mathbb{R}^{n\times m}\times\mathbb{R}^{p\times n}$ define the misfit by $\varepsilon(d,M):=0$ if M is a realization of d and $\varepsilon(d,M):=1$ otherwise. Moreover define the complexity of M by $c(M):=n$. Let $U:=\{-1,-2,-3,\dots\}\cup\{-\infty\}$. Define a utility by $u(n,1):=-\infty$ and $u(n,0):=-n$ for $n\in\mathbb{N}$. The procedure corresponding to this utility solves the minimal realization problem. The number n has the interpretation of the dimension of the state space. In case a solution exists, it is unique up to a choice of a basis in the state space. See e.g. Kalman, Falb and Arbib [7].

2.2.3. Minimum description length principle

As a final example of exact modelling we mention the minimum description length principle of Rissanen, see e.g. Rissanen [14]. In this case the data set D consists of finite sequences of (finite precision) real numbers. The model class $\mathbb{M}$ consists of finite sequences of binary digits. A model represents data exactly by means of an injective code $C:D\to\mathbb{M}$. It is assumed that C codes the data d by means of an auxiliary (countable) class $\mathbb{P}=\{P_\theta;\ \theta\in\Theta\}$ of probability distributions on D, in the following way. The binary sequence $C(d)$ consists of an initial part describing the parameter θ and a remaining part describing the data in a way which is optimal in P_θ (minimum mean description length code for P_θ).

The complexity of a model is defined as the length of the binary sequence. Given the class $\mathbb{P}$, the minimum description length principle corresponds to the procedure which consists of coding the data by means of the shortest possible binary string, i.e., by the model of least complexity. This minimum description length principle balances the desire for a small number of parameters (in θ) and a simple description of the data by means of P_θ (maximal likelihood). It is interesting to note that this approach gives a deterministic interpretation, in terms of exact modelling, of e.g. maximum likelihood estimation and modelling by means of minimizing prediction errors.

2.3. Minimal complexity, given tolerated misfit

Suppose that the complexity space C and the misfit space E both are totally ordered. We denote the orderings by $\leq$. A possible reconciliation between the objectives of low complexity and of low misfit is to specify a *maximal tolerated misfit* and to minimize the complexity under this constraint. Given $\varepsilon_{tol} \in E$, we define the utility $u_{\varepsilon_{tol}}$ as follows. Let $\underline{u} \notin C \times E$ and $U := (C \times E) \cup \{\underline{u}\}$. For $\varepsilon \geq \varepsilon_{tol}$ let $u_{\varepsilon_{tol}}(c,\varepsilon) := \underline{u}$, and for $\varepsilon < \varepsilon_{tol}$ $u_{\varepsilon_{tol}}(c,\varepsilon) := (c,\varepsilon)$. On U we impose the following total ordering: $\underline{u} < (c,\varepsilon)$ for all $(c,\varepsilon) \in C \times E$, and $(c_1,\varepsilon_1) < (c_2,\varepsilon_2)$ if $c_1 > c_2$ or if $c_1 = c_2$ and $\varepsilon_1 > \varepsilon_2$. So misfits of ε_{tol} or higher are not allowed. Further, models of low complexity are preferred, and for models of equal complexity low misfit is preferred. The procedure $P_{\varepsilon_{tol}}$ now is defined as the procedure corresponding to $u_{\varepsilon_{tol}}$.

Definition 2-3 $P_{\varepsilon_{tol}}(d) := \text{argmax}\{u(c(M),\ \varepsilon(d,M));\ M \in \mathrm{M}\}$, where $\{u(c_1,\varepsilon_1) = u(c_2,\varepsilon_2)\} :\Leftrightarrow \{\varepsilon_1, \varepsilon_2 \geq \varepsilon_{tol}$ or $(c_1,\varepsilon_1) = (c_2,\varepsilon_2)\}$ and $\{u(c_1,\varepsilon_1) < u(c_2,\varepsilon_2)\} :\Leftrightarrow \{\varepsilon_1 \geq \varepsilon_{tol} > \varepsilon_2$, or $\varepsilon_1, \varepsilon_2 < \varepsilon_{tol}$, $c_1 > c_2$, or $\varepsilon_1, \varepsilon_2 < \varepsilon_{tol}$, $c_1 = c_2$, $\varepsilon_1 > \varepsilon_2\}$.

Two of the procedures described in section 6 are of this type. These procedures are based upon the ones which will be presented in sections 2.6 and 2.7.

The procedure corresponding to the requirement $\varepsilon \leq \varepsilon_{tol}$ (instead of $\varepsilon < \varepsilon_{tol}$) will be denoted by $\bar{P}_{\varepsilon_{tol}}$.

Here we illustrate the approach by a simple geometric example.

Let D consist of the bounded convex subsets of $\mathbb{R}^2$ and M of the convex polyhedral subsets of $\mathbb{R}^2$. For $M \in \mathrm{M}$ define the complexity $c(M)$ as the number of extremal points of M. For $C \in D$ and $M \in \mathrm{M}$ define the misfit $\varepsilon(C,M)$ as the Lebesgue measure of the symmetric difference $(C \backslash M) \cup (M \backslash C)$. Let ε_{tol} be given. Then $P_{\varepsilon_{tol}}$ models C by means of the convex hull of a minimal number of points under the misfit restriction, and chooses among solutions those with minimal misfit. See figure 5 for an illustration.

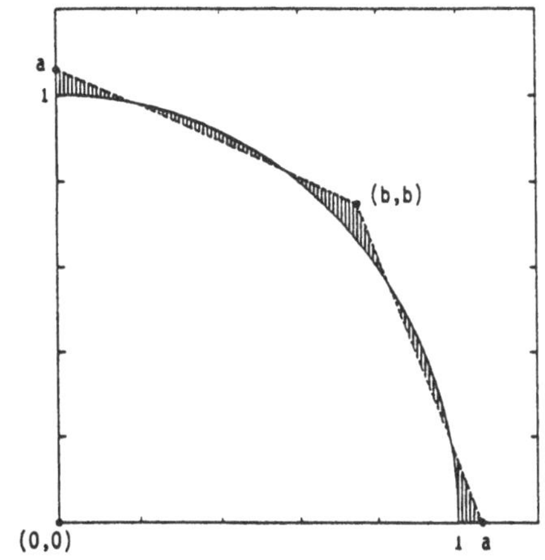

figure 5: $C=\{(x,y)\in\mathbb{R}^2;x^2+y^2\leq 1,\ x\geq 0,\ y\geq 0\}$, $\varepsilon_{tol}=0.05$; $P_{\varepsilon_{tol}}(C)$ is convex hull of $(0,0)$, $(0,a)$, $(a,0)$ and (b,b), with $a:=2(\alpha^2+1)^{1/2}/(4\alpha^2+1)^{1/2}$ and $b:=\alpha a/(1+\alpha)$, where $\alpha:=\tan(\frac{3}{8}\pi)$

Another example is speech processing. Let S denote the set of binary strings of finite length. The problem is to code, transmit and decode a signal $s\in S$ in the simplest way possible, given a tolerated misfit and an auxiliary class of models $\mathbb{M}_{aux}\subset S$. A coder is a map $f:S\to\mathbb{M}_{aux}\times S$ transforming a signal s into a transmitted signal $t\in S$. The signal t consists of an initial part describing the auxiliary model and a remaining part describing the signal s in an approximate way by means of the auxiliary model. A decoder is a map g: $\mathbb{M}_{aux}\times S\to S$ transforming a signal t into a decoded signal $\hat{s}$. See figure 6.

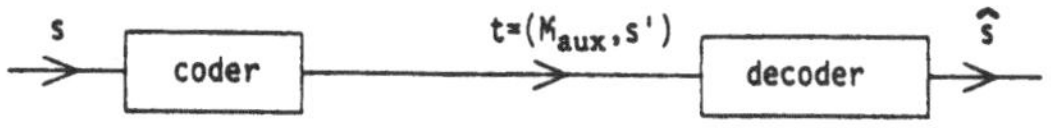

figure 6

For example, $\mathbb{M}_{aux}$ could be chosen to be the (set of parameters of the) class of autoregressive systems. The initial part of t then describes the order and the numerical values of the parameters of the auxiliary system. The remaining part of t could be used to describe the prediction errors of the estimates generated by the auxiliary system with respect to the signal

s. The decoder could construct a signal $\hat{s}$ based upon the estimates generated by the auxiliary system and the transmitted prediction errors. See e.g. Jayant and Noll [6].

Here the set of conceivable data is $D=S$ and the model class is $\mathbb{M}=\mathbb{M}_{aux}\times S$. Define the complexity of a model $t\in\mathbb{M}_{aux}\times S$ as the length of the string t. Let $\delta(s,\hat{s})$ denote a measure of the error of $\hat{s}$ with respect to s. Define the misfit of a model $t=(M_{aux},s')$ with respect to data s by $\varepsilon(s,(M_{aux},s')):=\ \delta(s,\hat{s})$ where $\hat{s}:=\ g(M_{aux},s')$. Given a tolerated misfit, one wants to minimize the complexity of the transmitted signal, i.e., of the model.

This approach resembles the minimum description length principle, though in speech processing it is not required that the data can be reconstructed exactly from the transmitted signal.

2.4. Minimal misfit, given tolerated complexity

Again suppose that C and E are totally ordered. Another possible reconciliation between the objectives of low complexity and of low misfit is to specify a *maximal tolerated complexity* and to minimize the misfit under this constraint. Given $c_{tol}\in C$, we define the utility $u_{c_{tol}}$ as follows. Let $\underline{u}\notin C\times E$ and $U:=\ (C\times E)\cup\{\underline{u}\}$. For $c>c_{tol}$ let $u_{c_{tol}}(c,\varepsilon):=\underline{u}$, and for $c\le c_{tol}$ define $u_{c_{tol}}(c,\varepsilon):=(c,\varepsilon)$. On U we impose the following total ordering: $\underline{u}<(c,\varepsilon)$ for all $(c,\varepsilon)\in C\times E$, and $(c_1,\varepsilon_1)<(c_2,\varepsilon_2)$ if $\varepsilon_1>\varepsilon_2$ or if $\varepsilon_1=\varepsilon_2$ and $c_1>c_2$. So a complexity above c_{tol} is not allowed. Further, models of low misfit are preferred, and for models of equal misfit low complexity is preferred. The procedure $P_{c_{tol}}$ now is defined as the procedure corresponding to $u_{c_{tol}}$.

Definition 2-4 $P_{c_{tol}}(d):=\operatorname{argmax}\{u(c(M),\varepsilon(d,M));\ M\in\mathbb{M}\}$, where $\{u(c_1,\varepsilon_1)=u(c_2,\varepsilon_2)\}:\Leftrightarrow\{c_1,c_2>c_{tol}$ or $(c_1,\varepsilon_1)=(c_2,\varepsilon_2)\}$ and $\{u(c_1,\varepsilon_1)<u(c_2,\varepsilon_2)\}:\Leftrightarrow\{c_1>c_{tol}\ge c_2,$ or $c_1,c_2\le c_{tol},\ \varepsilon_1>\varepsilon_2,$ or $c_1,c_2\le c_{tol},\ \varepsilon_1=\varepsilon_2,\ c_1>c_2\}$.

Again two of the procedures described in section 6 are of this type, along with procedures presented in sections 2.6 and 2.7.

Returning to the geometrical example of section 2.3, suppose c_{tol} is given. Then $P_{c_{tol}}$ models C by means of the convex hull of at most c_{tol} points in such a way that the resulting measure of the symmetric difference is minimal. Among solutions it chooses those with minimal number of extremal points. It can be shown that the last step in fact never will be invoked.

In the next section we give another example of modelling with given tolerated complexity.

2.5. Simultaneous equation models

We consider a modelling procedure which is sometimes followed in macro-econometrics and other disciplines dealing with complex dynamical phenomena. See e.g. Maddala [12].

Suppose one wants to describe the relationship between two groups of variables, one consisting of n_1 variables collected in $x \in \mathbb{R}^{n_1}$ and the other consisting of n_2 variables collected in $y \in \mathbb{R}^{n_2}$. For example, x could consist of the values of n_1 variables of interest at time t and y of values of these and possibly some other, auxiliary variables at times $s<t$.

Suppose one wants to use linear models. In general, no simple linear relationship will be exactly satisfied by the data. It is assumed that this misfit can be adequately modelled by means of a (Gaussian) disturbance term.

The model class of simultaneous equation models in this case can be parametrized by $\{(A,B,\Sigma);\ A \in \mathbb{R}^{n_1 \times n_1}$ nonsingular, $B \in \mathbb{R}^{n_1 \times n_2}, \Sigma \in \mathbb{R}^{n_1 \times n_1}, \Sigma = \Sigma^T \geq 0\}$. The parameter (A,B,Σ) corresponds to the model $Ax+By=\varepsilon$, where ε is a Gaussian random variable with mean zero and covariance matrix Σ.

Let data $\{(\tilde{x}_i,\tilde{y}_i);\ i=1,\dots,n\}$ be available. One possible approach to identify a model on the basis of these data, i.e., to estimate (A,B,Σ), is the following. Suppose the data are generated by a stochastic system $A_o x_i + B_o y_i = \varepsilon_i$, $i=1,\dots n$, where the ε_i are independent identically distributed zero mean Gaussian random variables with covariance matrix Σ_o. First estimate ($-A_o^{-1}B_o$, $A_o^{-1}\Sigma_o(A_o^{-1})^T$), e.g. by least squares (maximum likelihood). Denote the resulting estimates by $(\hat{\Pi},\hat{S})$. Impose restrictions on the parameter (A,B) in order to make the map $f:(A,B) \to -A^{-1}B$ injective. The injectivity of f is called identifiability in the literature. In this case the model could be estimated as $(\hat{A},\hat{B}) := f^{-1}(\hat{\Pi})$ and $\hat{\Sigma} := \hat{A}\hat{S}\hat{A}^T$.

We want to state some of the essential elements in this approach.

First, identifiability often is obtained by imposing prior restrictions on A and B, declaring certain elements of these matrices to be zero. The interpretation is that every equation corresponds to a part of the phenomenon which only incorporates certain variables. These zero restrictions are often inspired by theory. Imposing the restrictions resembles fixing the tolerated complexity, interpreted as the number of non-zero coefficients.

Second, it is not so much the least squares misfit as the variance of the estimated parameters which determines the confidence in the model. In a strict sense, every observation fits any model for which $\Sigma > 0$. However, inspection of the variability of the parameter estimates corresponds to some intuitive concept of misfit.

Finally, both the complexity and the "confidence" are defined in terms of parametrizations of models. In particular, every equation is investigated independent of the other ones. For example, declaring a parameter in a particular equation to be zero does not imply the absence of a direct relationship between the corresponding variables, as such a relationship can be due to the other equations.

In section 6 we decribe two modelling procedures for modelling dynamical phenomena which do not make use of stochastic assumptions. This in particular avoids the assumption of a stable distribution generating disturbances. Moreover, complexity and misfit measures are explicitly defined in terms of canonical parametrizations of dynamical models. These canonical forms are directly inspired by the objectives of modelling and do not depend on a theory concerning the phenomenon. The resulting measures have a clear interpretation in terms of model quality, as opposed to parameter quality. Moreover, the measures take the simultaneous nature of the model equations explicitly into account.

The procedures of section 6 for modelling dynamical phenomena make use of static modelling procedures. We will now describe these static procedures in sections 2.6 and 2.7.

2.6. Static descriptive modelling

Suppose we want to describe a finite number of points in $\mathbb{R}^n$ by means of a linear subspace. So D consists of the finite subsets of $\mathbb{R}^n$ and $\mathbb{M}$ consists of the linear subspaces of $\mathbb{R}^n$. A model M declares $x \in \mathbb{R}^n$ to be compatible with the phenomenon if and only if $x \in M$. As complexity we take $c^D : \mathbb{M} \to \{0, 1, \ldots, n\}$ defined as follows.

Definition 2-5 The *descriptive complexity* of a model $M \in \mathbb{M}$ is defined as its dimension, i.e., $c^D(M) := \dim(M)$.

So a simple model is one which excludes much.

Let $\mathbb{R}^n$ be equipped with e.g. the Euclidean inner product, denoted by $<\cdot,\cdot>$. To define a descriptive misfit, first consider models of codimension 1, i.e., there is $0 \neq a \in \mathbb{R}^n$ with $M = (\text{span}\{a\})^\perp$. Such a model claims the law $<x,a> = 0$ to hold true for the phenomenon. A measure of the quality of this law with respect to data $d = (\tilde{x}_1,\ldots,\tilde{x}_N) \in (\mathbb{R}^n)^N$ is $\varepsilon_1^D(d,M) := e^D(d,a)$, which is defined as follows.

Definition 2-6 For data $d = (\tilde{x}_1,\ldots,\tilde{x}_N) \in (\mathbb{R}^n)^N$ and $a \in \mathbb{R}^n$, the *descriptive misfit* of the law $<x,a> = 0$ with respect to d is defined as $e^D(d,a) := \{ \frac{1}{N} \sum_{i=1}^{N} <\tilde{x}_i,a>^2 / \|a\|^2 \}^{1/2}$.

If $\text{codim}(M) > 1$, then $\varepsilon_1^D(d,M)$ is defined as the descriptive misfit of the worst law claimed by M, i.e., $\varepsilon_1^D(d,M) := \max\{\varepsilon_1^D(d,M');\ M \subset M',\ \text{codim}(M') = 1\}$. Note that the model M claims that $\tilde{x}_i \in M$, so in particular $\tilde{x}_i \in M'$ for $M' \supset M$, $i = 1,\ldots,n$.

Definition 2-7 For $d \in (\mathbb{R}^n)^N$, $M \in \mathbb{M}$, the *first descriptive misfit* is $\varepsilon_1^D(d,M) := \max\{e^D(d,a);\ 0 \neq a \in M^\perp\}$.

Note that M claims that $<\tilde{x}_i,a> = 0$ for all $i = 1,\ldots,n$, $a \in M^\perp$. The *second* descriptive misfit is defined as the worst-but-one claimed law, i.e., if $\varepsilon_1^D(d,M) = e^D(d,a_1)$, $a_1 \in M^\perp$, then $\varepsilon_2^D(d,M) := \max\{e^D(d,a);\ 0 \neq a \in M^\perp \cap (\text{span}\{a_1\})^\perp\}$. So $\varepsilon_2^D(d,M)$ measures the quality of the laws claimed by M and orthogonal to the worst law a_1. For $k = 3,\ldots,n-c(M)$ the k-th descriptive misfit is inductively defined as follows: if for $j<k$ $\varepsilon_j^D(d,M) = e^D(d,a_j)$, $a_j \in M^\perp \cap (\text{span}\{a_1,\ldots,a_{j-1}\})^\perp$, then $\varepsilon_k^D(d,M) := \max\{e^D(d,a);\ 0 \neq a \in M^\perp \cap (\text{span}\{a_1,\ldots,a_{k-1}\})^\perp\}$. It can be shown that $\varepsilon_k^D(d,M)$ is well-defined this way, even if the a_j are not unique. For $k = n-c(M)+1,\ldots,n$ we define $\varepsilon_k^D(d,M) := 0$. In this way the misfit is a map ε^D: $D \times \mathbb{M} \to \mathbb{R}_+^n$.

On the complexity space $\{0,1,\ldots,n\}$ we take the natural ordering, as well as on $\mathbb{R}_+$. The misfit space $\mathbb{R}_+^n$ we order lexicographically, i.e., $(\varepsilon_1,\ldots,\varepsilon_n) \geq (\tilde{\varepsilon}_1,\ldots,\tilde{\varepsilon}_n)$ if and only if $\varepsilon_k = \tilde{\varepsilon}_k$ for all $k = 1,\ldots,n$ or if there is a k such that $\varepsilon_i = \tilde{\varepsilon}_i$ for $i<k$ and $\varepsilon_k > \tilde{\varepsilon}_k$.

We remark that complexity and misfit are defined on the level of models, not on the parameter level.

In the next propositions we give explicit algorithms for the procedures $P^D_{\varepsilon_{tol}}$ corresponding to minimizing complexity, given a tolerated misfit, and $P^D_{c_{tol}}$ corresponding to minimizing misfit, given a tolerated complexity, as described in sections 2.3 and 2.4 respectively.

For data $d=(\tilde{x}_1,\ldots,\tilde{x}_N)$ let $\frac{1}{N}\sum_{i=1}^{N}\tilde{x}_i\tilde{x}_i^T$ have singular value decomposition (S.V.D.) $\frac{1}{N}\sum_{i=1}^{N}\tilde{x}_i\tilde{x}_i^T=U\Sigma U^T$. Here U is orthogonal, i.e., $UU^T=U^TU=I_n$, the identity matrix in $\mathbb{R}^{n\times n}$. Σ is diagonal, $\Sigma=\text{diag}(\sigma_1,\ldots,\sigma_n)$ with $\sigma_1\geq\ldots\geq\sigma_n\geq 0$. Let $r:=\text{rank}(\frac{1}{N}\sum_{i=1}^{N}\tilde{x}_i\tilde{x}_i^T)$, then $\sigma_{r+1}=\ldots=\sigma_n=0$. Let u_j denote the j–th column of U. Define $M_k^*:=\text{span}\{u_1,\ldots,u_k\}$ and $M(\sigma):=\text{span}\{u_j;\ \sigma_j=\sigma\}$.

Proposition 2-8 For given data $d=(\tilde{x}_1,\ldots\tilde{x}_N)\in(\mathbb{R}^n)^N$ and tolerated complexity c_{tol}, $P^D_{c_{tol}}(d)$ is given by

(*i*) $P^D_{c_{tol}}(d)=\{0\}$ if $c_{tol}=0$;

(*ii*) $P^D_{c_{tol}}(d)=\text{span}\{\tilde{x}_1,\ldots,\tilde{x}_N\}$ if $c_{tol}\geq r$;

(*iii*) $P^D_{c_{tol}}(d)=M^*_{c_{tol}}$ if $0<c_{tol}<r$ and $\sigma_{c_{tol}}>\sigma_{c_{tol}+1}$;

(*iv*) if $\sigma_1\geq\ldots\geq\sigma_{c_1}>\sigma_{c_1+1}=\ldots=\sigma_{c_{tol}}=\sigma_{c_{tol}+1}\geq\sigma_{c_{tol}+2}\geq\ldots\geq\sigma_n$ then
$P^D_{c_{tol}}(d)=\{M^*_{c_1}+L;\ L\subset M(\sigma_{c_{tol}}),\ \dim(L)=c_{tol}-c_1\}$.

Proposition 2-9 Let data $d=(\tilde{x}_1,\ldots,\tilde{x}_N)\in(\mathbb{R}^n)^N$ be given. Assume moreover that a maximal misfit level is given with $\varepsilon_{tol}=\varepsilon_1^{tol}.(1,\ldots,1)$, so the misfit restriction concerns only the worst law claimed by a model. Then

(*i*) $P^D_{\varepsilon_{tol}}(d)=\{0\}$ if $\varepsilon_1^{tol}>\sigma_1$;

(*ii*) $P^D_{\varepsilon_{tol}}(d)=\text{span}\{\tilde{x}_1,\ldots,\tilde{x}_N\}$ if $\varepsilon_1^{tol}\leq\sigma_r$;

(*iii*) if $\sigma_r<\varepsilon_1^{tol}\leq\sigma_1$, then $P^D_{\varepsilon_{tol}}(d)=M_k^*$ with k such that $\sigma_k\geq\varepsilon_1^{tol}>\sigma_{k+1}$.

We also refer to Willems [15].

We finally remark that there is a close relationship between these procedures and total least squares. See e.g. Golub and Van Loan [1].Consider as a simple example the case $c_{tol}=n-1$. For $0\neq a\in\mathbb{R}^n$ let $M(a):=(\text{span}\{a\})^{\perp}:=\{x\in\mathbb{R}^n;\ \langle x,a\rangle=0\}$ and let π_a denote the orthogonal projection operator onto $M(a)$. For given data $d=(\tilde{x}_1,\ldots,\tilde{x}_N)\in(\mathbb{R}^n)^N$, in total least squares one determines a such that $\delta(d,a) := \frac{1}{N}\sum_{i=1}^{N} \|\tilde{x}_i - \pi_a \tilde{x}_i\|^2$ is minimal. See figure 7 for the case $n=2$.

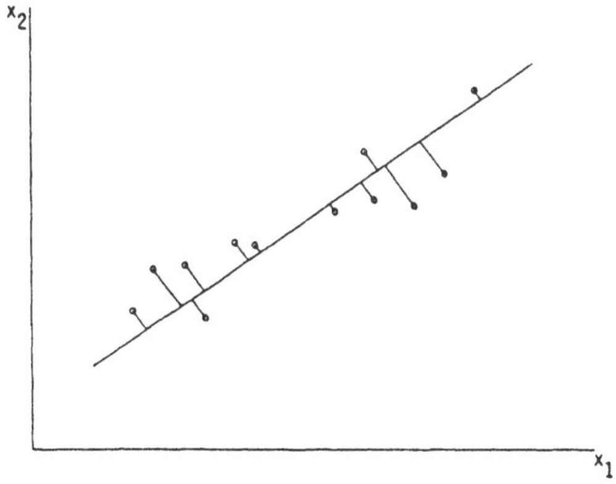

figure 7

It is easily shown that $\delta(d,a) = \{a^T(\frac{1}{N}\sum_{i=1}^{N} \tilde{x}_i\tilde{x}_i^T)a\}/\|a\|^2 = \{\varepsilon_1^D(d,M(a))\}^2$. So in this case of $c_{tol}=n-1$ the procedure $P^D_{c_{tol}}$ corresponds exactly to total least squares. Analogous results can be obtained for $c_{tol}<n-1$ and for $P^D_{\varepsilon_{tol}}$.

2.7. Static predictive modelling

Suppose we want to predict (or estimate) n_2 variables $y \in \mathbb{R}^{n_2}$ on the basis of n_1 other variables $x \in \mathbb{R}^{n_1}$ by means of a linear subspace of $\mathbb{R}^{n_1+n_2}$.

Let N observations $(\tilde{x}_i, \tilde{y}_i)$, $\tilde{x}_i \in \mathbb{R}^{n_1}$, $\tilde{y}_i \in \mathbb{R}^{n_2}$, $i=1,\ldots,N$ be available, so the data set is $D=(\mathbb{R}^{n_1+n_2})^N$.

Let M be a linear subspace of $\mathbb{R}^{n_1+n_2}$. The model M has the interpretation that, given x, it is predicted that y will belong to the set $M(x) := \{y \in \mathbb{R}^{n_2};\ (x,y) \in M\}$. Stated otherwise, let $x \in \mathbb{R}^{n_1}$ be observed. The model M amounts to predicting that the with x associated, but unobserved, y will be such that $<a_1,x> + <a_2,y> = 0$ for all $(a_1,a_2) \in M^{\perp}$, $a_1 \in \mathbb{R}^{n_1}$, $a_2 \in \mathbb{R}^{n_2}$. As model class $\mathbb{M}$ we will take the class of those linear subspaces M of $\mathbb{R}^{n_1+n_2}$ for which the projection on the x coordinate is surjective, i.e., $\{x;\ \exists y$ such that $(x,y) \in M\} = \mathbb{R}^{n_1}$. This means that prediction is possible for every $x \in \mathbb{R}^{n_1}$.

It is easily seen that $M(x) = y + M(0)$ for any $x \in \mathbb{R}^{n_1}$, $y \in M(x)$. So for given model $M \in \mathbb{M}$, the dimension of the (affine) predicted set is independent of the observation x. We define the predictive complexity $c^P: \mathbb{M} \to \{0,1,\ldots,n_2\}$ as follows.

Definition 2-10 The *predictive complexity* of a model $M \in \mathbb{M}$ is defined as the dimension of the affine predicted set, i.e., $c^P(M) := \dim(M(0))$.

So a simple model corresponds to predictions with few degrees of freedom.

To define a predictive misfit we again consider first models of codimension 1. Let $0 \neq a = (a_1, a_2) \in \mathbb{R}^{n_1} \times \mathbb{R}^{n_2}$ and $M = (\text{span}\{a\})^{\perp}$. Note that $M \in \mathbb{M}$ implies $a_2 \neq 0$. The model M predicts that, given x, y will satisfy $<a_2, y> = -<a_1, x>$. For data $d = \{(\tilde{x}_i, \tilde{y}_i);\ i = 1, \ldots, N\}$ the relative mean prediction error of this model is $\varepsilon_1^P(d, M) := e^P(d, a)$, which is defined as follows.

Definition 2-11 For data $d = \{(\tilde{x}_i, \tilde{y}_i);\ i = 1, \ldots, N\} \in (\mathbb{R}^{n_1} \times \mathbb{R}^{n_2})^N$ and $a = (a_1, a_2) \in \mathbb{R}^{n_1} \times \mathbb{R}^{n_2}$ with $a_2 \neq 0$, the *relative mean prediction error* is defined by $e^P(d, a) := [\{\frac{1}{N} \sum_{i=1}^{N} (<a_1, \tilde{x}_i> + <a_2, \tilde{y}_i>)^2\} / \{\frac{1}{N} \sum_{i=1}^{N} <a_2, \tilde{y}_i>^2\}]^{1/2}$.

If $\text{codim}(M) > 1$, then $\varepsilon^P(d, M)$ is defined in analogy with the misfit in section 2.6, i.e., $\varepsilon_1^P(d, M)$ measures the predictive misfit of the worst prediction made by M, $\varepsilon_2^P(d, M)$ the misfit of the prediction worst-but-one, and so on.

Formally, let $M_2^{\perp} := \{a_2; \exists a_1 \text{ such that } (a_1, a_2) \in M^{\perp}\}$, so $M_2^{\perp}$ consists of the space of predicted functionals on y. There holds $\dim(M_2^{\perp}) = n_2 - c(M)$. For $k = 1, \ldots, \dim(M_2^{\perp})$ we define $\varepsilon_k^P(d, M)$ inductively as follows.

Definition 2-12 For $d \in (\mathbb{R}^{n_1} \times \mathbb{R}^{n_2})^N$, $M \in \mathbb{M}$, the *first predictive misfit* is $\varepsilon_1^P(d, M) := \max\{e^P(d, a);\ a \in M^{\perp}\}$.

Further, if for $j = 1, \ldots, k-1$ $\varepsilon_j^P(d, M) = e^P(d, a^{(j)})$, $a_2^{(j)} \in M_2^{\perp} \cap (\text{span}\{a_2^{(1)}, \ldots, a_2^{(j-1)}\})^{\perp}$, then $\varepsilon_k^P(d, M) := \max\{e^P(d, a);\ a_2 \in M_2^{\perp} \cap (\text{span}\{a_2^{(1)}, \ldots, a_2^{(k-1)}\})^{\perp}\}$. For $k = \dim(M_2^{\perp}) + 1, \ldots, n_2$ we define $\varepsilon_k^P(d, M) := 0$. In this way the misfit $\varepsilon^P : D \times \mathbb{M} \to \mathbb{R}_+^{n_2}$ is well-defined, provided $N \geq n_2$ and provided that the data are generic in the sense that $\text{span}\{\tilde{y}_1, \ldots, \tilde{y}_N\} = \mathbb{R}^{n_2}$.

We order the complexity and misfit spaces as in section 2.6, i.e., naturally and lexicographically respectively.

Note that again complexity and misfit are defined on the level of models, not on the parameter level.

Next we will give explicit algorithms for the procedures $P^P_{\varepsilon_{tol}}$ corresponding to minimizing complexity, given a tolerated misfit, and $P^P_{c_{tol}}$ corresponding to minimizing predictive misfit, given a tolerated complexity.

Let the data be $d=\{(\tilde{x}_i,\tilde{y}_i);\ i=1,\ldots,N\}$. Suppose that $N\geq\max\{n_1,n_2\}$ and that the data are generic in the sense that $\text{span}\{\tilde{x}_1,\ldots,\tilde{x}_N\}=\mathbb{R}^{n_1}$ and $\text{span}\{\tilde{y}_1,\ldots,\tilde{y}_N\}=\mathbb{R}^{n_2}$. Let $\begin{pmatrix} S_{xx} & S_{xy} \\ S_{yx} & S_{yy}\end{pmatrix} := \frac{1}{N}\sum_{i=1}^{N}\begin{pmatrix}\tilde{x}_i\\ \tilde{y}_i\end{pmatrix}\begin{pmatrix}\tilde{x}_i\\ \tilde{y}_i\end{pmatrix}^T \in\mathbb{R}^{(n_1+n_2)\times(n_1+n_2)}$ and let $S_{xx}^{-1/2}S_{xy}S_{yy}^{-1/2}$ have S.V.D. $U\Lambda V^T$, with $U\in\mathbb{R}^{n_1\times n_1}$ and $V\in\mathbb{R}^{n_2\times n_2}$ both orthogonal matrices and $\Lambda=\begin{pmatrix}\Sigma & 0\\ 0 & 0\end{pmatrix}\in\mathbb{R}^{n_1\times n_2}$, $\Sigma=\text{diag}(\sigma_1,\ldots,\sigma_r)$, $\sigma_1\geq\ldots\geq\sigma_r>0$. There holds $\sigma_1\leq 1$ and $r=\text{rank}(S_{xy})$. Let r^* denote the number of singular values equal to 1. Denote the columns of $S_{xx}^{-1/2}U$ by $a_1^{(i)}$, $i=1,\ldots,n_1$, and those of $S_{yy}^{-1/2}V$ by $a_2^{(i)}$, $i=1,\ldots,n_2$. For $k=1,\ldots,r$ define $M_k^*:=\{(x,y);\ a_2^{(i)}y=\sigma_i\cdot a_1^{(i)}x,\ i=1,\ldots,k\}$. Then $c(M_k^*)=n_2-k$ and $\varepsilon^P(d,M_k^*)=((1-\sigma_k^2)^{1/2},\ldots,(1-\sigma_1^2)^{1/2},0,\ldots,0)$. Finally, let $M(\sigma):=\{(x,y);\ a_2^{(i)}y=\sigma a_1^{(i)}x$ for all i with $\sigma_i=\sigma\}$.

Proposition 2-13 For generic data $d=\{(\tilde{x}_i,\tilde{y}_i);\ i=1,\ldots,N\}$ and tolerated complexity c_{tol}, $P^P_{c_{tol}}$ is given by

(*i*) $P^P_{c_{tol}}(d)=\{M\in\mathbb{M};\ M\subset M_r^*,\ \dim(M_2^\perp)=n_2-c_{tol}\}$ if $c_{tol}<n_2-r$;

(*ii*) $P^P_{c_{tol}}(d)=M^*_{r^*}$ if $c_{tol}\geq n_2-r^*$;

(*iii*) $P^P_{c_{tol}}(d)=M^*_{n_2-c_{tol}}$ if $r^*<n_2-c_{tol}\leq r$ and $\sigma_{n_2-c_{tol}}>\sigma_{n_2-c_{tol}+1}$;

(*iv*) if $\sigma_1\geq\ldots\geq\sigma_{c_1}>\sigma_{c_1+1}=\ldots=\sigma_{n_2-c_{tol}}=\sigma_{n_2-c_{tol}+1}=\ldots=\sigma_{c_2}>\sigma_{c_2+1}\geq\ldots\geq\sigma_r>0$, then $P^P_{c_{tol}}(d)=\{M^*_{c_1}\cap L;\ L\supset M(\sigma_{n_2-c_{tol}+1}),\ c(L)=c_{tol}+c_1\}$.

Proposition 2-14 Let data $d=\{(\tilde{x}_i,\tilde{y}_i);\ i=1,\ldots,N\}$ be generic. Assume moreover that a maximal misfit level is given with $\varepsilon_{tol}=\varepsilon_1^{tol}.(1,\ldots,1)$, so the misfit restriction concerns only the worst prediction made by a model. Then

(*i*) $P^P_{\varepsilon_{tol}}(d)=M^*_{n_2}$ if $\varepsilon_1^{tol}>(1-\sigma^2_{n_2})^{1/2}$;

(*ii*) $P^P_{\varepsilon_{tol}}(d)=\mathbb{R}^{n_1+n_2}$ if $\varepsilon_1^{tol}\leq(1-\sigma_1^2)^{1/2}$;

(*iii*) $P^P_{\varepsilon_{tol}}(d)=M_r^*$ if $r<n_2$ and $(1-\sigma_r^2)^{1/2}<\varepsilon_1^{tol}\leq 1$;

(*iv*) if $(1-\sigma_1^2)^{1/2}<\varepsilon_1^{tol}\leq(1-\sigma_r^2)^{1/2}$, then $P^P_{\varepsilon_{tol}}(d)=M_k^*$ where k is such that $(1-\sigma_k^2)^{1/2}<\varepsilon_1^{tol}\leq(1-\sigma_{k+1}^2)^{1/2}$.

We also refer to Heij [4].

We remark that for $n_2 = 1$ and $c_{tol} = 0$ the procedure $P^P_{c_{tol}}$ reduces to ordinary least squares fitting. See figure 8.

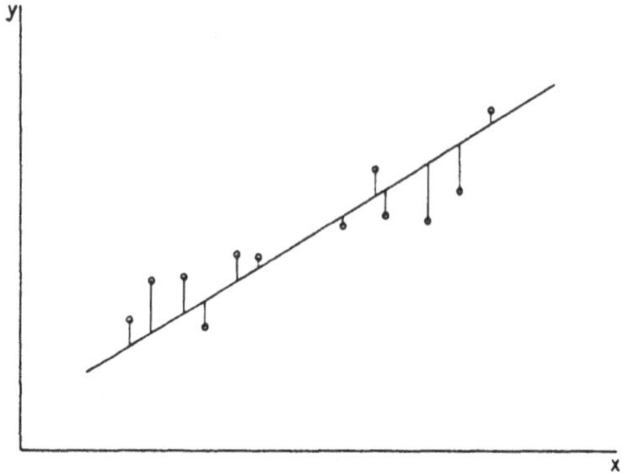

figure 8

The special (vertical) way of measuring the error in this case reflects the purpose of predicting y on the basis of x.

This concludes our section on approximate modelling. The procedures for static modelling in sections 2.6 and 2.7 are used for approximate modelling of time series by means of dynamical models in section 6. In order to do this, we introduce the concept of a dynamical system and a class of dynamical models in section 3. We define complexity and misfit in section 5 in terms of canonical parametrizations of these models. These canonical forms are described in section 4.

3. DYNAMICAL SYSTEMS

3.1. Definition of a dynamical system

Definition 3-1 A *dynamical system* is a triple (T, W, B) with $T \subset \mathbb{R}$ the time set, W the signal set and $B \subset W^T$ the behaviour of the system.

The behaviour B we will sometimes call a system or a model.

A dynamical system describes the relationships between variables of interest in the following way. Let W be the set in which the variables on every time instant take their values, and let T denote the time set under consideration. The behaviour B then consists of a set of time series $w:T\to W$ with the interpretation that time series $w\in B$ are compatible with the laws of the system, while time series $w\notin B$ are not compatible with these laws. This gives a deterministic description of the system.

For some illustrative examples we refer to Willems [15], [16].

3.2. AR-systems

In the sequel we will restrict attention to a special class of dynamical systems, namely those describable by a finite number of autoregressive equations. We will invariably consider discrete time systems with $T=\mathbb{Z}$ and with signal set $W=\mathbb{R}^q$. So there are q variables of interest which take on real values.

We will use the following notation. Let $R_k\in\mathbb{R}^{g\times q}$ for $k=d_1,d_1+1,\ldots,d_2$, where $d_1,d_2\in\mathbb{Z}$, $d_1\leq d_2$. Define $R\in\mathbb{R}^{g\times q}[s,s^{-1}]$ by $R(s,s^{-1}):=\sum_{k=d_1}^{d_2} R_k s^k$, so R is a finite Laurent series in s with coefficients in $\mathbb{R}^{g\times q}$. By a slight abuse of language we will call R a polynomial matrix in s and s^{-1}. By σ we denote left shift, i.e., if $w:\mathbb{Z}\to\mathbb{R}^q$ then $\sigma w:\mathbb{Z}\to\mathbb{R}^q$ is defined by $(\sigma w)(t):=w(t+1)$, $t\in\mathbb{Z}$. By σ^{-1} we denote the inverse of σ. The autoregressive system $B(R)$ then is defined as $\ker(R(\sigma,\sigma^{-1}))$, i.e., $B(R)$ is the set of those time series $w:\mathbb{Z}\to\mathbb{R}^q$ for which $R(\sigma,\sigma^{-1})w=0$, i.e., $\sum_{k=d_1}^{d_2} R_k w(t+k)=0$ for all $t\in\mathbb{Z}$.

Definition 3-2 Let $R\in\mathbb{R}^{g\times q}[s,s^{-1}]$. Then the *autoregressive system* (*AR-system*) $B(R)$ is defined by $B(R):=\{w\in(\mathbb{R}^q)^{\mathbb{Z}};\ R(\sigma,\sigma^{-1})w=0\}$.

We will denote the class of all AR-systems by $\mathbb{B}$, i.e., $\mathbb{B}:=\{B\subset(\mathbb{R}^q)^{\mathbb{Z}};\ \exists g\ \exists R\in\mathbb{R}^{g\times q}[s,s^{-1}]$ such that $B=B(R)\}$.

This class of systems is interesting for a number of reasons. First, it forms a class of models often used in practical modelling situations where one wants to describe linear relationships between the variables and their lagged values, as e.g. in econometrics, signal processing and linear control. Second, this class of systems includes some widely used systems as, for example, linear input/output systems with finite dimensional state space. Third, there exists a nice interpretation of AR-systems on the behavioural level of sets of time series, which we will now describe.

It can be shown that a system $B\subset(\mathbb{R}^q)^{\mathbb{Z}}$ is an AR-system, i.e., there is a polynomial matrix R such that $B = B(R)$, if and only is B is a *linear*, *time invariant*, *complete* system. B is called *linear* if it is a linear subspace of $(\mathbb{R}^q)^{\mathbb{Z}}$. It is called *time invariant* if $\sigma B = B$, i.e., shifted time series of the system also satisfy the laws of the system. This means that the laws of the system are time invariant. B is called *complete* if $\{w \in B\} \Leftrightarrow \{w|_{[t_0,t_1]} \in B|_{[t_0,t_1]}$ for all $-\infty < t_0 \leq t_1 < +\infty\}$. This means that in order to check whether a time series $w \in (\mathbb{R}^q)^{\mathbb{Z}}$ belongs to B or not it suffices to consider only windows $[t_0, t_1]$ of arbitrary finite length. Moreover it can be shown that if B is linear and time invariant, then B is complete if and only if there exists a $\Delta \geq 0$ such that $\{w \in B\} \Leftrightarrow \{w|_{[t,t+\Delta]} \in B|_{[0,\Delta]}$ for all $t \in \mathbb{Z}\}$. So in this case the laws which are imposed by B are local in time.

We finally mention that the class of AR-systems exactly consists of those subsets $B \subset (\mathbb{R}^q)^{\mathbb{Z}}$ which are linear, shift invariant and closed in the topology of pointwise convergence in $(\mathbb{R}^q)^{\mathbb{Z}}$. We will illustrate the use of this characterization by briefly returning to section 2.2.2 on undominated unfalsified modelling. Let $D = (\mathbb{R}^q)^{\mathbb{Z}}$, so the data consists of an infinite time series, and let $\mathbb{M} = \mathbb{B}$, so the model class consists of the AR-systems. The property of closedness of AR-systems implies that for every $\tilde{w} \in D$ there exists a unique $B^*(\tilde{w}) \in \mathbb{B}$ such that $\tilde{w} \in B^*(\tilde{w})$ and $\{ \tilde{w} \in B \in \mathbb{B} \} \Rightarrow \{ B^*(\tilde{w}) \subset B \}$. The procedure P_{uu} corresponding to undominated unfalsified modelling hence models $\tilde{w}$ by means of $B^*(\tilde{w})$. It is called the most powerful unfalsified model. In the sequel we will not consider exact modelling of an infinite time series, but approximate modelling of a finite time series.

3.3. Modelling a time series

Suppose we want to model a dynamical phenomenon. In terms of figure 1 in section 1.1, we assume that the objective is either description or prediction of the phenomenon. So we do not discuss control problems or objectives corresponding to theories concerning the phenomenon. Moreover, it is supposed that it is reasonable to model the phenomenon by means of a system which is linear, time invariant and complete. The interpretation is that the model gives a description of the phenomenon which is local, both in space (linearity) and in time (time invariance and completeness). The model class hence is $\mathbb{B}$. It is assumed that q real-valued variables have been specified which have to be included in the model and that data on these variables is available in the form of a finite time series. We denote the variables by $w := (w_1, \ldots, w_q)^T$, the time interval of observation by

$\mathcal{T}:= [t_0,t_1]$ for some $-\infty < t_0 \le t_1 < +\infty$, and the data by $\tilde{w}:= (\ \tilde{w}(t);\ t \in \mathcal{T}\ )$, an ordered sequence of observations. It is assumed that the data are directly related to the variables of interest and that there are no "missing observations".

In this case the data set is $D = \cup\{(\mathbb{R}^q)^n; n \in \mathbb{N}\}$, so the data consists of a time series of length n in $\mathbb{R}^q$. The model class is $\mathbb{M} = \mathbb{B} := \{B \subset (\mathbb{R}^q)^{\mathbb{Z}};\ B$ linear, time invariant, complete$\}$. The objective π is description or prediction. The modelling problem consists of choosing a procedure $P_\pi : D \to 2^{\mathbb{B}}$, corresponding to a utility u_π reflecting the purpose π of modelling. We will follow the approximate modelling approach described in section 2.1. Therefore we will define complexity maps $c_\pi : \mathbb{B} \to C_\pi$ and misfit maps ε_π: $D \times \mathbb{B} \to E_\pi$ and impose orderings on C_π and E_π. The resulting identification problem is depicted in figure 9.

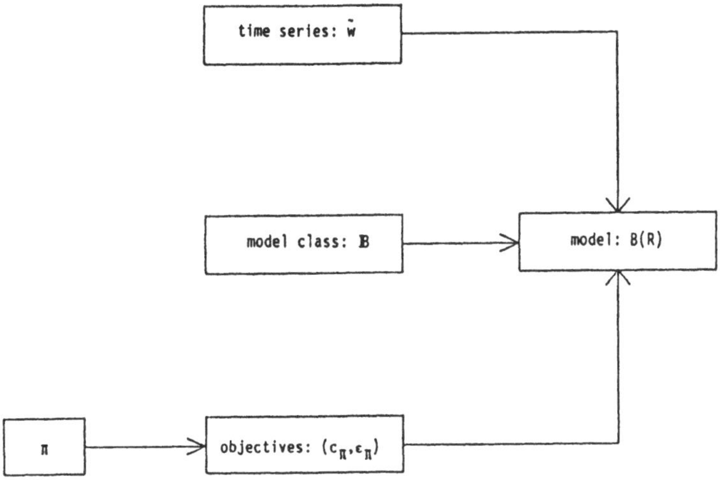

figure 9: modelling a time series

In order to implement procedures algorithmically it is desirable to express the utility not only in terms of the sets $B \subset (\mathbb{R}^q)^{\mathbb{Z}}$ but also in terms of a finite number of parameters parametrizing B, i.e., in terms of an AR-representation R such that $B = B(R)$. However, defining a utility in terms of R need not automatically be compatible with a utility in terms of B, as the map $f\colon \cup \{ \mathbb{R}^{g \times q}[s,s^{-1}];\ g \in \mathbb{N} \} \to \mathbb{B}$ with $f(R) := B(R)$ is not injective. The representation of B by means of R such that $B = B(R)$ is highly non-unique.

In section 4 we will describe the nature of the equivalence relation $\sim$ defined on $\cup\{\mathbb{R}^{g\times q}[s,s^{-1}];\ g\in\mathbb{N}\}$ by $\{R_1\sim R_2\}:\ \Leftrightarrow\ \{B(R_1)=B(R_2)\}$. Moreover we will define two canonical forms under this relation $\sim$, which are inspired by the objectives of modelling. In section 5 we will define complexity and misfit maps for the problem of modelling time series by means of AR-systems. These maps are defined in terms of the canonical forms, i.e., in terms of special AR-representations, and induce well-defined complexity and misfit measures for systems in $\mathbb{B}$. The corresponding modelling procedures defined in section 2.3 and 2.4 are described in section 6. In section 7 we give the resulting algorithms.

4. CANONICAL FORMS

4.1. Equivalent parametrizations

Let $\mathbb{B}$ denote the class of models $B\subset(\mathbb{R}^q)^{\mathbb{Z}}$ which are linear, time invariant and complete. As stated before, $B\in\mathbb{B}$ if and only if there exist $g\in\mathbb{N}$, $d_1,d_2\in\mathbb{Z}$, $d_1\leq d_2$, and a polynomial matrix $R=\sum_{k=d_1}^{d_2} R_k s^k\in\mathbb{R}^{g\times q}[s,s^{-1}]$ such that $B=B(R):=\{\,w\in(\mathbb{R}^q)^{\mathbb{Z}};\ R(\sigma,\sigma^{-1})w=0\,\}$.

We will use the following notation. R_1 is called equivalent to R_2, notation $R_1{\sim}R_2$, if $B(R_1)=B(R_2)$. For $B\in\mathbb{B}$ let $B^{\perp}$ denote the family of laws which are satisfied by the behaviour B, i.e., $B^{\perp}:=\{r\in\mathbb{R}^{1\times q}[s,s^{-1}];\ r(\sigma,\sigma^{-1})w=0$ for all $w\in B\}$. Let $R\in\mathbb{R}^{g\times q}[s,s^{-1}]$ have rows $r_i\in\mathbb{R}^{1\times q}[s,s^{-1}]$, $i=1,\ldots,g$, then the polynomial module generated by $r_1,\ldots,r_g$ is denoted by $M(R):=\{r\in\mathbb{R}^{1\times q}[s,s^{-1}];\ \exists p_i\in\mathbb{R}[s,s^{-1}],\ i=1,\ldots,g,$ such that $r=\sum_{i=1}^{g}p_ir_i\}$. Let $\mathbb{B}^{\perp}$ denote the class of these (finitely generated) submodules of $\mathbb{R}^{1\times q}[s,s^{-1}]$. By $\dim(M^{\perp})$ we denote the dimension of $M^{\perp}\in\mathbb{B}^{\perp}$ as a module, i.e., $\dim(M^{\perp})$ is the minimal number of elements of $M^{\perp}$ which generate $M^{\perp}$. Finally, $U\in\mathbb{R}^{g\times g}[s,s^{-1}]$ is called unimodular if it is invertible in $\mathbb{R}^{g\times g}[s,s^{-1}]$.

The next proposition summarizes some results on AR-representations of models in $\mathbb{B}$.

Proposition 4-1 (*i*) For every $B\in\mathbb{B}$, $B^{\perp}\in\mathbb{B}^{\perp}$; the map $f:\mathbb{B}\to\mathbb{B}^{\perp}:B\to B^{\perp}$ is a *bijection* of $\mathbb{B}$ onto $\mathbb{B}^{\perp}$; (*ii*) $\{B=B(R)\}\Leftrightarrow\{B^{\perp}=M(R)\}$; (*iii*) if $\dim(B^{\perp})=p$, then there exists $R\in\mathbb{R}^{p\times q}[s,s^{-1}]$ with $B=B(R)$; moreover, this R is unique up to left multiplication by a unimodular matrix.

This implies that the equivalence class of AR-parametrizations of a given model $B\in\mathbb{B}$ consists of those polynomials $R\in\mathbb{R}^{g\times q}[s,s^{-1}]$, for some $g\in\mathbb{N}$, for which the rows generate $B^{\perp}$. So the (autoregressive) laws which are satisfied for any time series in B consist of the rows of R and (polynomial) combinations of them.

We will use these results on equivalent parametrizations to define two canonical forms. A *canonical form* is defined as any subset $C\subset\cup\{\mathbb{R}^{g\times q}[s,s^{-1}];\ g\in\mathbb{N}\}$ which contains at least one element of every equivalence class, i.e., for any $g\in\mathbb{N}$ and $R\in\mathbb{R}^{g\times q}[s,s^{-1}]$ there exists an $R_c\in C$ such that $R\sim R_c$. C is called minimal if it contains exactly one element of every equivalence class, i.e., $R_1,R_2\in C$ with $R_1\sim R_2$ implies that $R_1=R_2$. The two canonical forms defined in sections 4.3 and 4.4 are not minimal. This non-minimality is rather intrinsic, i.e., forcing a reduction of the canonical form so that it would become minimal would require arguments which are not related to the objectives of modelling.

4.2. Preliminaries

In order to describe the canonical forms it is useful to introduce some vocabulary and notation.

For $r\in\mathbb{R}^{1\times q}[s,s^{-1}]$, $r=\sum_{k=-\infty}^{\infty} r_k s^k$, $r_k\in\mathbb{R}^{1\times q}$, define the order of r by $d(r):=\max\{k;\ r_k\neq 0\}-\min\{k;\ r_k\neq 0\}$. Let $R=\text{col}(r_1,\ldots,r_g)\in\mathbb{R}^{g\times q}[s,s^{-1}]$ denote the polynomial matrix with rows $r_1,\ldots,r_g$, then the order of R is defined as $d(R):=\max\{d(r_i);\ i=1,\ldots,g\}$. Suppose $r_i=\sum_{k=d_i'}^{d_i''} r_k^{(i)}s_k$ with $d_i''\geq d_i'$, $r_{d_i'}^{(i)}\neq 0\neq r_{d_i''}^{(i)}$, so $d(r_i)=d_i''-d_i'$. Let $L_+:=\text{col}(r_{d_i''}^{(i)};\ i=1,\ldots,g)$ and $L_-:=\text{col}(r_{d_i'}^{(i)};\ i=1,\ldots,g)$ be the leading and trailing coefficient matrices of R. Then R is called bilaterally row proper if L_+ and L_- both have full row rank g.

Let $R=\text{col}(r_1,\ldots,r_g)\in\mathbb{R}^{g\times q}[s,s^{-1}]$, then $(d(r_1),\ldots,d(r_g))$ is called the lag structure of R. In the sequel we will make use of the equation structure of R, which is defined in terms of the lag structure, as follows.

Definition 4-2 If $R\in\mathbb{R}^{g\times q}[s,s^{-1}]$ has lag structure $(d_1,\ldots,d_g)$, then the *equation structure* of R is defined as $e(R):=(e_t\,;\ t\geq 0)$, where $e_t:=\#\{i;d_i=t\}$ is the number of rows in R of order t.

For lag structures we define a total ordering by $\{(d_1',\ldots,d_{g'}') \leq (d_1'',\ldots,d_{g''}'')\} :\Leftrightarrow \{(d_1',\ldots,d_{g'}') = (d_1'',\ldots,d_{g''}'')$ or $g' < g''$ or there is a $g \leq g' = g''$ such that $d_g' < d_g''$ and $d_i' = d_i''$ for all $i < g\}$. So few equations and short lags are preferred. We order equation structures by $\{e' \leq e''\} :\Leftrightarrow \{e' = e''$ or $\sum_{t=0}^{\infty} e_t' < \sum_{t=0}^{\infty} e_t''$ or $\sum_{t=0}^{\infty} e_t' = \sum_{t=0}^{\infty} e_t''$ and there is a t_0 such that $e_{t_0}' > e_{t_0}''$ and $e_t' = e_t''$ for all $t < t_0\}$. For $B \in \mathbb{B}$ we call R a *shortest lag* or *tightest equation* representation of B if $B = B(R)$ and the lag or equation structure respectively is minimal in the class of AR-representations of B. Clearly, every $B \in \mathbb{B}$ has shortest lag and tightest equation representations. The following proposition characterizes these minimal descriptions.

Proposition 4-3 Let $B = B(R)$. Then the following statements are equivalent:

(*i*) R is bilaterally row proper;

(*ii*) R is a tightest equation representation of B;

(*iii*) there exists a permutation matrix Π such that ΠR is a shortest lag representation of B.

We will finally characterize shortest lag representations in terms of matrices. Let $B \in \mathbb{B}$ and $B^{\perp} := \{r \in \mathbb{R}^{1\times q}[s,s^{-1}];\ r(\sigma,\sigma^{-1})w = 0$ for all $w \in B\}$. Let $\mathbb{R}_t^{1\times q}[s]$ denote the class of polynomials in s of power at most t, i.e., $\mathbb{R}_t^{1\times q}[s] := \{r \in \mathbb{R}^{1\times q}[s];\ r = \sum_{k=-\infty}^{\infty} r_k s^k,\ r_k = 0$ for $k < 0$ and $k > t\}$. Let $B_t^{\perp} := B^{\perp} \cap \mathbb{R}_t^{1\times q}[s]$, then $B_t^{\perp}$ describes the family of laws of order at most t which are satisfied by the behaviour B. We will identify $B_t^{\perp}$ with a subspace of $(\mathbb{R}^{1\times q})^{t+1}$ as follows.

Definition 4.4 The bijection $v_t : \mathbb{R}_t^{1\times q}[s] \to (\mathbb{R}^{1\times q})^{t+1}$ is defined as follows. Let $r = \sum_{k=0}^{t} r_k s^k \in \mathbb{R}_t^{1\times q}[s]$, then $v_t(r) \in (\mathbb{R}^{1\times q})^{t+1}$ is defined by $v_t(r) := (r_0, r_1, \ldots, r_t)$.

It can be shown that $v_t(B_t^{\perp})$ is the (Euclidean) orthogonal complement in $(\mathbb{R}^q)^{t+1}$ of $B_t := B|_{[-t,0]} = B|_{[s,s+t]}$ for any $s \in \mathbb{Z}$, i.e., the behaviour on an interval of length $t+1$.

Next we define spaces $L_t \subset B^{\perp}$ as follows. Let $L_0 := B_0^{\perp}$ consist of the zero order laws for B. Define $V_0 := v_0(L_0)$. Observe that $B_0^\perp + sB_0^\perp \subset B_1^\perp$. We will say that the first order laws in $B_0^\perp + sB_0^\perp$ are *implied* by zero order laws. *Truly* first order laws for B, collected in $L_1 \subset B_1^\perp$, are required to be independent of those implied laws. Formally, let V_1 be a *complementary space* of $v_1(B_0^\perp + sB_0^\perp)$ in $v_1(B_1^\perp)$, i.e., $V_1 \cap v_1(B_0^\perp + sB_0^\perp) = \{0\}$ and $V_1 + v_1(B_0^\perp + sB_0^\perp) = v_1(B_1^\perp)$. Then $L_1 := v_1^{-1}(V_1)$. Analogously, the t-th order laws in $B_{t-1}^\perp + sB_{t-1}^\perp \subset B_t^\perp$ are implied by lower order laws. Truly t-th order laws are collected in $L_t \subset B_t^\perp$, defined as $L_t := v_t^{-1}(V_t)$ for a complementary space V_t of $v_t(B_{t-1}^\perp + sB_{t-1}^\perp)$ in $v_t(B_t^\perp)$, i.e., $V_t \cap v_t(B_{t-1}^\perp + sB_{t-1}^\perp) = \{0\}$ and $V_t + v_t(B_{t-1}^\perp + sB_{t-1}^\perp) = v_t(B_t^\perp)$.

Clearly, the spaces V_t and L_t in general are not uniquely defined. Let $n_t := \dim(V_t)$ and let $\{ v_1^{(t)}, \ldots, v_{n_t}^{(t)} \}$ be an arbitrary basis of V_t. Moreover define $r_i^{(t)} := v_t^{-1}(v_i^{(t)})$, $i = 1, \ldots, n_t$. The following proposition establishes the relationship between the sets L_t and shortest lag representations of a model $B \in \mathbb{B}$.

Proposition 4-5 Let $B \in \mathbb{B}$. Then there exists a d such that $n_d \neq 0$ and $n_t = 0$ for all $t > d$. Any tightest equation representation R of B has equation structure $e(R) = (n_0, \ldots, n_d, 0, 0, \ldots)$. Finally, R is a tightest equation representation of B if and only if there exists a choice of the complementary spaces V_t, of bases $\{v_i^{(t)};\ i = 1, \ldots, n_t\}$ of V_t, and of numbers $k_i(t) \in \mathbb{Z}$ for $i = 1, \ldots, n_t$, $t = 0, \ldots, d$, such that the rows of R consist of $\{\sigma^{k_i(t)} \cdot r_i^{(t)};\ i = 1, \ldots, n_t,\ t = 0, \ldots, d\}$.

The canonical forms will correspond to a special choice of the complementary spaces V_t, which we will describe in the next two sections.

4.3. Canonical descriptive form

In section 5 we will define the descriptive complexity and misfit of models in terms of tightest equation representations of a special type. Note that proposition 4-5 characterizes the non-unicity of tightest equation representations in terms of the choice of the complementary spaces V_t and bases of these spaces. The canonical descriptive form selects particular complementary spaces, but the choice of bases is left arbitrary. The complexity and misfit in section 5 will be defined independent of this choice of bases.

We choose truly t-th order laws of B such that they are (Euclidean) orthogonal to the t-th order laws which are implied by lower order ones. Formally, we define $L_t^D \subset B_t^\perp$ as follows. $L_0^D := B_0^\perp$, and $L_t^D := v_t^{-1}\{ [v_t(B_{t-1}^\perp + sB_{t-1}^\perp)]^\perp \cap [v_t(B_t^\perp)] \}$. So, intuitively, the laws $r \in L_t^D$ are orthogonal to those in $B_{t-1}^\perp + sB_{t-1}^\perp$. The orthogonality is imposed to ensure that laws in L_t^D are "far" from being implied by laws of lower order. Of course, in some cases it could be sensible to choose other inner products than the Euclidean one.

Now R is defined to be in canonical descriptive form if it is itself a tightest equation description of the corresponding behaviour $B(R)$ and if the laws of truly order t are contained in L_t^D. We will then say that laws of different order are orthogonal.

Definition 4-6 R is in *canonical descriptive form* (CDF) if

(i) R is a tightest equation representation of $B(R)$;

(ii) laws of different order are orthogonal.

Proposition 4-7 (CDF) is a canonical form.

Note that for R in (CDF) $R \in \mathbb{R}^{g\times q}[s]$, i.e., R is a polynomial matrix in s.

We will describe (CDF) in terms of matrices as follows. Let $R \in \mathbb{R}^{g\times q}[s]$ and let $R^{(t)} := \text{col}(r_i^{(t)};\ i=1,\dots,n_t)$ consist of the rows of R of order t, $t \geq 0$, $n_t \geq 0$, $\sum_{t=0}^{\infty} n_t = g$. Let d be the highest power of s in R and for $t \geq 0$ let $N_t := \text{col}(v_d(r_i^{(t)});\ i=1,\dots,n_t) \in \mathbb{R}^{n_t \times (d+1)q}$ correspond to the t-th order laws in R. Let $N_t = [R_0^{(t)} \dots R_d^{(t)}]$ with $R_i^{(t)} \in \mathbb{R}^{n_t \times q}$, $i=0,\dots,d$. Let $k_t := \max\{i;\ R_i^{(t)} \neq 0\}$. Let $L_- := \text{col}(R_0^{(0)},\dots,R_0^{(d)}) \in \mathbb{R}^{g\times q}$ and $L_+ := \text{col}(R_{k_0}^{(0)},\dots,R_{k_d}^{(d)}) \in \mathbb{R}^{g\times q}$. Define s: $\mathbb{R}^{1\times(d+1)q} \to \mathbb{R}^{1\times(d+1)q}$ as follows. If $v = (v_0,\dots,v_{d-1},v_d)$ with $v_i \in \mathbb{R}^{1\times q}$, $i=0,\dots,d$, then $s(v) := (0,v_0,\dots,v_{d-1})$. Let $V_0 := N_0$ and define $\bar{V}_t$ for $t=1,\dots,d$ inductively by $\bar{V}_t := \text{col}(\bar{V}_{t-1}, s\bar{V}_{t-1}, N_t)$. Finally, for matrices A_1 and A_2 let $A_1 \perp A_2$ denote that every row of A_1 is orthogonal to any row of A_2.

Proposition 4-8 R is in canonical descriptive form if and only if

(i) L_+ and L_- have full row rank; (this implies $k_t = t$)

(ii) $N_t \perp \text{col}(\bar{V}_{t-1}, s\bar{V}_{t-1})$ for all $t=1,\dots,d$.

So, whether R is in (CDF) or not can be checked by means of proposition 4-8 in terms of matrices which can be easily calculated from R. These algebraic conditions will play a role in the algorithms of section 7.

The next proposition describes the non-unicity of (CDF) representations of systems $B \in \mathbb{B}$.

Proposition 4-9 Let $B \in \mathbb{B}$, $B = B(R)$ with $d(R) = d$ and R in (CDF). Let the rows of R be ordered with increasing degree. Then $B = B(R')$ with R' in (CDF) if and only if there exists a permutation matrix Π and a blockdiagonal matrix $A = \text{diag}(A_{00}, \ldots, A_{dd})$ with $A_{tt} \in \mathbb{R}^{n_t \times n_t}$ nonsingular such that $R' = \Pi A R$.

4.4. Canonical predictive form

The canonical predictive form also corresponds to a particular tightest equation representation of the AR-equations describing a behaviour. Again, the complementary spaces V_t of section 4.2 are chosen in a particular way and the choice of bases is left arbitrary. The spaces are intimately connected with the purpose of prediction and corresponding complexity and misfit maps, which will be defined in section 5.

To define the canonical predictive form, we consider the (forward) predictive interpretation of a law $r \in \mathbb{R}^{1 \times q}[s]$. Let $d(r) = d$, $r = \sum_{k=-\infty}^{\infty} r_k s^k$ with $r_k = 0$ for $k < 0$ and $k > d$. The law r corresponding to $r(\sigma)w = 0$ predicts that, given $w(s)$ for $s = t-d, \ldots, t-1$, $w(t)$ will be such that $r_d w(t) = -\sum_{k=0}^{d-1} r_k w(t-d+k)$, $t \in \mathbb{Z}$. We call r a predictive law of order d, r_d a predicted functional of order d, and $-\sum_{k=0}^{d-1} r_k s^k$ a prediction polynomial of order d. Intuitively speaking, we will choose the complementary spaces V_t such that the predicted functionals of different order are orthogonal and such that prediction polynomials of a certain order are orthogonal to predictive laws of lower order. This ensures that predictive laws of different order are "far" from each other.

Formally, for $B \in \mathbb{B}$ define $L_t^P \subset B_t^\perp$ as follows. Let $F_t := \{\tilde{r} \in \mathbb{R}^{1 \times q};\ \exists r \in B_t^\perp,\ r = \sum_{k=0}^{t} r_k s^k$, such that $r_t = \tilde{r}\}$ denote the set of predicted functionals of order at most t. Then $L_0^P := B_0^\perp$ and $L_t^P := v_t^{-1}\{\, [v_t(F_{t-1} \cdot s^t) + v_t(B_{t-1}^\perp)]^\perp \cap [v_t(B_t^\perp)] \,\}$. R is said to be in canonical predictive form if it is itself a tightest equation representation of the corresponding behaviour $B(R)$ and if the predictive laws of order t are contained in L_t^P. We will then say that predicted functionals of different order are orthogonal, corresponding to $v_t(L_t^P) \perp v_t(F_{t-1}.s^t)$, and that the prediction polynomials are orthogonal to predictive laws of lower order, corresponding to $v_t(L_t^P) \perp v_t(B_{t-1}^\perp)$.

Definition 4-10 R is in *canonical predictive form* (CPF) if

(*i*) R is a tightest equation representation of $B(R)$;

(*ii*) predicted functionals of different orders are orthogonal;

(*iii*) prediction polynomials are orthogonal to predictive laws of lower order.

Proposition 4-11 (CPF) is a canonical form.

Using the notation of section 4.3, proposition 4-12 gives simple algebraic conditions for R to be in (CPF). These conditions will be used in the algorithms of section 7.

Proposition 4-12 R is in canonical predictive form if and only if

(*i*) L_+ and L_- have full row rank; (this implies $k_t = t$)

(*ii*) $R_t^{(t)} \perp R_s^{(s)}$ for all $t \neq s$, $t, s = 0, \ldots, d$;

(*iii*) $N_t \perp \bar{V}_{t-1}$ for all $t = 1, \ldots, d$.

The non-unicity of (CPF) representations is exactly of the same kind as described for (CDF) in proposition 4-9, i.e., the representation is unique up to a permutation of the rows and a choice of bases in the spaces L_t^P.

We conclude this section by giving a simple example illustrating the canonical forms (CDF) and (CPF). Consider $B \in \mathbb{B}$ defined by $B := \{w \in (\mathbb{R}^3)^{\mathbb{Z}};\ w_1(t) + w_2(t-1) = 0,\ w_1(t) + w_3(t) + w_2(t-2) = 0,\ t \in \mathbb{Z}\}$. Then $B = B(R)$ with $R := \begin{bmatrix} 0 & 1 & 0 \\ 0 & 1 & 0 \end{bmatrix} + \begin{bmatrix} 1 & 0 & 0 \\ 0 & 0 & 0 \end{bmatrix} . s + \begin{bmatrix} 0 & 0 & 0 \\ 1 & 0 & 1 \end{bmatrix} . s^2$. R is neither in (CDF) nor in (CPF). Let $U_1 := \begin{bmatrix} 1 & 0 \\ -1/2 & 1 \end{bmatrix} + \begin{bmatrix} 0 & 0 \\ -1/2 & 0 \end{bmatrix} . s$, $U_2 := \begin{bmatrix} 1 & 0 \\ -1/2 & 1 \end{bmatrix} + \begin{bmatrix} 0 & 0 \\ -1 & 0 \end{bmatrix} . s$, $R_1 := U_1 \cdot R$ and $R_2 := U_2 \cdot R$. Then $B = B(R_1) = B(R_2)$, $R_1 = \begin{bmatrix} 0 & 1 & 0 \\ 0 & 1/2 & 0 \end{bmatrix} + \begin{bmatrix} 1 & 0 & 0 \\ -1/2 & -1/2 & 0 \end{bmatrix} . s + \begin{bmatrix} 0 & 0 & 0 \\ 1/2 & 0 & 1 \end{bmatrix} . s^2$ is in (CDF) and $R_2 = \begin{bmatrix} 0 & 1 & 0 \\ 0 & 1/2 & 0 \end{bmatrix} + \begin{bmatrix} 1 & 0 & 0 \\ -1/2 & -1 & 0 \end{bmatrix} . s + \begin{bmatrix} 0 & 0 & 0 \\ 0 & 0 & 1 \end{bmatrix} . s^2$ is in (CPF).

5. COMPLEXITY AND MISFIT

5.1. Complexity

As before, let $\mathbb{B}$ denote the class of linear, time invariant, complete systems in $(\mathbb{R}^q)^{\mathbb{Z}}$. Intuitively, a system is more complex if more time series are compatible with the system, i.e., if the system imposes less restrictions on the behaviour. A simple system is one with a few degree of freedom. In particular, if $B_1, B_2 \in \mathbb{B}$ and $B_1 \subset B_2$, $B_1 \neq B_2$, then we call B_1 less complex than B_2. More general, we will call B_1 less complex than B_2 if it allows less time series. The complexity of a system will express the magnitude of the set of time series compatible with the system. For $B \in \mathbb{B}$, let $B_t := B|_{[0,t]}$ denote the space of time series of length $t+1$ which are compatible with the system. By $\mathbb{Z}_+$ we denote the set $\mathbb{Z}_+ := \{0,1,2,3,...\}$. We now define the complexity as a sequence of numbers $c_t(B)$, $t \in \mathbb{Z}_+$, where $c_t(B)$ measures the magnitude of B_t.

Definition 5-1 The *complexity* of dynamical systems is defined by $c: \mathbb{B} \to (\mathbb{R}_+)^{\mathbb{Z}_+}$, $c(B) := (c_t(B);\ t \in \mathbb{Z}_+)$, where $c_t(B) := \frac{1}{t+1} \cdot \dim(B_t)$.

It can be shown that the limits $\lim_{t\to\infty} c_t(B) =: m$ and $\lim_{t\to\infty} t.\{c_t(B) - m\} =: n$ exist and that m is the number of inputs in B and n the (minimal) number of state variables.

A natural ordering of complexities is the partial ordering defined by $\{c^{(1)} \geq c^{(2)}\} :\Leftrightarrow \{c_t^{(1)} \geq c_t^{(2)}$ for all $t \in \mathbb{Z}_+\}$. This ordering is related to tightest equation representations. For $B \in \mathbb{B}$ let $e^* = (e_t^*;\ t \geq 0)$ denote the equation structure of a tightest equation representation of B. If $B_1, B_2 \in \mathbb{B}$ with equation structures $e^{*(1)}$ and $e^{*(2)}$ respectively, then $\dim(B_i|_{[0,t]}) = (t+1)q - \sum_{k=0}^{t} (t+1-k) e_k^{*(i)}$, so $c(B_1) \geq c(B_2)$ if and only if for all $t \in \mathbb{Z}_+$ $\sum_{k=0}^{t} (t+1-k) e_k^{*(1)} \leq \sum_{k=0}^{t} (t+1-k) e_k^{*(2)}$. So systems are complex if their behaviour is restricted by few laws which are of high order.

In the approximate modelling procedures of section 6 we will use utility functions involving the complexity. These utility functions will be based on a total (lexicographic) ordering of complexities which is a refinement of the natural ordering, and which is defined by $\{c^{(1)} \geq c^{(2)}\} :\Leftrightarrow \{c^{(1)} = c^{(2)}$ or there is a $t_0 \in \mathbb{Z}_+$ such that $c_{t_0}^{(1)} > c_{t_0}^{(2)}$ and $c_t^{(1)} = c_t^{(2)}$ for all $t < t_0\}$.

We want to make some remarks on this ordering.

First, in assessing the complexity of a system the number of short lag equations is decisive. Indeed, as $c_t = q - \frac{1}{t+1} \cdot \sum_{k=0}^{t} (t+1-k) e_k^*$, it follows that $\{c^{(1)} \geq c^{(2)}\} \Leftrightarrow \{e^{*(1)} = e^{*(2)}$ or there is a $t_0 \in \mathbb{Z}_+$ such that $e_{t_0}^{*(1)} < e_{t_0}^{*(2)}$ and $e_t^{*(1)} = e_t^{*(2)}$ for all $t < t_0\}$. Note that this ordering of equation structures differs from the one described in section 4.2.

Second, it can be shown that for a system $B \in \mathbb{B}$ there holds $m = q - \sum_{t=0}^{\infty} e_t^*$ and $n = \sum_{t=0}^{\infty} t . e_t^*$, where m denotes the number of inputs or unrestricted variables, n the number of states and $(e_t^*;\ t \in \mathbb{Z}_+)$ the tightest equation structure of B. A simple model is one which leaves little unrestricted, i.e., for which the total number of laws $\sum_{t=0}^{\infty} e_t^*$ is large, and which has small memory, i.e., for which $\sum_{t=0}^{\infty} t . e_t^*$ is small. This amounts to preference of many equations and of short lag, i.e., of small values of $c_t(B)$ for t small. This is reflected by the lexicographic ordering of complexities. Note that the complexity is related to the system considered as a set of trajectories and not to the number of parameters needed to represent the system.

Third, this lexicographic ordering allows for simple recursive algorithms, as will be seen in section 7.

Finally, the reverse lexicographic ordering defined by $\{c^{(1)} \geq c^{(2)}\} : \Leftrightarrow \{c^{(1)} = c^{(2)}$ or there is a $t_0 \in \mathbb{Z}_+$ such that $c_{t_0}^{(1)} > c_{t_0}^{(2)}$ and $c_t^{(1)} \geq c_t^{(2)}$ for all $t > t_0\}$ seems more appealing. It is directly connected with m and n, as for this ordering $\{m_1 > m_2\} \Rightarrow \{c^{(1)} > c^{(2)}\}$ and $\{m_1 = m_2, n_1 > n_2\} \Rightarrow \{c^{(1)} > c^{(2)}\}$. This does not hold true for the lexicographic ordering. However, the construction of algorithms for modelling procedures based on the reverse lexicographic ordering seems to be difficult.

We conclude this section by defining the (total) complexity ordering which we will use in the sequel and by expressing this ordering in terms of equation structures.

Definition 5-2 The *ordering* of complexities of systems in $\mathbb{B}$ is defined by $\{c(B_1) \geq c(B_2)\} : \Leftrightarrow \{c(B_1) = c(B_2)$ or there is a $t_0 \in \mathbb{Z}_+$ such that $c_{t_0}(B_1) > c_{t_0}(B_2)$ and $c_t(B_1) = c_t(B_2)$ for all $t < t_0\}$.

Proposition 5-3 Let $B_i \in \mathbb{B}$ have tightest equation structure $e^*(B_i) := (e_t^*(B_i);\ t \in \mathbb{Z}_+)$, $i = 1,2$. Then $c(B_1) \geq c(B_2)$ if and only if $e^*(B_1) \leq e^*(B_2)$ in the lexicographic ordering, i.e., $e^*(B_1) = e^*(B_2)$ or there is a $t_0 \in \mathbb{Z}_+$ such that $e^*_{t_0}(B_1) < e^*_{t_0}(B_2)$ and $e^*_t(B_1) = e^*_t(B_2)$ for all $t < t_0$.

The complexity ordering can easily be characterized in terms of the canonical forms of sections 4.3 and 4.4 by using proposition 4-3.

Corollary 5-4 Let $B_i \in \mathbb{B}$, $B_i = B(R_d^{(i)}) = B(R_p^{(i)})$ with $R_d^{(i)}$ in (CDF) and $R_p^{(i)}$ in (CPF), $i = 1,2$. Let $e_d^{(i)}$ and $e_p^{(i)}$ denote the equation structure of $R_d^{(i)}$ and $R_p^{(i)}$ respectively, $i = 1,2$. Then $\{c(B_1) \geq c(B_2)\} \Leftrightarrow \{e_p^{(1)} = e_d^{(1)} \leq e_d^{(2)} = e_p^{(2)}$ in lexicographic ordering$\}$.

5.2. Descriptive misfit

In this section we define the misfit of a model $B \in \mathbb{B}$ in describing data consisting of a finite time series $\tilde{w} := (\tilde{w}(t);\ t \in \mathcal{T})$ on an interval $\mathcal{T} = [t_0, t_1]$. As in section 2.6 we first consider the case where B imposes one restriction, in the sense that $B = B(r)$ for some $r \in \mathbb{R}^{1 \times q}[s, s^{-1}]$.

As descriptive misfit we consider the average equation error. Let $n \in \mathbb{Z}$, $d \in \mathbb{Z}_+$, $r = \sum_{k=n}^{n+d} r_k s^k$ with $r_k \in \mathbb{R}^{1 \times q}$, $r_n \neq 0 \neq r_{n+d}$. We define $\|r\|^2 := \sum_{k=n}^{n+d} \|r_k\|^2$ and $\|r\tilde{w}\|^2 := \frac{1}{t_1 - t_0 - d + 1} \cdot \sum_{t=t_0-n}^{t_1-n-d} \{ \sum_{k=n}^{n+d} r_k \tilde{w}(t+k) \}^2$. So $\|r\tilde{w}\|$ measures in how far $\tilde{w}$ satisfies the restriction imposed by $B(r)$ that $(r\tilde{w})(t) = 0$ for $t = t_0 - n, \ldots, t_1 - n - d$. It is assumed that $d(r) = d \leq t_1 - t_0$.

Definition 5-5 The *descriptive misfit* of $r \in \mathbb{R}^{1 \times q}[s, s^{-1}]$ with respect to data $\tilde{w} \in (\mathbb{R}^q)^{\mathcal{T}}$ is defined as the mean *equation error*, i.e., $e^D(\tilde{w}, r) := \|r\tilde{w}\| / \|r\|$.

We define the misfit of $B(r)$ by $\varepsilon^D_{d,1}(\tilde{w}, B(r)) := e^D(\tilde{w}, r)$.

Next let $\dim(B^\perp) \geq 2$. For $r \in B^\perp$ we measure the descriptive misfit by $e^D(\tilde{w}, r)$. The problem is to define the misfit of B, which imposes an infinite number of laws on the phenomenon. We will define the misfit of B by choosing a canonical basis in $B^\perp$, using the canonical descriptive form (CDF). The idea is to define a sequence of misfits, measuring the quality of laws of different order claimed by B. Note that using (CDF) guarantees that laws of different order are orthogonal, so loosely speaking these quality measures become more or less *independent*. By this we mean that e.g.

a first order law should not be judged as being of small misfit if this is due to the fact that this first order law is ("near" to being) implied by good zero order laws. This is made explicit by the orthogonality conditions in (CDF) as stated in section 4.3 and will be illustrated by means of examples in section 9.

To define $\varepsilon^D(\tilde{w},B)$, consider the spaces L_t^D of truly t-th order decriptive laws as defined in section 4.3. Let $n_t := \dim(v_t(L_t^D))$, then $n_t = e_t$ where $(e_t;\ t\in\mathbb{Z}_+)$ is the tightest equation structure of AR-representation of B. For $n_t>0$ define $\varepsilon_{t,1}^D(\tilde{w},B)$ as the worst fit of the truly t-th order laws claimed by B, i.e. $\varepsilon_{t,1}^D(\tilde{w},B) := \max\{e^D(\tilde{w},r);\ r\in L_t^D\}$.

Definition 5-6 For $B\in\mathbb{B}$, let L_t^D denote the space of truly t-th order descriptive laws of B. For data $\tilde{w}\in(\mathbb{R}^q)^{\mathcal{T}}$, the *main* t-th *descriptive misfit* is defined by $\varepsilon_{t,1}^D(\tilde{w},B) := \max\{e^D(\tilde{w},r);\ r\in L_t^D\}$ if $\dim(v_t(L_t^D))>0$, else $\varepsilon_{t,1}^D(\tilde{w},B) := 0$.

If $n_t>1$, then we define $\varepsilon_{t,2}^D(\tilde{w},B)$ as the misfit of the worst-but-one t-th order law, i.e., if $\varepsilon_{t,1}(\tilde{w},B) = e^D(\tilde{w},r_1)$, $r_1\in L_t^D$, then $\varepsilon_{t,2}^D(\tilde{w},B) := \max\{e^D(\tilde{w},r);\ r\in v_t^{-1}\{\,v_t(L_t^D)\cap[v_t(r_1)]^{\perp}\}\}$. For $k=2,\ldots,n_t$, $\varepsilon_{t,k}^D(\tilde{w},B)$ is inductively defined as the worst-but-$(k-1)$ t-th order misfit, as follows. If $\varepsilon_{t,j}^D(\tilde{w},B) = e^D(\tilde{w},r_j)$, $r_j\in v_t^{-1}\{\,v_t(L_t^D)\cap[\mathrm{span}(v_t(r_1),\ldots,v_t(r_{j-1}))]^{\perp}\}$ for $j=1,2,\ldots,k-1$, then $\varepsilon_{t,k}^D(\tilde{w},B) := \max\{e^D(\tilde{w},r);\ r\in v_t^{-1}\{\,v_t(L_t^D)\cap[\mathrm{span}(v_t(r_1),\ldots,v_t(r_{k-1}))]^{\perp}\}\}$. For $k=n_t+1,\ldots,q$, $\varepsilon_{t,k}^D(\tilde{w},B) := 0$. It can be shown that $\varepsilon_{t,k}^D$ is well-defined in this way, i.e., independent of the maximizing arguments r_j.

Definition 5-7 The *descriptive misfit* is a map $\varepsilon^D : (\mathbb{R}^q)^{\mathcal{T}}\times\mathbb{B} \to (\mathbb{R}_+^{1\times q})^{\mathbb{Z}_+}$, where $\varepsilon_{t,k}^D(\tilde{w},B)$ is the descriptive misfit of the worst-but-$(k-1)$ law of the truly t-th order descriptive laws in L_t^D claimed by B, $t\in\mathbb{Z}_+$, $k=1,\ldots,q$.

We remark that both the complexity and the descriptive misfit are defined in terms of the spaces L_t^D, hence in terms of (CDF), but independent of a choice of basis in L_t^D. A convenient basis for L_t^D could be $\{r_1,\ldots,r_{n_t}\}$ as defined above.

Note that there are at most $\sum_{t=0}^{\infty} e_t = q-m \le q$ misfit numbers unequal to zero. These numbers give the equation error of a suitably chosen basis of all the equations which are claimed by the model. The numbers $\{\varepsilon^D_{t,k}\ ;\ k=1,\ldots,q\}$ measure the quality of the t-th order equations, which are orthogonal to the lower order ones.

We will impose the following lexicographic ordering on misfits.

Definition 5-8 $\{\varepsilon'=(\varepsilon'_{t,k}) \geq \varepsilon''=(\varepsilon''_{t,k})\}: \Leftrightarrow \{\varepsilon'=\varepsilon''$; or there exists $t_0\in\mathbb{Z}_+$, $k_0\leq q$ such that $\varepsilon'_{t_0,k_0}>\varepsilon''_{t_0,k_0}$ and $\varepsilon'_{t,k}=\varepsilon''_{t,k}$ for all $t<t_0$, $k=1,\ldots,q$ and for $t=t_0$, $k=1,\ldots,k_0-1$; or there exists $t_0\in\mathbb{Z}_+$ such that $\varepsilon'_{t_0,1}>\varepsilon''_{t_0,1}$ and $\varepsilon'_{t,k}=\varepsilon''_{t,k}$ for all $t<t_0$, $k=1,\ldots,q\}$.

Note that if B_1 has lower order laws than B_2, then the misfit of B_1 in general will be larger than that of B_2. On the other hand the complexity of B_1 is smaller than that of B_2. In section 6 we will describe two procedures to balance the desires for low misfit and low complexity by fixing a maximal tolerated level for one of the objectives and optimizing with respect to the other one. These procedures correspond to the utilities defined in sections 2.3 and 2.4. We will do the same for predictive misfit, defined in the next section.

5.3. Predictive misfit

The one-step-ahead predictive misfit of a dynamical system in predicting a time series is based on the prediction error defined in section 2.7 for static prediction. Now the data consists of a finite time series $\tilde{w}=(\tilde{w}(t);\ t\in\mathcal{T}=[t_0,t_1])$ and the model class consists of the class of linear, time invariant, complete systems $\mathbb{B}$.

Again we first consider the case where $B=B(r)$ with $r\in\mathbb{R}^{1\times q}[s,s^{-1}]$. Let $n\in\mathbb{Z}$, $d\in\mathbb{Z}_+$, $r=\sum_{k=n}^{n+d} r_k s^k$ with $r_k\in\mathbb{R}^{1\times q}$, $r_n\neq 0\neq r_{n+d}$. Then $B(r)$ predicts that $r_{n+d}w(t+n+d) = -\sum_{k=n}^{n+d-1} r_k w(t+k)$. Let $r_{n+d}\tilde{w}(t+n+d) = -\sum_{k=n}^{n+d-1} r_k\tilde{w}(t+k)+e(t+n+d)$ for $t=t_0-n,\ldots,t_1-n-d$. So $e(t)$ is the error made at time t in the prediction of $r_{n+d}w(t)$. Let $\|e\|^2 := \frac{1}{t_1-t_0-d+1}\sum_{t=t_0+d}^{t_1} e^2(t)$ denote the average prediction error and let $\|r_{n+d}\tilde{w}\|^2_d := \frac{1}{t_1-t_0-d+1}\sum_{t_1=t_0+d}^{t_1}\{r_{n+d}\tilde{w}(t)\}^2$ denote the average magnitude of the predicted functional. It is assumed that $d\leq t_1-t_0$.

> **Definition 5-9** The *predictive misfit* of $r \in \mathbb{R}^{1\times q}[s,s^{-1}]$, with $1 \le d(r) \le t_1 - t_0$ and with leading coefficient vector $r^* \in \mathbb{R}^{1\times q}$, with respect to data $\tilde{w} \in (\mathbb{R}^q)^{\mathcal{T}}$ is defined as the *relative mean prediction error*, i.e., $e^P(\tilde{w},r) := \|r\tilde{w}\| \,/\, \|r^*\tilde{w}\|_d = \|e\| \,/\, \|r^*\tilde{w}\|_d$.

We define the predictive misfit of $B(r)$ by $\varepsilon^P_{d,1}(\tilde{w},B(r)) := e^P(\tilde{w},r)$.

Next we define the misfit for models with $\dim(B^{\perp}) \ge 2$. Again we will measure the predictive quality of a model by means of a sequence of numbers which measure the quality of predictive laws of different order. The quality assessment for laws of different orders is made independently by using the canonical predictive form (CPF). First of all we require the t-th order laws to be truly t-th order, i.e., the t-th order laws should not be implied by lower order ones. Second, we require predicted functionals of different order to be orthogonal. This is essential to guarantee that good quality of one predictive law is not due to good quality of another predictive law. This is made explicit by the orthogonality conditions of (CPF) in section 4.4 and will be illustrated by means of examples in section 9.

To define $\varepsilon^P(\tilde{w},B)$, consider the spaces L^P_t defined in section 4.4 and let $n_t := \dim(v_t(L^P_t)) = e_t$. We give the definition of predictive misfit in analogy with the definition of descriptive misfit in section 5.2 and with the same motivation. For $t=0$ we define $\varepsilon^P_{t,k}(\tilde{w},B) := \varepsilon^D_{t,k}(\tilde{w},B)$, as for $d(r)=0$ $e^P(\tilde{w},r)=1$ for any $\tilde{w}$, so the predictive misfit makes no sense for these static laws. In this case we measure the misfit simply by $\|e\|/\|r\|$.

> **Definition 5-10** For $B \in \mathbb{B}$, let L^P_t denote the space of truly t-th order predictive laws of B. For data $\tilde{w} \in (\mathbb{R}^q)^{\mathcal{T}}$, the *main t-th order predictive misfit* for $t \ge 1$ is defined by $\varepsilon^P_{t,1}(\tilde{w},B) := \max\{e^P(\tilde{w},r);\ r \in L^P_t\}$ if $\dim(v_t(L^P_t)) > 0$, else $\varepsilon^P_{t,1}(\tilde{w},B) := 0$.

Moreover, $\varepsilon^P_{t,k}$ measures the predictive misfit of the worst-but-$(k-1)$ law of the truly t-th order predictive laws in L^P_t claimed by B. If $t \ge 1$ and $n_t > 1$, then $\varepsilon^P_{t,k}(\tilde{w},B)$ for $k = 2,\dots,n_t$ is inductively defined as follows. If $\varepsilon^P_{t,j}(\tilde{w},B) = e^P(\tilde{w},r_j)$ with $r_j \in v_t^{-1}\{\, v_t(L^P_t) \cap [\mathrm{span}(v_t(r_1),\dots,v_t(r_{j-1}))]^{\perp}\}$ for $j = 1,\dots,k-1$, then $\varepsilon^P_{t,k}(\tilde{w},B) := \max\{e^P(\tilde{w},r);\ r \in v_t^{-1}\{\, v_t(L^P_t) \cap [\mathrm{span}(v_t(r_1),\dots, v_t(r_{k-1}))]^{\perp}\}\,\}$. For $k = n_t+1,\dots,q$ we define $\varepsilon^P_{t,k}(\tilde{w},B) := 0$. It can be shown that $\varepsilon^P_{t,k}(\tilde{w},B)$ is well-defined.

Definition 5-11 The *predictive misfit* is a map $\varepsilon^P:(\mathbb{R}^q)^{\mathcal{T}}\times\mathbb{B}\to(\mathbb{R}_+^{1\times q})^{\mathbb{Z}_+}$ where $\varepsilon^P_{0,k}(\tilde{w},B):=\varepsilon^D_{0,k}(\tilde{w},B)$ and for $t\geq 1$ $\varepsilon^P_{t,k}(\tilde{w},B)$ is the predictive misfit of the worst-but-$(k-1)$ law of the truly t-th order predictive laws in L^P_t claimed by B, $k=1,\dots,q$.

We order the predictive misfit sequences in the same way as the descriptive misfit sequences, i.e., lexicographically. Corresponding modelling procedures are described in the next section.

6. MODELLING PROCEDURES

6.1. Introduction

In this section we describe four modelling procedures. Both for the purpose of description and for that of prediction we define two utility functions, corresponding to fixing the tolerated misfit or the tolerated complexity and optimizing complexity and misfit respectively. The corresponding procedures lead to relatively simple algorithms, the details of which are given in section 7.

6.2. Deterministic descriptive modelling procedures

Let $\mathbb{B}$ consist of the class of AR-systems $B\subset(\mathbb{R}^q)^{\mathbb{Z}}$ and let the set of conceivable data be $D:=\cup\{(\mathbb{R}^q)^n;\ n\in\mathbb{N}\}$, so the data consists of a finite time series $\tilde{w}\in(\mathbb{R}^q)^{\mathcal{T}}$ for some $\mathcal{T}=[t_0,t_1]$.

First consider the case that a maximal tolerated complexity $c_{tol}:=(c_t^{tol};\ t\in\mathbb{Z}_+)$ is given. Fixing c_{tol} is interpreted as requiring that allowable models should satisfy $c_t(B)\leq c_t^{tol}$ for all $t\in\mathbb{Z}_+$. As $c_t=q-\frac{1}{t+1}\sum_{k=0}^{t}(t+1-k)e_k^*$ this amounts to requiring $\sum_{k=0}^{t}(t+1-k)e_k^*(B)\geq(q-c_t^{tol}).(t+1)$ for all $t\in\mathbb{Z}_+$, where $(e_t^*(B);\ t\in\mathbb{Z}_+)$ is the equation structure of a tighest equation representation of B. So a maximal tolerated complexity amounts to requiring that B imposes a minimal tolerated number of (truly) t-th order restrictions. Under this requirement the descriptive misfit will be minimized. The misfit of B is the sequence $\varepsilon^D(\tilde{w},B)\in(\mathbb{R}_+^{1\times q})^{\mathbb{Z}_+}$ with lexicographic ordering as defined in section 5.2. The procedure $P^D_{c_{tol}}:D\to 2^{\mathbb{B}}$ then is defined as in section 2.4, i.e., as follows.

Definition 6-1 For $\tilde{w}\in D$, $P^D_{c_{tol}}(\tilde{w}):=$ argmax$\{ u_{c_{tol}}(c(B),\varepsilon^D(\tilde{w},B));\ B\in\mathbb{B} \}$, where the ordering for $u:= u_{c_{tol}}$ is defined by

(*i*) $\{ u(c^{(1)},\varepsilon^{(1)})=u(c^{(2)},\varepsilon^{(2)})\}: \Leftrightarrow \{ \exists t_i\in\mathbb{Z}_+\ c^{(i)}_{t_i} > c^{tol}_{t_i},\ i=1,2;$ or $(c^{(1)},\varepsilon^{(1)}) = (c^{(2)},\varepsilon^{(2)}) \}$;

(*ii*) $\{u(c^{(1)},\varepsilon^{(1)})<u(c^{(2)},\varepsilon^{(2)})\}: \Leftrightarrow \{ \exists t_0\in\mathbb{Z}_+\ c^{(1)}_{t_0}>c^{tol}_{t_0}$ and $\forall t\in\mathbb{Z}_+\ c^{(2)}_t \leq c^{tol}_t$; or $\forall t\in\mathbb{Z}_+\ c^{(1)}_t,c^{(2)}_t\leq c^{tol}_t$ and $\exists t_0\in\mathbb{Z}_+$ such that $\varepsilon^{(1)}_{t_0}>\varepsilon^{(2)}_{t_0}$ and $\varepsilon^{(1)}_t=\varepsilon^{(2)}_t$ for all $t<t_0$; or $\forall t\in\mathbb{Z}_+\ c^{(1)}_t,c^{(2)}_t\leq c^{tol}_t, \varepsilon^{(1)}=\varepsilon^{(2)}$ and $\exists t_0\in\mathbb{Z}_+$ such that $c^{(1)}_{t_0}>c^{(2)}_{t_0}$ and $c^{(1)}_t=c^{(2)}_t$ for all $t<t_0\}$. Here the vectors $\varepsilon_t\in\mathbb{R}^{1\times q}$ are ordered lexicographically.

Note that the requirement $c(B)\leq c_{tol}$ is not interpreted in the lexicographic ordering, but in the pointwise ordering, i.e., $c(B)\leq c_{tol}$ if and only if $c_t(B)\leq c^{tol}_t$ for all $t\in\mathbb{Z}_+$.

Next suppose that a maximal tolerated misfit $\varepsilon_{tol}:= (\varepsilon^{tol}_t;\ t\in\mathbb{Z}_+)\in(\mathbb{R}^{1\times q})^{\mathbb{Z}_+}$ is given. We will invariably assume that $\varepsilon^{tol}_t=\bar{\varepsilon}^{tol}_t.(1,\ldots,1)$ with $\bar{\varepsilon}^{tol}_t\in\mathbb{R}$. The requirement $\varepsilon^D(\tilde{w},B)<\varepsilon_{tol}$ also is not interpreted in the lexicographical sense, but pointwise. As $\varepsilon^D_{t,k}(\tilde{w},B)\leq\varepsilon^D_{t,l}(\tilde{w},B)$ for $k\geq l$, this means that a model $B\in\mathbb{B}$ is tolerated if and only if $\varepsilon^D_{t,1}(\tilde{w},B)<\bar{\varepsilon}^{tol}_t$ for all $t\in\mathbb{Z}_+$. So fixing ε_{tol} amounts to requiring that the misfit of (truly) t-th order laws should be smaller than $\bar{\varepsilon}^{tol}_t$. One can impose an upper bound L on the order of equations by taking $\bar{\varepsilon}^{tol}_t<0$ for $t>L$.

Under the requirement $\varepsilon^D_{t,1}(\tilde{w},B)<\bar{\varepsilon}^{tol}_t$ the complexity has to be minimized. The complexity of a system is $c(B)\in(\mathbb{R}_+)^{\mathbb{Z}_+}$ with lexicographic ordering, as defined in section 5.1. Equivalently, under the misfit restriction the equation structure $(e^*_t(B);\ t\geq 0)$ has to be maximized lexicographically. So the purpose is to find as many relationships of small order as possible.

The procedure $P^{*D}_{\varepsilon_{tol}}: D\rightarrow 2^{\mathbb{B}}$ corresponding to the one described in section 2.3 for minimizing complexity given a misfit restriction is defined as follows. For $\tilde{w}\in D$, $P^{*D}_{\varepsilon_{tol}}(\tilde{w}):=$ argmax$\{ u(c(B),\varepsilon^D(\tilde{w},B));\ B\in\mathbb{B} \}$, with the ordering $\{u(c^{(1)},\varepsilon^{(1)}) = u(c^{(2)},\varepsilon^{(2)})\}: \Leftrightarrow \{ \exists t_i\in\mathbb{Z}_+\ \varepsilon^{(i)}_{t_i,1}\geq\bar{\varepsilon}^{tol}_{t_i},\quad i=1,2;$ or $(c^{(1)},\varepsilon^{(1)}) = (c^{(2)},\varepsilon^{(2)}) \}$, and $\{u(c^{(1)},\varepsilon^{(1)})<u(c^{(2)},\varepsilon^{(2)})\}: \Leftrightarrow \{ \exists t_0\in\mathbb{Z}_+\ \varepsilon^{(1)}_{t_0,1}\geq\bar{\varepsilon}^{tol}_{t_0}$ and $\forall t\in\mathbb{Z}_+\ \varepsilon^{(2)}_{t,1}<\bar{\varepsilon}^{tol}_t$; or $\forall t\in\mathbb{Z}_+\ \varepsilon^{(1)}_{t,1},\ \varepsilon^{(2)}_{t,1}<\varepsilon^{tol}_t$ and $\exists t_0\in\mathbb{Z}_+$ such that $c^{(1)}_{t_0}>c^{(2)}_{t_0}$ and $c^{(1)}_t=c^{(2)}_t$ for all $t<t_0$; or $\forall t\in\mathbb{Z}_+\ \varepsilon^{(1)}_{t,1},\ \varepsilon^{(2)}_{t,1}<\bar{\varepsilon}^{tol}_t$, $c^{(1)}=c^{(2)}$ and $\varepsilon^{(1)}>\varepsilon^{(2)}$ in lexicographic ordering$\}$.

However, $P^{*D}_{\varepsilon_{tol}}$ is difficult to implement algorithmically. We will consider a slight variation $P^{D}_{\varepsilon_{tol}}$ of $P^{*D}_{\varepsilon_{tol}}$. We will illustrate the difference between these two procedures by means of a simple example in section 9. The procedure $P^{D}_{\varepsilon_{tol}}$ allows for a relatively simple algorithm, described in section 7.

We now first define $P^{D}_{\varepsilon_{tol}}$ and subsequently give an interpretation.

Definition 6-2 For $\tilde{w}\in D$, $P^{D}_{\varepsilon_{tol}}(\tilde{w}):=\operatorname{argmax}\{ u_{\varepsilon_{tol}}(c(B),\varepsilon^{D}(\tilde{w},B));\ B\in\mathbb{B}\}$ where the ordering for $u:=u_{\varepsilon_{tol}}$ is defined by

(*i*) $\{u(c^{(1)},\varepsilon^{(1)})=u(c^{(2)},\varepsilon^{(2)})\}:\Leftrightarrow\{\exists t_i\in\mathbb{Z}_+\ \varepsilon^{(i)}_{t_i,1}\geq\bar{\varepsilon}^{tol}_{t_i},\ i=1,2;$ or $(c^{(1)},\varepsilon^{(1)})=(c^{(2)},\varepsilon^{(2)})\}$;

(*ii*) $\{u(c^{(1)},\varepsilon^{(1)})<u(c^{(2)},\varepsilon^{(2)})\}:\Leftrightarrow\{\exists\, t_0\in\mathbb{Z}_+\ \varepsilon^{(1)}_{t_0,1}\geq\bar{\varepsilon}^{tol}_{t_0}$ and $\forall t\in\mathbb{Z}_+\ \varepsilon^{(2)}_{t,1}<\bar{\varepsilon}^{tol}_{t}$; or $\forall t\in\mathbb{Z}_+\ \varepsilon^{(1)}_{t,1},\ \varepsilon^{(2)}_{t,1}<\bar{\varepsilon}^{tol}_{t}$ and $(c_0^{(1)},\varepsilon^{(1)}_{0,1},\ldots,\varepsilon^{(1)}_{0,e_0^{(1)}},\ c_1^{(1)},\ \varepsilon^{(1)}_{1,1},\ldots,\varepsilon^{(1)}_{1,e_1^{(1)}},\ c_2^{(1)},\ \varepsilon^{(1)}_{2,1},\ldots,\varepsilon^{(1)}_{2,e_2^{(1)}},\ c_3^{(1)},\ldots)>(c_0^{(2)},\ \varepsilon^{(2)}_{0,1},\ldots,\varepsilon^{(2)}_{0,e_0^{(2)}},\ c_1^{(2)},\ \varepsilon^{(2)}_{1,1},\ldots,\varepsilon^{(2)}_{1,e_1^{(2)}},\ c_2^{(2)},\ \varepsilon^{(2)}_{2,1},\ldots,\varepsilon^{(2)}_{2,e_2^{(2)}},\ c_3^{(2)},\ldots)$ in the lexicographic ordering, where $e^{(i)}$ is the tightest equation structure corresponding to $c^{(i)}$, $i=1,2\}$.

This means that $P^{D}_{\varepsilon_{tol}}$ maximizes the number of zero order relations under the misfit constraint. Among solutions, which in general are highly non-unique, it chooses the one with minimal misfit. Subsequently the number of first order relations is maximized, and then the first order misfit is minimized, and so on. Note that these first order relations should be orthogonal to the zero order ones, as the utility is defined in terms of $\varepsilon^{D}(\tilde{w},B)$ which involves (CDF). The resulting model is optimal with respect to the utility $u_{\varepsilon_{tol}}$. Proposition 5-3 indicates a close relationship between $P^{D}_{\varepsilon_{tol}}$ and $P^{*D}_{\varepsilon_{tol}}$. However, $P^{D}_{\varepsilon_{tol}}$ need not always minimize the complexity with respect to the lexicographic ordering on $(c_t(B);\ t\in\mathbb{Z}_+)$, as will be illustrated by means of an example in section 9. This is due to the auxiliary minimization of misfits, which is essential for obtaining simple (recursive) algorithms.

Proposition 6-3 The procedures $P^{D}_{c_{tol}}$ and $P^{D}_{\varepsilon_{tol}}$ are well-defined maps from D into $2^{\mathbb{B}}$.

Finally, by $\bar{P}^{D}_{\varepsilon_{tol}}(\tilde{w})$ we denote the procedure which is defined in analogy with $P^{D}_{\varepsilon_{tol}}$, but requiring $\varepsilon^{D}_{t,1}(\tilde{w},B)\leq\bar{\varepsilon}^{tol}_{t}$ in contrast with $P^{D}_{\varepsilon_{tol}}$ which requires $\varepsilon^D_{t,1}(\tilde{w},B) < \bar{\varepsilon}^{tol}_t$.

6.3. Two deterministic predictive modelling procedures

In this section we briefly describe two predictive procedures, corresponding to fixing a maximal tolerated complexity or misfit and minimizing misfit and complexity respectively. These procedures are analogues of the descriptive procedures defined in section 6.2 and are obtained by replacing the descriptive misfit ε^D by the predictive misfit ε^P.

Again, fixing a maximal tolerated complexity amounts to requiring of an allowable model B that it imposes a minimal tolerated number of (truly) t-th order restrictions on the phenomenon, $t \in \mathbb{Z}_+$. Under this requirement the relative mean prediction error ε^P is minimized lexicographically. So first the misfit of the zero order laws (in L^P_0) is minimized, then the misfit of the truly first order laws (in L^P_1, hence orthogonal to the zero order laws), and so on.

On the other hand, one can fix a maximal tolerated relative mean prediction error $\bar{\varepsilon}^{tol}_t \in \mathbb{R}$ for predictive laws of (truly) order t. The procedure $P^{*P}_{\varepsilon_{tol}}(\tilde{w})$ corresponding to minimizing the complexity lexicographically under the constraint $\varepsilon^P_{t,1}(\tilde{w},B) < \bar{\varepsilon}^{tol}_t$, $t \in \mathbb{Z}_+$, again is difficult to implement algorithmically. Therefore we will consider a slightly different procedure $P^P_{\varepsilon_{tol}}$, in analogy with $P^D_{\varepsilon_{tol}}$. This procedure corresponds to first finding a maximal number of zero order relations, then minimizing the misfit of these, subsequently maximizing the number of first order relations and minimizing their predictive misfit, and so on. Due to proposition 5-3 there is a close relationship between $P^P_{\varepsilon_{tol}}$ and $P^{*P}_{\varepsilon_{tol}}$. However, they are not equivalent, due to the auxiliary minimization of the misfit.

We define $\bar{P}^P_{\varepsilon_{tol}}$ in analogy with $P^P_{\varepsilon_{tol}}$, replacing the constraints $\varepsilon^P_{t,1}(\tilde{w},B) < \bar{\varepsilon}^{tol}_t$ by $\varepsilon^P_{t,1}(\tilde{w},B) \le \bar{\varepsilon}^{tol}_t$.

For completeness we define $P^P_{c_{tol}}$ and $P^P_{\varepsilon_{tol}}$ explicitly.

Definition 6-4 For given $c_{tol} \in (\mathbb{R}_+)^{\mathbb{Z}_+}, \varepsilon_{tol} \in (\mathbb{R}^{1\times q})^{\mathbb{Z}_+}$ with $\varepsilon^{tol}_t = \bar{\varepsilon}^{tol}_t.(1,\dots,1)$, $\bar{\varepsilon}^{tol}_t \in \mathbb{R}$, the procedures $P^P_{c_{tol}}: D \to 2^{\mathbb{B}}$ and $P^P_{\varepsilon_{tol}}: D \to 2^{\mathbb{B}}$ are defined as follows. For $\tilde{w} \in D$, $P^P_{c_{tol}}(\tilde{w}) := \operatorname{argmax}\{ u_{c_{tol}}(c(B), \varepsilon^P(\tilde{w},B));\ B \in \mathbb{B} \}$ and $P^P_{\varepsilon_{tol}}(\tilde{w}) := \operatorname{argmax}\{ u_{\varepsilon_{tol}}(c(B), \varepsilon^P(\tilde{w},B));\ B \in \mathbb{B} \}$, with the orderings for $u_{c_{tol}}$ and $u_{\varepsilon_{tol}}$ defined as in the definition of $P^D_{c_{tol}}$ and $P^D_{\varepsilon_{tol}}$ respectively.

We finally remark that for univariate time series, i.e., $q=1$, the descriptive and predictive procedures are equivalent. That is, for $\tilde{w} \in \mathbb{R}^{\mathcal{T}}$ $P^D_{c_{tol}}(\tilde{w}) = P^P_{c_{tol}}(\tilde{w})$ for all c_{tol}, and $P^D_{\varepsilon_{tol}}(\tilde{w}) = P^P_{\varepsilon_{tol}}(\tilde{w})$, $\bar{P}^D_{\varepsilon_{tol}}(\tilde{w}) = \bar{P}^P_{\varepsilon_{tol}}(\tilde{w})$ for all ε_{tol}.

7. ALGORITHMS

7.1. Introduction

In this section we describe algorithms for the four deterministic approximate modelling procedures of section 6. These algorithms basically consist of sequential application of the results stated in propositions 2–8 and 2–9 in section 2.6 and propositions 2–13 and 2–14 in section 2.7. Before giving a detailed description of the algorithms we first introduce some concepts and notation and illustrate the approach by describing $P^D_{c_{tol}}$ in general terms.

Let the data consist of a finite time series $\tilde{w} \in (\mathbb{R}^q)^{\mathcal{T}}$ with $\mathcal{T} = [t_0, t_1]$. Let $0 \leq d \leq t_1 - t_0$ and $\tau(\mathcal{T}, d) := t_1 - t_0 - d + 1$, then for $r \in \mathbb{R}^{1\times q}[s, s^{-1}]$, $r = \sum_{k=n}^{n+d} r_k s^k$, $r_k \in \mathbb{R}^{1\times q}$, $r_n \neq 0 \neq r_{n+d}$, there holds $\|r\tilde{w}\|^2 := \frac{1}{\tau(\mathcal{T},d)} \cdot \sum_{t=t_0-n}^{t_1-n-d} \{ \sum_{k=n}^{n+d} r_k \tilde{w}(t+k) \}^2 = v_d(r).S(w,d).v_d(r)^T$ where $S(\tilde{w},d) := \frac{1}{\tau(\mathcal{T},d)} \cdot \sum_{t=t_0}^{t_1-d} (\tilde{w}(t)^T, \ldots, \tilde{w}(t+d)^T)^T . (\tilde{w}(t)^T, \ldots, \tilde{w}(t+d)^T)$ is the empirical covariance matrix of order d.

The algorithms consist of constructing complementary spaces $\{V_t;\, t \in \mathbb{Z}_+\}$. The corresponding models $B \in \mathbb{B}$ are then defined in terms of $L_t := v_t^{-1}(V_t)$ by $B := \{w \in (\mathbb{R}^q)^{\mathbb{Z}};\ r(\sigma)w = 0$ for all $r \in L_t,\ t \in \mathbb{Z}_+\}$. Here $L_t = \{0\}$ for t sufficiently large.

The models identified by the algorithms coincide with the models corresponding to the procedures of section 6 for specifications of c_{tol} and ε_{tol} which are in accordance with the number of observations and for generic data. In general terms, one should not allow laws for which the order is too large in comparison with the number of data. Moreover, the algorithms generate optimal models for λ-generic data, i.e., non-optimality only can arise in a subset N of $(\mathbb{R}^q)^{\mathcal{T}}$ for which $(\mathbb{R}^q)^{\mathcal{T}}\backslash N$ contains an open set of full Lebesgue measure in $(\mathbb{R}^q)^{\mathcal{T}}$.

We will illustrate the foregoing by considering $P^D_{c_{tol}}$. We will make a sensibility assumption on c_{tol} which is related to the number of observations. Moreover we will make some generic assumptions on the data.

First, in order that the descriptive misfit $e^D(\tilde{w},r):=\|r\tilde{w}\|/\|r\|$ is well-defined, it is required that $d:=d(r)\leq t_1-t_0$. Moreover, $\{\varepsilon^D(\tilde{w},r)\}^2=\|r\|^{-2}.v_d(r).S(\tilde{w},d).v_d(r)^T$, with $\text{rank}(S(\tilde{w},d))\leq\min\{t_1-t_0-d+1,\ q(d+1)\}$. If $t_1-t_0-d+1<q(d+1)$, then for any $\tilde{w}\in(\mathbb{R}^q)^{\mathcal{T}}$ there exists an r with $d(r)\leq d$ and $e^D(\tilde{w},r)=0$. To prevent overparametrization it is reasonable at least to require $t_1-t_0-d+1\geq q(d+1)$,i.e., $d\leq\bar{d}(\mathcal{T}):=(t_1-t_0+1-q)/(q+1)$. This restricts the set of laws for which the quality can be reasonably assessed, and implies restrictions on the requirements in c_{tol} to be sensible. In order to state this exactly as well as some generic assumptions on the data, we consider for given $c_{tol}\in(\mathbb{R}_+)^{\mathbb{Z}_+}$ the class of allowable models $B\in\mathbb{B}$ for which $c_t(B)\leq c_t^{tol}$ for all $t\in\mathbb{Z}_+$ and the corresponding class of tightest equation structures $E(c_{tol}):=\{(e_t^*;\ t\in\mathbb{Z}_+);\ \exists B\in\mathbb{B},\ c_t(B)\leq c_t^{tol}$ for all $t\in\mathbb{Z}_+$, such that $(e_t^*;\ t\in\mathbb{Z}_+)$ is the tightest equation structure of $B\}$. Equip $E(c_{tol})$ with the lexicographic ordering, and let $e(c_{tol})$ be the corresponding minimal element of $E(c_{tol})$.

Definition 7-1 For given tolerated complexity c_{tol}, the *equation structure corresponding to* c_{tol} is defined as the minimal achievable tightest equation structure of tolerated models in $\mathbb{B}$ with respect to the lexicographic ordering.

We will now first state the assumptions and then comment on them.

Assumption 7-2 Let $c_{tol}\in(\mathbb{R}_+)^{\mathbb{Z}_+}$ and $\tilde{w}\in(\mathbb{R}^q)^{\mathcal{T}}$ be given.

(i) $\max\{t;\ e_t(c_{tol})\neq 0\}<\bar{d}(\mathcal{T}):=(t_1-t_0+1-q)/(q+1)$;

(ii) $P^D_{c_{tol}}(\tilde{w})=\{B\}$, i.e., a singleton;

(iii) B has tightest equation structure $e(c_{tol})$.

Proposition 7-3 Given (i), then (ii) and (iii) hold true for generic data $\tilde{w}$.

Assumption 7-2(i) expresses a sensibility requirement for c_{tol}, as equations of order more than $\bar{d}(\mathcal{T})$ are not sensible. Assumption 7-2(iii) also expresses a sensibility requirement which we only illustrate in detail for $e_0^*(B)=e_0(c_{tol})$, as the other requirements have a similar interpretation. The condition $c_0(B)\le c_0^{tol}$ implies that at least $q-c_0^{tol}$ zero order laws need to be accepted. Let n_0 denote the number of independent equations of order zero which are exactly satisfied by the data $\tilde{w}$. It is reasonable to suppose that $q-c_0^{tol}\ge n_0$. In this case any optimal model B has a tightest equation structure $(e_t^*(B);\, t\in\mathbb{Z}_+)$ with $e_0^*(B)=q-c_0^{tol}$, which is minimal in view of the requirement $c_0(B)\le c_0^{tol}$. That $e_0^*(B)=q-c_0^{tol}$ for optimal models B is seen as follows. Let $e_0^*(B)>q-c_0^{tol}\ge n_0$. It follows from the definition of ε^D in section 5.2 that $\varepsilon^D_{0,q-c_0^{tol}-n_0+1}(\tilde{w},B)>0$. As the ordering on ε^D is lexicographic, an optimal model should satisfy $e_0^*(B)=q-c_0^{tol}$, because models with $e_0^*(B)<q-c_0^{tol}$ are not allowed and models with $e_0^*(B)>q-c_0^{tol}$ can be improved by deleting an equation. Similarly, once $B_{t-1}^{\perp}$ has been identified, the requirements in c_{tol} imply a minimal required number e_t of truly t-th order laws in the space $v_t^{-1}\{\,[v_t(B_{t-1}^{\perp}+sB_{t-1}^{\perp})]^{\perp}\,\}$. Let n_t denote the number of independent t-th order equations in this space which are exactly satisfied by the data. Under the reasonable assumption that $e_t\ge n_t$ it follows that for optimal models $e_t^*(B)=e_t$. Roughly stated, due to the lexicographic ordering it is preferable to accept as few low order equations as possible, given the complexity constraint.

It can be shown that for generic data $\tilde{w}$ there holds $n_t=0$ for all $t\le\bar{d}(\mathcal{T})$. So in this case assumption (iii) is satisfied

Under assumption 7-2, due to the lexicographic ordering on ε^D we first have to identify $e_0(c_{tol})$ zero order equations of minimal misfit. In the following section it will be assumed that this problem has a unique solution. This holds true for generic data. Let the solution be L_0 and define $B_0^{\perp}:=L_0$, $V_0:=v_0(L_0)$. Next we have to identify $e_1(c_{tol})$ equations of first order and minimal misfit, under the restriction that the equations are truly first order, i.e., orthogonal to $B_0^{\perp}+sB_0^{\perp}$. A second (generically satisfied) assumption is that this problem also has a unique solution, say L_1. Let $V_1:=v_1(L_1)\perp v_1(B_0^{\perp}+sB_0^{\perp})$ and $B_1^{\perp}:=B_0^{\perp}+sB_0^{\perp}+L_1$. In the same way we identify $e_t(c_{tol})$ equations of truly t-th order of minimal misfit. It is assumed that this problem has a unique solution L_t. Let $V_t:=v_t(L_t)$ and $B_t^{\perp}:=B_{t-1}^{\perp}+sB_{t-1}^{\perp}+L_t$. The resulting model is then defined by $B:=\{\,w\in(\mathbb{R}^q)^{\mathbb{Z}}\,;\ r(\sigma)w=0 \text{ for all } r\in\bigcup_{t\ge0}B_t^{\perp}\,\}$. For this B there holds $L_t=L_t^D$ of (CDF). Moreover, for generic data $\tilde{w}$ the model B is uniquely defined by $\tilde{w}$

and gives the optimal model $P^D_{c_{tol}}(\tilde{w})$.

Note that the foregoing consists of sequential optimal choice of $e_t(c_{tol})$ descriptive equations of minimal misfit. Every step of this sequential optimization will be solved by means of an algorithm corresponding to proposition 2-8.

In the next sections we describe computational details of this algorithm and the other ones. We specify input, initialization, recursive part, termination and output of the algorithms. Moreover, we state the optimality properties of the resulting models in terms of assumptions on the data which are generically satisfied. We refer also to Willems [15] and Heij [4].

In the algorithms we will use the notation $A=\text{col}(A_1,\ldots,A_n)$ to indicate the matrix $A\in\mathbb{R}^{l\times n}$ with blockrows $A_i\in\mathbb{R}^{l_i\times m}$, $i=1,\ldots,n$, where $l:=\sum_{i=1}^{n} l_i$.

7.2. Descriptive modelling, given tolerated complexity

In this section we describe an algorithm which for generic data $\tilde{w}\in(\mathbb{R}^q)^{\mathcal{T}}$ and sensible tolerated complexity c_{tol} generates the model $\{B\}=P^D_{c_{tol}}(\tilde{w})$ as defined in section 6.2. We first give the algorithm and subsequently state the generic conditions on the data.

Algorithm for $P^D_{c_{tol}}$.

1. *Input.*

1.1. Data $\tilde{w}=(\tilde{w}(t);\ t\in\mathcal{T}=[t_0,t_1])\in(\mathbb{R}^q)^{\mathcal{T}}$.

1.2. Tolerated complexity $c_{tol}=(c_t^{tol};\ t\in\mathbb{Z}_+)\in(\mathbb{R}_+)^{\mathbb{Z}_+}$.

Let $e_{tol}:=e(c_{tol})$ denote the equation structure corresponding to c_{tol}.

2. *Initialization* (*step* 0).

2.1. Let $S(\tilde{w},0):=\frac{1}{t_1-t_0+1}\cdot\sum_{t=t_0}^{t_1}\tilde{w}(t)\tilde{w}(t)^T$, the empirical covariance matrix of order 0, have singular value decomposition (SVD) $S(\tilde{w},0)=U_0\Sigma_0U_0^T$, $\Sigma_0=\text{diag}(\sigma_1^{(0)},\ldots,\sigma_q^{(0)})$, $\sigma_1^{(0)}\geq\ldots\geq\sigma^{(0)}_{q-e_0^{tol}}\geq\sigma^{(0)}_{q-e_0^{tol}+1}\geq\ldots\geq\sigma_q^{(0)}\geq 0$.

2.2. If $U_0=(u_1^{(0)},\ldots,u_q^{(0)})$, $u_k^{(0)}\in\mathbb{R}^q$, $k=1,\ldots,q$, then define $V_0:=\text{span}\{u_k^{(0)T};\ k\geq q-e_0^{tol}+1\}$ and $B_0^{\perp}:=v_0^{-1}(V_0)$.

2.3. Define $p_1:=2e_0^{tol}$ and let $\{v_k^{(1)T};\ k=1,\ldots,p_1\}$ be an orthonormal basis of $v_1(B_0^\perp + sB_0^\perp) \subset \mathbb{R}^{1\times 2q}$, e.g., $v_k^{(1)T}$ is the k-th row of $\begin{bmatrix} \bar{U}_0 & 0 \\ 0 & \bar{U}_0 \end{bmatrix}$ where $\bar{U}_0 := \mathrm{col}(u_k^{(0)T};\ k = q - e_0^{tol} + 1, \ldots, q)$.

3. *Recursion (step t).*

3.0. Input from step $t-1$: an orthonormal basis $\{v_k^{(t)T};\ k=1,\ldots,p_t\}$ of $v_t(B_{t-1}^\perp + sB_{t-1}^\perp) \subset \mathbb{R}^{1\times q(t+1)}$, where $p_t = \dim(v_t(B_{t-1}^\perp + sB_{t-1}^\perp)) = \sum_{k=0}^{t-1} (t+1-k).\ e_k^{tol}$.

SVD: $\sum_{k=1}^{p_t} v_k^{(t)} v_k^{(t)T} = \bar{V}_t \bar{\Sigma}_t \bar{V}_t^T$, $\bar{\Sigma}_t = \mathrm{diag}(\bar{\sigma}_1^{(t)}, \ldots, \bar{\sigma}_{q(t+1)}^{(t)})$, $1 = \bar{\sigma}_1^{(t)} = \ldots = \bar{\sigma}_{p_t}^{(t)} > \bar{\sigma}_{p_t+1}^{(t)} = \ldots = \bar{\sigma}_{q(t+1)}^{(t)} = 0$, $\bar{V}_t = (v_1^{(t)}, \ldots, v_{p_t}^{(t)}, v_{p_t+1}^{(t)}, \ldots, v_{q(t+1)}^{(t)})$. Let $q_t := q(t+1) - p_t$ and define $P_t := \mathrm{col}(v_k^{(t)T}; k = p_t+1, \ldots, q(t+1)) \in \mathbb{R}^{q_t \times q(t+1)}$. So the rows of P_t form an orthonormal basis for $[v_t(B_{t-1}^\perp + sB_{t-1}^\perp)]^\perp \subset \mathbb{R}^{1\times q(t+1)}$.

3.1. Let $S(\tilde{w},t) := \frac{1}{t_1 - t_0 - t + 1} . \sum_{k=t_0}^{t_1 - t} (\tilde{w}(k)^T, \ldots, \tilde{w}(k+t)^T)^T . (\tilde{w}(k)^T, \ldots, \tilde{w}(k+t)^T)$, the empirical covariance matrix of order t, and let $P_t S(\tilde{w},t) P_t^T$ have SVD $P_t S(\tilde{w},t) P_t^T = U_t \Sigma_t U_t^T$, $\Sigma_t = \mathrm{diag}(\sigma_1^{(t)}, \ldots, \sigma_{q_t}^{(t)})$, $\sigma_1^{(t)} \geq \ldots \geq \sigma_{q_t - e_t^{tol}}^{(t)} \geq \sigma_{q_t - e_t^{tol} + 1}^{(t)} \geq \ldots \geq \sigma_{q_t}^{(t)} \geq 0$.

3.2. If $U_t = (u_1^{(t)}, \ldots, u_{q_t}^{(t)})$, $u_k^{(t)} \in \mathbb{R}^{q_t}$, $k = 1, \ldots, q_t$, then define $V_t := \mathrm{span}\{u_k^{(t)T}.P_t;\ k \geq q_t - e_t^{tol} + 1\}$, $L_t := v_t^{-1}(V_t) \subset \{r \in \mathbb{R}^{1\times q}[s];\ r = \sum_{k=0}^{t} r_k s^k,\ r_k \in \mathbb{R}^{1\times q},\ k = 0, \ldots, t\}$ and $B_t^\perp := B_{t-1}^\perp + sB_{t-1}^\perp + L_t$.

3.3. Output to step $t+1$: an orthonormal basis $\{v_k^{(t+1)T};\ k=1,\ldots,p_{t+1}\}$ of $v_{t+1}(B_t^\perp + sB_t^\perp)$, $p_{t+1} := \sum_{k=0}^{t} (t+2-k).e_k^{tol}$.

Note that $O_t := \{v_k^{(t)T};\ k = 1, \ldots, p_t\} \cup \{u_k^{(t)T}.P_t;\ k = q_t - e_t^{tol} + 1, \ldots, q_t\}$ forms an orthonormal basis of $v_t(B_t^\perp)$, with $\dim(O_t) = \sum_{k=0}^{t} (t+1-k) e_k^{tol}$. Let $O_t^0 := \{(v,0);\ v \in O_t,\ 0 \in \mathbb{R}^{1\times q}\}$ and ${}^0O_t := \{(0,v);\ 0 \in \mathbb{R}^{1\times q},\ v \in O_t\}$, then it suffices to choose $\sum_{k=0}^{t} e_k^{tol}$ orthonormal vectors in span ${}^0O_t$, orthogonal to O_t^0.

4. *Termination (at step t^*).*

Either at $t^* = \bar{d}(\mathcal{T}) := (t_1 - t_0 + 1 - q)/(q+1)$, or at $t^* < \bar{d}(\mathcal{T})$ when $\sum_{t=0}^{t^*} e_t^{tol} = q$.

5. *Output.*

Bases for V_t, $t \le t^*$, and $B^{\perp}_{t^*}$. Define $B := \{w \in (\mathbb{R}^q)^{\mathbb{Z}};\ r(\sigma)w = 0, r \in B^{\perp}_{t^*}\}$.

We remark that the algorithm basically consists of *sequential* application of proposition 2–8 in section 2.6. In the initialization the data is $x_i := \tilde{w}(t_0+i)$, $i = 0,\ldots,t_1-t_0$. In step t of the recursion the data consists of $x_i := P_t.\mathrm{col}(\tilde{w}(t_0+i),\ldots,\tilde{w}(t_0+i+t))$, $i = 0,\ldots,t_1-t_0-t$. The operators P_t take care of the requirement that the new laws should be orthogonal to the old ones. Concerning step 3.1 note that for laws r with $d(r) = t$ and $v_t(r) \in [v_t(B^{\perp}_{t-1}+sB^{\perp}_{t-1})]^{\perp}$ there holds $\|r\tilde{w}\|^2 = v_t(r).P_t.S(\tilde{w},t).P^T_t.v_t(r)^T$.

Next we state the assumptions on $\tilde{w}$ and c_{tol}.

Assumption 7–4 ($P^D_{c_{tol}}$). Let $c_{tol} \in (\mathbb{R}_+)^{\mathbb{Z}_+}$ and $\tilde{w} \in (\mathbb{R}^q)^{\mathcal{T}}$ be given.

(*i*) assumption 7–2(i);

(*ii*) $\sigma^{(0)}_{q-e_0^{tol}} > \sigma^{(0)}_{q-e_0^{tol}+1}$; in step t $\sigma^{(t)}_{q_t-e_t^{tol}} > \sigma^{(t)}_{q_t-e_t^{tol}+1}$;

(*iii*) for step t, let $u_k^{(t)T}.P_t = (u_{k,0},\ldots,u_{k,t})$, $u_{k,j} \in \mathbb{R}^{1\times q}$, and $U_0 :=$ $\mathrm{col}\{u_{k,0};\ k \ge q_t - e_t^{tol}+1\}$, $U_t := \mathrm{col}\{\ u_{k,t};\ k \ge q_t - e_t^{tol}+1\}$; assume $\mathrm{rank}(U_0) = \mathrm{rank}(U_t) = e_t^{tol}$.

Assumption (i) expresses a sensibility requirement for c_{tol}. Assumption (ii) is satisfied for generic data and guarantees the existence of a unique solution for the problem of optimal choice of e_t^{tol} equations of order t, orthogonal to $B^{\perp}_{t-1}+sB^{\perp}_{t-1}$. Assumption 7–4(ii) implies assumption 7–2(ii) and (iii). Assumption 7–4(iii) is satisfied for generic data and corresponds to requiring that the laws, identified in step t, really have order t, i.e., $\{0 \ne r \in L_t\} \Rightarrow \{d(r) = t\}$.

Theorem 7–5 Suppose assumption 7–4 is satisfied, then

(*i*) $P^D_{c_{tol}}(\tilde{w}) = \{B\}$, the model generated by the algorithm;

(*ii*) $e^*(B) = e_{tol}$;

(*iii*) $\varepsilon^D_{t,k}(\tilde{w},B) = \{\sigma^{(t)}_{q_t-e_t^{tol}+k}\}^{1/2}$, $k = 1,\ldots,e_t^{tol}$;

(*iv*) $L_t = L_t^D$ for B, so the algorithm gives a CDF representation of B.

Optimality of the model generated by the algorithm follows from proposition 2–8, due to the lexicographic ordering on ε^D and assumption 7–4(ii).

It can be shown that the algorithm always generates an allowable model, i.e., $c_t(B) \le c_t^{tol}$ for all $t \in \mathbb{Z}$. However, the generated model may be suboptimal in case assumption 7-4 is not satisfied, i.e., for non-generic data.

7.3. Descriptive modelling, given tolerated misfit

Next we describe an algorithm which for generic data $\tilde{w} \in (\mathbb{R}^q)^{\mathcal{T}}$ and sensible tolerated misfit generates the model $P^D_{\varepsilon_{tol}}(\tilde{w})$ as defined in section 6.2. The algorithm basically consists of sequential application of proposition 2-9. The (generic) optimality of the model generated by the algorithm is a consequence of proposition 2-9 and the special utility $u_{\varepsilon_{tol}}$ as defined in definition 6-2.

Algorithm for $P^D_{\varepsilon_{tol}}$.

1. *Input.*

1.1. Data $\tilde{w} = (\tilde{w}(t);\ t \in \mathcal{T} = [t_0, t_1]) \in (\mathbb{R}^q)^{\mathcal{T}}$.

1.2. Tolerated misfit $\varepsilon_{tol} = (\varepsilon_t^{tol};\ t \in \mathbb{Z}_+)$, $\varepsilon_t^{tol} = \bar{\varepsilon}_t^{tol}.(1,\ldots,1) \in \mathbb{R}^{1\times q}$, $\bar{\varepsilon}_t^{tol} \in \mathbb{R}$.

2. *Initialization* (*step* 0).

2.1. SVD: $S(\tilde{w},0) = U_0 \Sigma_0 U_0^T$, $\Sigma_0 = \text{diag}(\sigma_1^{(0)},\ldots,\sigma_q^{(0)})$, $\sigma_1^{(0)} \ge \ldots \ge \sigma_{q-e_0}^{(0)} \ge (\bar{\varepsilon}_0^{tol})^2 > \sigma_{q-e_0+1}^{(0)} \ge \ldots \ge \sigma_q^{(0)} \ge 0$.

2.2. If $U_0 = (u_1^{(0)},\ldots,u_q^{(0)})$, $u_k^{(0)} \in \mathbb{R}^q$, $k=1,\ldots,q$, then define $V_0 := \text{span}\{u_k^{(0)T}; k \ge q-e_0+1\}$ and $B_0^{\perp} := v_0^{-1}(V_0)$.

2.3. Define $p_1 := 2e_0$ and let $\{v_k^{(1)T};\ k=1,\ldots,p_1\}$ be an orthonormal basis of $v_1(B_0^{\perp} + sB_0^{\perp}) \subset \mathbb{R}^{1\times 2q}$, e.g., $v_k^{(1)T}$ is the k-th row of $\begin{bmatrix} \bar{U}_0 & 0 \\ 0 & \bar{U}_0 \end{bmatrix}$ where $\bar{U}_0 := \text{col}(u_k^{(0)T};\ k = q-e_0+1,\ldots,q)$.

3. *Recursion* (*step* t).

3.0. Input from step $t-1$: an orthonormal basis $\{v_k^{(t)T};\ k=1,\ldots,p_t\}$ of $v_t(B_{t-1}^{\perp} + sB_{t-1}^{\perp}) \subset \mathbb{R}^{1\times q(t+1)}$, where $p_t = \dim(v_t(B_{t-1}^{\perp} + sB_{t-1}^{\perp})) = \sum_{k=0}^{t-1}(t+1-k).e_k$, where e_k is the number of accepted k-th order laws. Let $q_t := q(t+1) - p_t$, $e_t' := q - \sum_{k=0}^{t-1} e_k$ and define P_t as in step 3.0 of the algorithm for $P^D_{c_{tol}}$.

3.1. SVD: $P_t S(\tilde{w},t) P_t^T = U_t \Sigma_t U_t^T$, $\Sigma_t = \text{diag}(\sigma_1^{(t)},\ldots,\sigma_{q_t}^{(t)})$, $\sigma_1^{(t)} \ge \ldots \ge \sigma_{q_t-e_t''}^{(t)} \ge (\bar{\varepsilon}_t^{tol})^2 > \sigma_{q_t-e_t''+1}^{(t)} \ge \ldots \ge \sigma_{q_t}^{(t)} \ge 0$.

3.2. If $U_t = (u_1^{(t)},\ldots,u_{q_t}^{(t)})$, $u_k^{(t)} \in \mathbb{R}^{q_t}$, $k=1,\ldots,q_t$, then with $e_t := \min\{e_t', e_t''\}$ define $V_t := \text{span}\{u_k^{(t)T}.P_t;\ k \ge q_t - e_t + 1\}$, $L_t := v_t^{-1}(V_t)$ and $B_t^{\perp} :=$

$B^{\perp}_{t-1}+sB^{\perp}_{t-1}+L_t$.

3.3. Output to step $t+1$: an orthonormal basis $\{v_k^{(t+1)T};\ k=1,\ldots,p_{t+1}\}$ of $v_{t+1}(B^{\perp}_t+sB^{\perp}_t)$, $p_{t+1}:=\sum_{k=0}^{t}(t+2-k).e_k$. See also step 3.3 of the algorithm for $P^D_{c_{tol}}$.

4. *Termination (at step t^*).*

Either at $t^*=\bar{d}(\mathcal{T})$, or at $t^*<\bar{d}(\mathcal{T})$ when $\sum_{t=0}^{t^*}e_t=q$ or $\bar{\varepsilon}_t^{tol}\leq 0$ for $t>t^*$.

5. *Output.*

Bases for V_t, $t\leq t^*$, and $B^{\perp}_{t^*}$. Define $B:=\{w\in(\mathbb{R}^q)^{\mathbb{Z}};\ r(\sigma)w=0,\ r\in B^{\perp}_{t^*}\}$.

We will make the following assumptions on $\tilde{w}$ and ε_{tol}.

Assumption 7-6 ($P^D_{\varepsilon_{tol}}$). Let $(\bar{\varepsilon}_t^{tol};\ t\in\mathbb{Z}_+)\in\mathbb{R}^{\mathbb{Z}_+}$ and $\tilde{w}\in(\mathbb{R}^q)^{\mathcal{T}}$ be given.

(*i*) $\bar{\varepsilon}_t^{tol}\leq 0$ for all $t>\bar{d}(\mathcal{T})$;

(*ii*) if at t^* $e''_{t^*}>e'_{t^*}(>0)$, then assume $\sigma^{(t^*)}_{q_t-e'_{t^*}}>\sigma^{(t^*)}_{q_t-e'_{t^*}+1}$;

(*iii*) assumption 7-4(iii), with e_t^{tol} replaced by e_t.

Here (i) expresses a sensibility requirement for ε_{tol}, (ii) is satisfied for generic data and guarantees the uniqueness of $P^D_{\varepsilon_{tol}}(\tilde{w})$, and (iii) is satisfied for generic data and amounts to requiring that the laws, identified in step t, really have order t.

Theorem 7-7 Suppose assumption 7-6 is satisfied, then

(*i*) $P^D_{\varepsilon_{tol}}(\tilde{w})=\{B\}$, the model generated by the algorithm;

(*ii*) $e^*(B)=(e_t;\ t\in\mathbb{Z}_+)$;

(*iii*) $\varepsilon^D_{t,k}(\tilde{w},B)=\{\sigma^{(t)}_{q_t-e_t+k}\}^{1/2}$, $k=1,\ldots,e_t$;

(*iv*) $L_t=L_t^D$ for B, so the algorithm gives a CDF representation of B.

7.4. Predictive modelling, given tolerated complexity

In this section we give an algorithm which for generic data $\tilde{w}\in(\mathbb{R}^q)^{\mathcal{T}}$ and sensible tolerated complexity c_{tol} generates the model $\{B\}=P^P_{c_{tol}}(\tilde{w})$ as defined in section 6.3. We first give the algorithm and subsequently state the generic conditions on the data.

Algorithm for $P^P_{c_{tol}}$.

1. *Input.*

As for $P^D_{c_{tol}}$.

2. *Initialization* (*step* 0).

2.1. As for $P^D_{c_{tol}}$.

2.2. As for $P^D_{c_{tol}}$.

2.3. Define $p_0 := e_0^{tol}$, $n_0 := e_0^{tol}$ and let $\{v_k^{(0)T};\ k \geq q - e_0^{tol} + 1\}$, $v_k^{(0)} := u_k^{(0)}$, $k \geq q - e_0^{tol} + 1$, be an orthonormal basis of $v_0(B_0^\perp)$ and $F_0 = v_0(B_0^\perp)$, where F_0 is as defined in section 4.4.

3. *Recursion* (*step* t).

3.0. Input from step $t-1$: an orthonormal basis $\{v_k^{(t-1)T};\ k = 1, \ldots, p_{t-1}\}$, $p_{t-1} := \sum_{k=0}^{t-1} (t-k) e_k^{tol}$, of $v_{t-1}(B_{t-1}^\perp) \subset \mathbb{R}^{1 \times qt}$, and an orthonormal basis $\{f_k^{(t-1)T};\ k = 1, \ldots, n_{t-1}\}$, $n_{t-1} := \sum_{k=0}^{t-1} e_k^{tol}$, of $F_{t-1} := \{\tilde{r} \in \mathbb{R}^{1 \times q};\ \exists r \in B_{t-1}^\perp,\ r = \sum_{k=0}^{t-1} r_k s^k$, such that $r_{t-1} = \tilde{r}\}$.

SVD: $\sum_{k=1}^{p_{t-1}} v_k^{(t-1)} v_k^{(t-1)T} = \bar{V}_{t-1} \bar{\Sigma}_{t-1} \bar{V}_{t-1}^T$, $\bar{\Sigma}_{t-1} = \mathrm{diag}(\bar{\sigma}_1^{(t-1)}, \ldots, \bar{\sigma}_{q.t}^{(t-1)})$, $1 = \bar{\sigma}_1^{(t-1)} = \ldots = \bar{\sigma}_{p_{t-1}}^{(t-1)} > \bar{\sigma}_{p_{t-1}+1}^{(t-1)} = \ldots = \bar{\sigma}_{q.t}^{(t-1)} = 0$, $\bar{V}_{t-1} = (v_1^{(t-1)}, \ldots, v_{p_{t-1}}^{(t-1)}, v_{p_{t-1}+1}^{(t-1)}, \ldots, v_{q.t}^{(t-1)})$. Let $q_t := q.t - p_{t-1}$ and define $P_{1t} := \mathrm{col}(v_k^{(t-1)T};\ k = p_{t-1}+1, \ldots, q{\cdot}t) \in \mathbb{R}^{q_t \times q \cdot t}$.

Similarly, SVD: $\sum_{k=1}^{t-1} f_k^{(t-1)} f_k^{(t-1)T} = \bar{\bar{V}}_{t-1} \bar{\bar{\Sigma}}_{t-1} \bar{\bar{V}}_{t-1}^T$, $\bar{\bar{\Sigma}}_{t-1} = \mathrm{diag}(\bar{\bar{\sigma}}_1^{(t-1)}, \ldots, \bar{\bar{\sigma}}_q^{(t-1)})$, $1 = \bar{\bar{\sigma}}_1^{(t-1)} = \ldots = \bar{\bar{\sigma}}_{n_{t-1}}^{(t-1)} > \bar{\bar{\sigma}}_{n_{t-1}+1}^{(t-1)} = \ldots = \bar{\bar{\sigma}}_q^{(t-1)} = 0$, $\bar{\bar{V}}_{t-1} = (f_1^{(t-1)}, \ldots, f_q^{(t-1)})$. Define $P_{2t} := \mathrm{col}(f_k^{(t-1)T};\ k = n_{t-1}+1, \ldots, q) \in \mathbb{R}^{(q - n_{t-1}) \times q}$.

Finally let $P_t := \begin{bmatrix} P_{1t} & 0 \\ 0 & P_{2t} \end{bmatrix}$. Then the rows of P_t form an orthonormal basis for $[v_t(F_{t-1}.s^t) + v_t(B_{t-1}^\perp)]^\perp \subset \mathbb{R}^{1 \times q(t+1)}$.

3.1 Let $P_t S(\tilde{w}, t) P_t^T = \begin{bmatrix} S_-^{(t)} & S_{-+}^{(t)} \\ S_{+-}^{(t)} & S_+^{(t)} \end{bmatrix}$ with $S_-^{(t)} \in \mathbb{R}^{q_t \times q_t}$,

$S_{+}^{(t)} \in \mathbb{R}^{(q-n_{t-1})\times(q-n_{t-1})}$, $S_{-+}^{(t)} = S_{+-}^{(t)T} \in \mathbb{R}^{q_t \times (q-n_{t-1})}$.

SVD: $(S_{-}^{(t)})^{-1/2}.S_{-+}^{(t)}.(S_{+}^{(t)})^{-1/2} = U_t^{-}\Lambda_t U_t^{+T}$, $\Lambda_t = \begin{bmatrix} \Sigma_t \\ 0 \end{bmatrix} \in \mathbb{R}^{q_t \times (q-n_{t-1})}$,

$\Sigma_t = \operatorname{diag}(\sigma_1^{(t)},\ldots,\sigma_{q-n_{t-1}}^{(t)})$, $\sigma_1^{(t)} \geq \ldots \geq \sigma_{e_t^{tol}}^{(t)} \geq \sigma_{e_t^{tol}+1}^{(t)} \geq \ldots \geq \sigma_{q-n_{t-1}}^{(t)} \geq 0$.

3.2. If $(S_{-}^{(t)})^{-1/2}.U_t^{-} = (\bar{u}_1^{(t)},\ldots,\bar{u}_{q_t}^{(t)})$ and $(S_{+}^{(t)})^{-1/2}.U_t^{+} = (\bar{\bar{u}}_1^{(t)},\ldots,\bar{\bar{u}}_{q-n_t}^{(t)})$, then for $k \leq e_t^{tol}$ let $u_k^{(t)T} := (-\sigma_k^{(t)}.\bar{u}_k^{(t)T}, \bar{\bar{u}}_k^{(t)T}).P_t \in \mathbb{R}^{1\times q(t+1)}$.

Define $V_t := \operatorname{span}\{u_k^{(t)T};\ k \leq e_t^{tol}\}$, $L_t := v_t^{-1}(V_t)$ and $B_t^{\perp} := B_{t-1}^{\perp} + sB_{t-1}^{\perp} + L_t$.

3.3. Output to step $t+1$: orthonormal bases $\{v_k^{(t)};\ k=1,\ldots,p_t\}$ of $v_t(B_t^{\perp})$ and $\{f_k^{(t)T};\ k=1,\ldots,n_t\}$ of F_t. Here $p_t := p_{t-1} + \sum_{k=0}^{t} e_k^{tol}$ and $n_t := n_{t-1} + e_t^{tol}$.

Note that a basis for F_t is $\{f_k^{(t-1)T};\ k=1,\ldots,n_{t-1}\} \cup \{\bar{\bar{u}}_k^{(t)T}.P_{2t};\ k \leq e_t^{tol}\}$. Further, let $O_{t-1} := \{v_k^{(k-1)T};\ k=1,\ldots,p_{t-1}\}$, $O_{t-1}^{0} := \{(v,0);\ v \in O_{t-1},\ 0 \in \mathbb{R}^{1\times q}\}$ and ${}^{0}O_{t-1} := \{(0,v);\ 0 \in \mathbb{R}^{1\times q},\ v \in O_{t-1}\}$. For $v_t(B_t^{\perp})$ it then suffices to take O_{t-1}^{0}, V_t, and n_{t-1} orthonormal vectors in span ${}^{0}O_{t-1}$, orthogonal to $O_{t-1}^{0} + V_t$.

4. *Termination* (*at step* t^{*}).
   As for $P_{c_{tol}}^{D}$.

5. *Output*.
   Bases for V_t, $t \leq t^{*}$, and $B_{t^{*}}^{\perp}$. Define $B := \{w \in (\mathbb{R}^q)^{\mathbb{Z}};\ r(\sigma)w = 0,\ r \in B_{t^{*}}^{\perp}\}$.

We remark that the algorithm basically consists of *sequential* application of proposition 2–13 of section 2.7. As a rough outline, $P_{c_{tol}}^{P}$ models data by successively minimizing the misfit of a required number e_0^{tol} of zero order laws, then minimizing the predictive misfit of a required number e_1^{tol} of first order laws, and so on. In order to measure the misfit more or less independently, as made precise in section 5.3, the newly identified laws r of order t have to be elements of the space $[v_t(F_{t-1}.s^{t}) + v_t(B_{t-1}^{\perp})]^{\perp}$, see section 4.4. The operator P_t takes care of this requirement. The resulting optimization problem of step t of the recursion is of a static nature as described in section 2.7. The data consists of (x_i, y_i), $i = 0,\ldots,t_1-t_0-t$, with $y_i := P_{2t}\tilde{w}(t_0+t+i)$ and $x_i := P_{1t}.\operatorname{col}(\tilde{w}(t_0+i),\ldots,\tilde{w}(t_0+t-1+i))$.

Next we state the assumption on $\tilde{w}$ and c_{tol}.

Assumption 7-8 ($P_{c_{tol}}^{P}$). Let $c_{tol} \in (\mathbb{R}_{+})^{\mathbb{Z}_{+}}$ and $\tilde{w} \in (\mathbb{R}^q)^{\mathcal{T}}$ be given.

(*i*) assumption 7–2(i);

(*ii*) $\sigma^{(0)}_{q-e_0^{tol}} > \sigma^{(0)}_{q-e_0^{tol}+1}$; in step t $\sigma^{(t)}_{e_t^{tol}} > \sigma^{(t)}_{e_t^{tol}+1}$;

(*iii*) for step t, let $u_k^{(t)T} = (u_{k,0}, \ldots, u_{k,t})$, $u_{k,j} \in \mathbb{R}^{1\times q}$, and $U_0 := \text{col}\{u_{k,0};\ k \le e_t^{tol}\}$, $U_t := \text{col}\{u_{k,t};\ k \le e_t^{tol}\}$; assume $\text{rank}(U_0) = \text{rank}(U_t) = e_t^{tol}$;

(*iv*) for step t, $S_-^{(t)}$ and $S_+^{(t)}$ have full rank.

Here (i) is a sensibility requirement for c_{tol}. Assumption (ii) is satisfied for generic data and implies assumption 7–2(ii) and (iii). Assumption (iii) also is satisfied for generic data and corresponds to requiring that the laws, identified in step t, really have order t, i.e., $\{0 \ne r \in L_t\} \Rightarrow \{d(r) = t\}$. Also, given assumption (i), assumption (iv) is satisfied for generic data, which is seen as follows. For step t, the number of data is $t_1 - t_0 - t + 1$ and $S_-^{(t)} \in \mathbb{R}^{q_t \times q_t}$, $S_+^{(t)} \in \mathbb{R}^{(q-n_{t-1})\times(q-n_{t-1})}$. As $q_t \le q.t$, $q - n_{t-1} \le q.t$, $S_-^{(t)}$ and $S_+^{(t)}$ generically have full rank if $t_1 - t_0 - t + 1 \ge q.t$, i.e., $t \le (t_1 - t_0 + 1)/(q+1)$, which is implied by assumption (i).

The following theorem is a consequence of proposition 2–13 and the lexicographic ordering of ε^P.

Theorem 7–9 Suppose assumption 7–8 is satisfied, then

(*i*) $P^P_{c_{tol}}(\tilde{w}) = \{B\}$, the model generated by the algorithm;

(*ii*) $e^*(B) = e_{tol}$;

(*iii*) $\varepsilon^P_{t,k}(\tilde{w}, B) = \{1 - (\sigma^{(t)}_{e_t^{tol}-k+1})^2\}^{1/2}$, $k = 1, \ldots, e_t^{tol}$;

(*iv*) $L_t = L_t^P$ for B, so the algorithm gives a CPF representation of B.

7.5. Predictive modelling, given tolerated misfit

Finally we give an algorithm which for generic data $\tilde{w} \in (\mathbb{R}^q)^{\mathcal{T}}$ and sensible ε_{tol} generates the model $P^P_{\varepsilon_{tol}}(\tilde{w})$ as defined in section 6.3. The algorithm basically consists of sequential application of proposition 2–14 of section 2.7. The (generic) optimality of the model generated by the algorithm is a consequence of proposition 2–14 and the special utility $u_{\varepsilon_{tol}}$ as defined in definition 6–2.

Algorithm for $P^P_{\varepsilon_{tol}}$.

1. *Input.*

   As for $P^D_{\varepsilon_{tol}}$

2. *Initialization* (*step* 0).

2.1. As for $P^D_{\varepsilon_{tol}}$.

2.2. As for $P^D_{\varepsilon_{tol}}$.

2.3. As for $P^P_{c_{tol}}$, with e_0^{tol} replaced by e_0.

3. *Recursion* (*step* t).

3.0. As for $P^P_{c_{tol}}$, with e_k^{tol} replaced e_k, $k \leq t-1$; let $e_t' := q - \sum_{k=0}^{t-1} e_k$.

3.1. As for $P^P_{c_{tol}}$. Let $0 \leq 1-(\sigma_1^{(t)})^2 \leq \ldots \leq 1-(\sigma_{e_t''}^{(t)})^2 < (\bar{\varepsilon}_t^{tol})^2 \leq 1-(\sigma_{e_t''+1}^{(t)})^2 \leq \ldots \leq 1-(\sigma_{q-n_{t-1}}^{(t)})^2 \leq 1$.

3.2. As for $P^P_{c_{tol}}$, with e_t^{tol} replaced by $e_t := \min\{e_t', e_t''\}$.

3.3. As for $P^P_{c_{tol}}$, with e_t^{tol} replaced by e_t.

4. *Termination* (*at step* t^*).

As for $P^D_{\varepsilon_{tol}}$.

5. *Output*.

Bases for V_t, $t \leq t^*$, and $B^{\perp}_{t^*}$. Define $B := \{w \in (\mathbb{R}^q)^{\mathbb{Z}};\ r(\sigma)w = 0,\ r \in B^{\perp}_{t^*}\}$.

Assumption 7-10 ($P^P_{\varepsilon_{tol}}$).

(*i*) assumption 7–6(i);

(*ii*) assumption 7–6(ii);

(*iii*) assumption 7–8(iii) with e_t^{tol} replaced by e_t;

(*iv*) assumption 7–8(iv).

Again (i) is a sensibility requirement for ε_{tol}. Given (i), the assumptions (ii), (iii) and (iv) are satisfied for generic data.

Theorem 7-11 Suppose assumption 7–10 is satisfied, then

(*i*) $P^P_{\varepsilon_{tol}}(\tilde{w}) = \{B\}$, the model generated by the algorithm;

(*ii*) $e^*(B) = (e_t;\ t \in \mathbb{Z}_+)$;

(*iii*) $\varepsilon^P_{t,k}(\tilde{w}, B) = \{1-(\sigma^{(t)}_{e_t-k+1})^2\}^{1/2}$, $k = 1, \ldots, e_t$;

(*iv*) $L_t = L_t^P$ for B, so the algorithm gives a CPF representation of B.

7.6. Comments

The algorithms described in the foregoing sections allow for a simple numerical implementation of the procedures of section 6. The computational complexity is mainly determined by singular value analysis of empirical covariance matrices and, in the case of predictive modelling, determination of the square root of positive definite matrices. The algorithms have been numerically implemented and employed, e.g., for the simulations described in section 9.

The essential part of the algorithms is the construction of the complementary spaces V_t, either generating a canonical descriptive form or a canonical predictive form. The operators P_t guarantee that newly identified laws are "far" from being implied by the already identified laws. In this way the misfit is measured according to the principles of section 5. This perhaps is one of the main contributions of the paper. In assessing the quality of a model, the simultaneous nature of AR-equations representing a system is fully taken into account. The quality is measured by means of canonical parametrizations, which are not determined by (scientific) theory, but which are based upon the purpose of modelling, i.e. here, description or prediction.

The identified models may be rather sensitive for changes in c_{tol}. For changes in ε_{tol} the identified models only change at discrete critical values. This indicates that fixing the complexity (the structural form) leads to non-robust identified models. Minimizing misfit of a given parametrized model hence often leads to models which are less robust than models obtained by minimizing complexity under the constraint of a maximal tolerated misfit. So in cases where one has no strong reasons to postulate the structure of a phenomenon, it seems preferable to infer approximate structure from the data by imposing a pragmatic requirement of fit.

8. CONSISTENCY

8.1. Definition of consistency

The procedures of section 6 have a clear optimality property as *data modelling* procedures. The identified models are optimal with respect to the utility $u_{c_{tol}}$ or $u_{\varepsilon_{tol}}$. The procedures give a solution for the identification problem, i.e., given data and the model class $\mathbb{B}$, a model is chosen from the model class which is optimal in view of a criterion, based on the objective of modelling. It need not be assumed that the data are generated by a phenomenon of a certain structure. This pure data modelling is of interest e.g. in data compression, speech processing, econometrics, and so on.

However, in other cases one wants to construct a good model of the *phenomenon* which generates the data. The identified model then should not only be good with respect to the particular data, but it should be good with respect to the generating system.

In this section we will define a general concept of consistency, reflecting the purpose of constructing models which approximate the generating system in an optimal way. The approach is inspired by Ljung [9], [10]. We also refer to Heij and Willems [5].

Intuitively, a procedure is called *consistent* if the model, identified by the procedure, converges to an *optimal approximation of the generating system* when the number of observations tends to infinity. So in the limit a consistent procedure identifies a model which, within the given model class, is as close as possible to the phenomenon. In this sense a consistent procedure gives a good model of the phenomenon, provided the number of observations is large enough.

To define consistency we introduce some additional concepts. Let the set of conceivable data be $D:=\cup\{(\mathbb{R}^q)^n;\ n\in\mathbb{N}\}$, so data $\tilde{w}\in D$ consists of a finite time series $\tilde{w}=(\tilde{w}(t);\ t\in\mathcal{T}=[t_0,t_1])$ in q variables. Let $\#(\mathcal{T}):=t_1-t_0+1$ denote the number of observations. Let $\mathbb{M}$ be a class of models and $\mathbb{G}$ a class of generating systems. It is assumed that the phenomenon generating the data corresponds to a system $G\in\mathbb{G}$. This means that there is a time series $w\in(\mathbb{R}^q)^{\mathbb{Z}}$ compatible with G from which we observe $\tilde{w}=w|_{\mathcal{T}}$.

Suppose that the objectives π have been used to construct a procedure $P:D\to 2^{\mathbb{M}}$. Moreover, assume that π induces an optimal approximation map $A:\mathbb{G}\to 2^{\mathbb{M}}$. This means that, with respect to π, $A(G)$ is the set of optimal approximations within the class $\mathbb{M}$ of the system $G\in\mathbb{G}$. Often $A(G)$ will consist of a singleton. Further, let $\to$ be a concept of convergence in $2^{\mathbb{M}}$, possibly also related to π. Finally, let n.a. denote a concept of "nearly always" for systems $G\in\mathbb{G}$. Such a concept is crucial, as optimal properties of procedures can fail to hold true for nasty data which nearly never occur.

Consistency now is defined as follows.

Definition 8-1 P is called *consistent* if for all $G \in \mathbb{G}$, n.a. in $w \in G$, $P(w|_{\mathcal{T}}) \to A(G)$ if $\#(\mathcal{T}) \to \infty$.

This means that, if the length of the observed time series tends to infinity, the set of models identified by a consistent procedure converges "nearly always" to the set of optimal approximations within $\mathbb{M}$ of the generating system G.

In this paper, $A(\mathbb{G})$ will consist of singleton , i.e., for $G \in \mathbb{G}$ there exists a unique approximation $a(G) \in \mathbb{M}$, so $A(G) = \{a(G)\}$. In this case, let $\to$ be a concept of convergence in $\mathbb{M}$. Then $P: D \to 2^{\mathbb{M}}$ is called *consistent* if for all $G \in \mathbb{G}$, n.a. in $w \in G$, $P(w|_{\mathcal{T}}) = \{M(w|_{\mathcal{T}})\}$, i.e., a singleton, for $\#(\mathcal{T})$ sufficiently large, and $M(w|_{\mathcal{T}}) \to a(G)$ for $\#(\mathcal{T}) \to \infty$. By slight abuse of notation we will indicatate this by $P(w|_{\mathcal{T}}) \to A(G)$.

The consistency problem is depicted in figure 10.

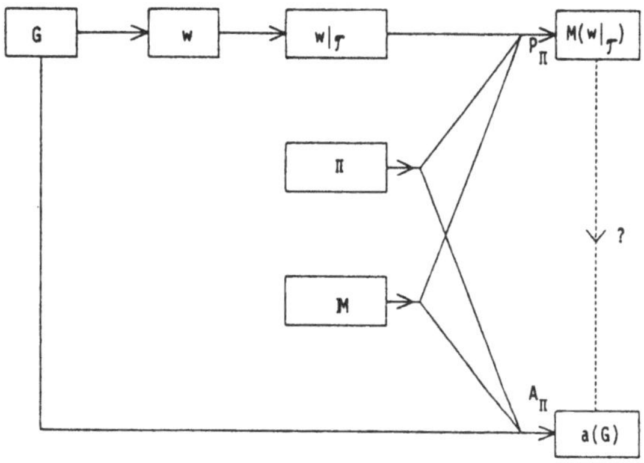

figure 10: consistency

This concept of model consistency differs in some important aspects from the concept of parameter consistency in statistics, see e.g. Kendall and Stuart [8]. In the latter case $\mathbb{M} = \mathbb{G} = \{M(\theta);\ \theta \in \Theta\}$ for some parametrized class of models (probability distributions). The data modelling problem is formulated as an estimation problem, and a modelling procedure is a map $E: D \to \Theta$. The procedure is called consistent if (n.a.) $E(w|_{\mathcal{T}}) \to \theta$ when $\#(\mathcal{T}) \to \infty$, where θ parametrizes the generating system. Model consistency differs in four main respects from this parameter consistency. First, it need not be assumed that $\mathbb{M}=\mathbb{G}$, i.e., that the generating system belongs to the model class. Second, convergence is defined in terms of models, not in terms of parametrizations. Third, parameter consistency raises problems in case of non-unique parametrizations, model consistency avoids these problems. Fourth, the models need not be stochastic.

For the case of time series analysis, see e.g. Hannan, Dunsmuir and Deistler [3] for parameter consistency and e.g. Ljung and Caines [11] for model consistency.

In the next two sections we investigate consistency of some of the procedures of section 6 for certain classes of generating systems $\mathbb{G}$. In section 8.2. we suppose $\mathbb{G}=\mathbb{B}$, i.e. the phenomenon itself is a linear, time invariant, complete (deterministic) dynamical system. In section 8.3 we consider the case where $\mathbb{G}$ consists of stochastic ARMA models and the purpose π is prediction. For this case we define optimal deterministic approximations of stochastic systems.

8.2. Deterministic generating AR-systems

Let the model class $\mathbb{M}$ again consist of the AR-models, i.e., $\mathbb{M}:=\mathbb{B}$. Suppose that the data are generated by a system $G\in\mathbb{G}=\mathbb{B}$, i.e., the generating system itself is an AR-system, so there exists an exact model of the phenomenon in the model class. In this case it is assumed that there is a system $B\in\mathbb{B}$ such that the data $\tilde{w}\in(\mathbb{R}^q)^{\mathcal{T}}$ is a finite observation of a time series $w\in(\mathbb{R}^q)^{\mathbb{Z}}$ generated by B, i.e., there is $w\in B$ with $\tilde{w}=w|_{\mathcal{T}}$. We restrict attention to so-called controllable systems B, cf. Willems [15].

Let $D:=\cup\{(\mathbb{R}^q)^n;\ n\in\mathbb{N}\}$ and $P:D\to 2^{\mathbb{M}}$ a procedure. To define consistency we specify an optimal approximation map $A:\mathbb{G}\to\mathbb{B}$ and a concept of convergence on $\mathbb{B}$. As $\mathbb{G}=\mathbb{B}$, an obvious choice for A is the identity map. Moreover, we take the discrete topology on $\mathbb{B}$. A procedure P then is consistent if for all $B\in\mathbb{B}$, n.a. in $w\in B$, there holds $P(w|_{\mathcal{T}})=\{B\}$ for $\#(\mathcal{T})$ sufficiently large. In this case, nearly always after observing a sufficiently large finite part of the time series the procedure identifies the generating system exactly.

To define n.a., we use the concept of genericity. Let $V\subset(\mathbb{R}^q)^{\mathcal{T}}$ be a linear subspace. A subset $V'\subset V$ is called generic in V if there is a polynomial $p:V\to\mathbb{R}$, $p\neq 0$ such that the complement of V' in V is contained in $p^{-1}(0)$. For $B\in\mathbb{B}$ we call $B'\subset B$ generic in B if $B'|_{\mathcal{T}}\subset B|_{\mathcal{T}}$ is generic in $B|_{\mathcal{T}}$ for $\#(\mathcal{T})$ sufficiently large. A property now is said to hold true n.a. for B if the set of points $w \in B$ where the property holds true is generic in B.

In this setting of consistency we first consider the exact modelling procedure P_{uu} as described in section 2.2.2, i.e., the procedure corresponding to undominated unfalsified modelling. So $P_{uu}: D \to 2^{\mathbb{B}}$, where for $\tilde{w} \in (\mathbb{R}^q)^{\mathcal{T}}$ $B \in P_{uu}(\tilde{w})$ if and only if $B \in \mathbb{B}$, B is unfalsified, i.e., $\tilde{w} \in B|_{\mathcal{T}}$, and B is undominated, i.e., $\{\tilde{w} \in B'|_{\mathcal{T}}, B' \in \mathbb{B}, B' \subset B\} \Rightarrow \{B' = B\}$.

Proposition 8-2 P_{uu} is not consistent.

As a simple example, take $B = (\mathbb{R}^q)^{\mathbb{Z}}$. For any $w \in B$ and any $\mathcal{T}$ of finite length there exist $B' \in \mathbb{B}$ such that $w|_{\mathcal{T}} \in B'|_{\mathcal{T}}$ and $\dim(B') \leq q.\#(\mathcal{T})$, hence $B \notin P_{uu}(w|_{\mathcal{T}})$.

Next we consider the procedures described in section 6. We define two exact and sensible modelling procedures as follows. For $k \in \mathbb{Z}_+$ let $\bar{\varepsilon}_{tol}(k) = (\bar{\varepsilon}_t^{tol}(k);\ t \in \mathbb{Z}_+) \in \mathbb{R}^{\mathbb{Z}_+}$ be defined by $\bar{\varepsilon}_t^{tol}(k) := 0$ for $0 \leq t \leq \bar{d}(k) := (k-q)/(q+1)$ and $\bar{\varepsilon}_t^{tol}(k) := -1$ for $t > \bar{d}(k)$. Let $\varepsilon_{tol}(k) := (\varepsilon_t^{tol}(k);\ t \in \mathbb{Z}_+)$ with $\varepsilon_t^{tol}(k) := \bar{\varepsilon}_t^{tol}(k).(1, \ldots, 1)$. The procedures $\bar{\bar{P}}^D_{\varepsilon_{tol}(k)}$ and $\bar{\bar{P}}^P_{\varepsilon_{tol}(k)}$ as defined in sections 6.2 and 6.3 correspond to accepting only exact laws of order at most $\bar{d}(k)$. Now define $P^D(w|_{\mathcal{T}}) := \bar{\bar{P}}^D_{\varepsilon_{tol}(\#(\mathcal{T}))}(w|_{\mathcal{T}})$ and $P^P(w|_{\mathcal{T}}) := \bar{\bar{P}}^P_{\varepsilon_{tol}(\#(\mathcal{T}))}(w|_{\mathcal{T}})$. So P^D and P^P accept the exact laws which are significant, given the number of data.

Proposition 8-3 P^D and P^P are consistent on controllable systems.

For fixed c_{tol} or ε_{tol}, i.e., independent of the number of data, the procedures $P^D_{c_{tol}}$, $P^D_{\varepsilon_{tol}}$, $P^P_{c_{tol}}$, $P^P_{\varepsilon_{tol}}$, $\bar{\bar{P}}^D_{\varepsilon_{tol}}$ and $\bar{\bar{P}}^P_{\varepsilon_{tol}}$ are not consistent, in the strict sense of exact identification for generic finite time series. We illustrate this for $P^D_{c_{tol}}$ and $P^D_{\varepsilon_{tol}}$. Similar arguments hold true for the other procedures. First suppose c_{tol} is given. Let $e_{tol} := e(c_{tol})$ be the equation structure corresponding to c_{tol}. If $e_{tol} = 0$, then $P^D_{c_{tol}}$ is not consistent for the same reasons as given for P_{uu}. If there is $t \in \mathbb{Z}_+$ with $e_t^{tol} \geq 1$, then $B \in \mathbb{B}$ with $e_t^*(B) = 0$ cannot be exactly identified, hence $P^D_{c_{tol}}$ is not consistent. Next suppose ε_{tol} is given. If $\varepsilon_{t,1}^{tol} \leq 0$ for some $t \in \mathbb{Z}_+$, then exact identification of $B \in \mathbb{B}$ with $e_t^*(B) \geq 1$ is impossible. If $\varepsilon_{t,1}^{tol} > 0$ for all $t \in \mathbb{Z}_+$, then ε_{tol} does not satisfy the sensibility assumption 7-6(i) for any $\mathcal{T}$. Moreover, as $\varepsilon_{0,1}^{tol} > 0$ $P^D_{\varepsilon_{tol}}(w|_{\mathcal{T}})$ will accept laws of order 0 for $w|_{\mathcal{T}} \in (\mathbb{R}^q)^{\mathcal{T}}$ of sufficiently small norm. Not having this sufficiently small norm is not a generic property for any $B \in \mathbb{B}$ with $B \neq \{0\}$. If $B \in \mathbb{B}$ with $e_0^*(B) = 0$, then $P^D_{\varepsilon_{tol}}$

in this case cannot exactly identify B generically, hence $P^D_{\varepsilon_{tol}}$ is not consistent.

An interesting question is the relationship between consistency of $P^D_{\varepsilon_{tol}}$ and $P^P_{\varepsilon_{tol}}$ and a definition of n.a. in terms of *"sufficient excitation"*. Without going into details, the procedures are consistent for the class of controllable systems if n.a. is defined in terms of sufficient excitation of the inputs with respect to ε_{tol}. Exact identification then is guaranteed provided the inputs are sufficiently rich with respect to ε_{tol}.

8.3. Stochastic generating ARMA-systems

8.3.1. Introduction

In this section we will consider the predictive procedures $P^P_{c_{tol}}$ and $P^P_{\varepsilon_{tol}}$ in case the data consist of a finite part of a realization of a stochastic process. In section 8.3.4 we will define the optimal approximation of a stochastic process by a deterministic system, given c_{tol} or ε_{tol}. Roughly speaking, the optimal deterministic approximation is described by the predictive relationships corresponding to c_{tol} or ε_{tol} in case the stochastic process were known. Note that both deterministic and stochastic systems generally can be given an interpretation in terms of (optimal) one-step-ahead prediction by means of deterministic equations.

A similar exposition could be given for the descriptive procedures $P^D_{c_{tol}}$ and $P^D_{\varepsilon_{tol}}$. However, in general it seems difficult to give an interpretation of stochastic systems in terms of deterministic descriptive relationships. Therefore we restrict attention to $P^P_{c_{tol}}$ and $P^P_{\varepsilon_{tol}}$.

In the following we introduce a concept of convergence on $\mathbb{B}$, describe a class of generating ARMA-systems, define optimal approximation maps $A^P_{c_{tol}}$ and $A^P_{\varepsilon_{tol}}$ and state consistency results.

8.3.2. Convergence

Let $B_k \in \mathbb{B}$, $k \in \mathbb{N}$, and $B_\infty \in \mathbb{B}$. Then B_k is defined to converge to B_∞ for $k \to \infty$ if there exist parametrizations $B_k = B(R_k)$, $k \in \mathbb{N}$, and $B_\infty = B(R_\infty)$ with the following properties. R_∞ has full row rank over the polynomials, $\{d(R_k);\ k \in \mathbb{N}\}$ is bounded, and $R_k \to R_\infty$ for $k \to \infty$ in Euclidean sense. By this we mean that for k sufficiently large R_k has as many rows as R_∞, and if $R_k =$

$\sum_{j=-\infty}^{\infty} R_j^{(k)} s^j$, $R_j^{(k)} = (r_{lm}^{jk}) \in \mathbb{R}^{p\times q}$, $k \in \mathbb{N} \cup \{\infty\}$, then $\sum_{j=-\infty}^{\infty} \sum_{l=1}^{p} \sum_{m=1}^{q} (r_{lm}^{jk} - r_{lm}^{j\infty})^2 \to 0$ if $k \to \infty$.

This concept of convergence is analysed by Nieuwenhuis and Willems [13]. There it is shown that this convergence in terms of parametrizations is equivalent to a natural concept of convergence of systems, considered as subsets of $(\mathbb{R}^q)^{\mathbb{Z}}$.

8.3.3. Generating stochastic systems

We assume that the generating system belongs to the class $\mathbb{G}$ of stochastic processes $w = \{w(t);\ t \in \mathbb{Z}\}$ which satisfy the following assumption.

Assumption 8-4 (*i*) w is second order stationary with for all $t \in \mathbb{Z}$ $Ew(t) = 0$, $C_k := Ew(t)w(t+k)^T$; (*ii*) almost surely for realizations w_r of w there holds for all $k \in \mathbb{Z}_+$ $\frac{1}{t_1 - t_0 + 1} \cdot \sum_{t=t_0}^{t_1-k} w_r(t) w_r(t+k)^T \to C_k$ if $|t_1 - t_0| \to \infty$.

A sufficient condition for the assumption to be satisfied is that w is strictly stationary and ergodic, e.g., that w is Gaussian with a spectral distribution Φ which is continuous on the unit circle. We refer to Hannan [2]. This especially holds true for Gaussian ARMA-processes, in which case $\Phi(z) := \sum_{k=-\infty}^{\infty} C_k z^{-k}$ is a rational function with no poles on the unit circle. The process w then has a representation of the following form. There exist $m \in \mathbb{N}$, polynomial matrices $N \in \mathbb{R}^{q\times m}[s]$ and $M \in \mathbb{R}^{q\times q}[s]$ with $\det(M(s)) \neq 0$ on $|s| \leq 1$, and an m-dimensional Gaussian white noise process n, i.e., $En(t) = 0$ and $En(t)n(s)^T = 0$ for $t \neq s$, such that $M(\sigma^{-1})w = N(\sigma^{-1})n$.

The consistency result stated in section 8.3.5 is in terms of generic subclasses of $\mathbb{G}$ which we will define in section 8.3.4. Here genericity is defined as follows. Define $\mathbb{C} \subset (\mathbb{R}^{q\times q})^{\mathbb{Z}}$ as the collection of $(C_k;\ k \in \mathbb{Z})$ for which there exist $w \in \mathbb{G}$ with $C_k = Ew(t)w(t+k)^T$, $k \in \mathbb{Z}$. A subset $\mathbb{C}' \subset \mathbb{C}$ is called generic if for all $-\infty < t_0 \leq t_1 < +\infty$ $\mathbb{C}'|_{[t_0, t_1]}$ is a λ-generic set in $\mathbb{C}|_{[t_0, t_1]}$, i.e., it contains an open subset of full Lebesgue measure in $\mathbb{C}|_{[t_0, t_1]}$. A class of stochastic systems $\mathbb{G}' \subset \mathbb{G}$ is called generic if $\mathbb{C}' := \{(C_k;\ k \in \mathbb{Z});$ $\exists w \in \mathbb{G}'$ with $C_k = Ew(t)w(t+k)^T$ for all $k \in \mathbb{Z}\}$ is generic, i.e., if the set of covariance sequences in $\mathbb{G}'$ is λ-generic.

The classes $\mathbb{G}_{c_{tol}}$ and $\mathbb{G}_{\varepsilon_{tol}}$ of section 8.3.4 are generic. Moreover, the Gaussian ARMA-processes in $\mathbb{G}_{c_{tol}}$ and $\mathbb{G}_{\varepsilon_{tol}}$ are generic in the class of all Gaussian ARMA-processes in $\mathbb{G}$. So the consistency results of section 8.3.5 in particular hold true for generic ARMA-processes.

8.3.4. Approximation maps and the classes $\mathbb{G}_{c_{tol}}$, $\mathbb{G}_{\varepsilon_{tol}}$

In this section we construct for a given stochastic process w optimal approximations in $\mathbb{B}$. The optimality has to be understood in the sense of a utility corresponding to the purpose of modelling. For w we define the optimal approximations $A^P_{c_{tol}}(w)$ and $A^P_{\varepsilon_{tol}}(w)$ as the models of optimal prediction of w for c_{tol} and ε_{tol} respectively in case the generating system w were known.

The foregoing is made precise as follows. For $r \in \mathbb{R}^{1\times q}[s,s^{-1}]$ with $d(r)>0$ define the relative expected prediction error in analogy with section 5.3 as $e^P(w,r) := \{ (E\|rw\|^2) / (E\|r^*w\|^2) \}^{1/2}$, where r^* is the leading coefficient vector of r and $E\|rw\|^2 := E\{(r(\sigma,\sigma^{-1})w)(t)\}^2$ which does not depend on t due to stationarity. If $d(r)=0$ then define $e^P(w,r) := \{ E\|rw\|^2 / \|r\|^2 \}^{1/2}$. For $B \in \mathbb{B}$ we define $\varepsilon^P(w,B) \in (\mathbb{R}_+^{1\times q})^{\mathbb{Z}_+}$ exactly analogous to $\varepsilon^P(\tilde{w},B)$ in section 5.3. Hence $\varepsilon^P_{t,1}(w,B)$ measures the largest relative expected prediction error of the truly t-th order predictive laws claimed by B, $t \in \mathbb{Z}_+$, and so on. We now define $A^P_{c_{tol}}(w)$ and $A^P_{\varepsilon_{tol}}(w)$ as the predictive models which are optimal for c_{tol} and ε_{tol} respectively, in case w were known.

Definition 8-5 For $w \in \mathbb{G}$, $A^P_{c_{tol}}(w) := \text{argmax}\{ u_{c_{tol}}(c(B), \varepsilon^P(w,B));\ B \in \mathbb{B} \}$ and $A^P_{\varepsilon_{tol}}(w) := \text{argmax}\{ u_{\varepsilon_{tol}}(c(B), \varepsilon^P(w,B));\ B \in \mathbb{B} \}$.

So $A^P_{c_{tol}}$ and $A^P_{\varepsilon_{tol}}$ give deterministic approximations of stochastic processes which are optimal in terms of a utility on complexity and predictive quality of models described by (deterministic) autoregressive equations.

In the sequel we will restrict attention to subclasses of $\mathbb{G}$ for which $A^P_{c_{tol}}$ and $A^P_{\varepsilon_{tol}}$ consist of singletons. For $w \in \mathbb{G}$ define $S(w,t) := E[\text{col}(w(t),\ldots,w(t+k)).\text{col}(w(t),\ldots,w(t+k))^T]$, $t \in \mathbb{Z}_+$. Now consider the algorithms of sections 7.4 and 7.5 with $S(\tilde{w},t)$ replaced by $S(w,t)$. Note that any c_{tol} satisfies assumption 7-2(i) for $\#(\mathcal{T})$ sufficiently large. Suppose that ε_{tol} is such that there is a t such that $\varepsilon^{tol}_{s,1} \leq 0$ for $s>t$.

Definition 8-6 $\mathbb{G}_{c_{tol}} := \{ w \in \mathbb{G};$ assumption 7-8(ii), (iii), (iv) is satisfied$\}$; $\mathbb{G}_{\varepsilon_{tol}} := \{ w \in \mathbb{G};$ assumption 7-10(ii), (iii), (iv) is satisfied and $\sigma^{(0)}_{q-e_0+1} < (\bar{\varepsilon}^{tol}_0)^2 < \sigma^{(0)}_{q-e_0}$, $1-(\sigma^{(t)}_{e'_t})^2 < (\bar{\varepsilon}^{tol}_t)^2 < 1-(\sigma^{(t)}_{e'_t+1})^2$ }.

Proposition 8-7 (*i*) $\mathbb{G}_{c_{tol}}$ and $\mathbb{G}_{\varepsilon_{tol}}$ are generic in $\mathbb{G}$;

(*ii*) for $w \in \mathbb{G}_{c_{tol}}$ $A^P_{c_{tol}}(w)$ is a singleton, generated by the algorithm of section 7.4 with $S(\tilde{w},t)$ replaced by $S(w,t)$;

(*iii*) for $w \in \mathbb{G}_{\varepsilon_{tol}}$ $A^P_{\varepsilon_{tol}}(w)$ is a singleton, generated by the algorithm of section 7.5 with $S(\tilde{w},t)$ replaced by $S(w,t)$.

Moreover, the Gaussian ARMA-processes in $\mathbb{G}_{c_{tol}}$ and $\mathbb{G}_{\varepsilon_{tol}}$ are generic in the class of all Gaussian ARMA-processes in $\mathbb{G}$.

8.3.5. Consistency results

Assume that the data $\tilde{w}$ consist of a (finite) observation on $\mathcal{T}$ of a realization $w_r \in (\mathbb{R}^q)^{\mathbb{Z}}$ of a stochastic process w. As definition of n.a. in w we take a.s., i.e., "almost sure" with respect to the process . The next theorem states consistency results for $P^P_{c_{tol}}$ and $P^P_{\varepsilon_{tol}}$, with the approximation maps as in section 8.3.4 and the concept of convergence as defined in section 8.3.2. It is assumed that for ε_{tol} there is a t such that $\varepsilon^{tol}_{s,1} \leq 0$ for $s > t$, in which case we call ε_{tol} finite.

Theorem 8-8 For every c_{tol}, $P^P_{c_{tol}}$ is consistent on $\mathbb{G}_{c_{tol}}$. For every finite ε_{tol}, $P^P_{\varepsilon_{tol}}$ is consistent on $\mathbb{G}_{\varepsilon_{tol}}$.

This means the following. Let w_r be a realization of a stochastic process $w \in \mathbb{G}_{c_{tol}}$ and let $\tilde{w} = w_r|_{\mathcal{T}}$. Let $A^P_{c_{tol}}(w) = B \in \mathbb{B}$ with corresponding predictive spaces $V^P_t := v_t(L^P_t)$, where L^P_t is as defined in section 4.4. Then almost sure $P^P_{c_{tol}}(\tilde{w})$ is a singleton for $\#(\mathcal{T})$ sufficiently large. Denote the corresponding predictive spaces by $V^P_t(\mathcal{T})$, the complexity by $c(\mathcal{T})$ and the predictive misfit by $\varepsilon(\mathcal{T})$. Then for $\#(\mathcal{T}) \to \infty$ there holds a.s. that $c_t(\mathcal{T}) \to c_t(B)$, $V^P_t(\mathcal{T}) \to V^P_t$ in the Grassmannian topology (i.e., there exist choices of bases of $V^P_t(\mathcal{T})$ which converge to a basis of V^P_t), and $\varepsilon_{t,k}(\mathcal{T}) \to \varepsilon^P_{t,k}(w,B)$, $k = 1,\ldots,q$, $t \in \mathbb{Z}_+$. A similar result holds true for $P^P_{\varepsilon_{tol}}$. The convergence $V^P_t(\mathcal{T}) \to V^P_t$ implies convergence of AR-relations and of the corresponding models. So if the number of observations tends to infinity, the identified model a.s. converges to the optimal (prediction) model B which would be identified in case w were known.

Proof of the theorem consists of using the ergodic properties of w and establishing continuity properties of the steps of the algorithms in sections 7.4 and 7.5 with respect to changes in $S(\tilde{w},t)$, $t\in\mathbb{Z}_+$.

We remark that also the procedure $\bar{P}^P_{\varepsilon_{tol}}$ is consistent on $\mathbb{G}_{\varepsilon_{tol}}$. Moreover, $P^P_{\varepsilon_{tol}}$ is not consistent if ε_{tol} is not finite. Note that such ε_{tol} is not sensible.

We conclude this section by commenting on the optimality. Consider e.g. $P^P_{\varepsilon_{tol}}$ and suppose that $w\in\mathbb{G}_{\varepsilon_{tol}}$ is such that $B:=A^P_{\varepsilon_{tol}}(w)$ satisfies $\sum_{t=0}^{\infty} e^*_t(B)=q$. Then use of B leads to one-step-ahead pointpredictions, which we indicate by $\hat{w}^*$. In this case a.s. and for $\#(\mathcal{T})$ sufficiently large $P^P_{\varepsilon_{tol}}(\tilde{w})$ also leads to pointpredictions, indicated by $\hat{w}(\mathcal{T})$. There holds $E\|\hat{w}^*-\hat{w}(\mathcal{T})\|\to 0$ if $\#(\mathcal{T})\to\infty$. In this sense the one-step-ahead predictions converge to the optimal ones. However, if $q>1$ in general there does not exist a choice of ε_{tol} such that $\hat{w}^*$ (and hence $\hat{w}(\mathcal{T})$) is close to the least squares (causal) predictor for w. So the optimality has to be interpreted in terms of $u_{\varepsilon_{tol}}$, not in terms of minimal mean square prediction error. It is not unreasonable to be slightly non-optimal in accuracy if the predictions can be made by much simpler models.

9. SIMULATIONS

9.1. Introduction

In this section we will illustrate the modelling procedures of section 6 by means of four simple numerical examples.

In section 9.2 we consider exact modelling. In this case only exactly satisfied laws are accepted. This corresponds to applying the procedures $\bar{P}^D_{\varepsilon_{tol}}$ and $\bar{P}^P_{\varepsilon_{tol}}$ with $\varepsilon_{tol}=0$. The data consists of an exact observation of a time series generated by an AR-system.

Section 9.3 gives an example of descriptive modelling of a time series, given a maximal tolerated complexity, i.e., of the procedure $P^D_{c_{tol}}$. The data consists of a noisy observation of a signal generated by an AR-system. We will compare the (non-causal) impulse response of the generating system with that of the identified model.

In section 9.4 we illustrate the difference between descriptive and predictive modelling. For a given time series we compare the models identified by the procedures $P^D_{\varepsilon_{tol}}$ and $P^P_{\varepsilon_{tol}}$.

Finally section 9.5 contains a simulation illustrating the fact that the procedures for modelling, given a maximal tolerated misfit, need not generate models of minimal complexity. This indicates the difference between the procedures $P^D_{\varepsilon_{tol}}$ $(P^P_{\varepsilon_{tol}})$ and $P^{*D}_{\varepsilon_{tol}}$ $(P^{*P}_{\varepsilon_{tol}})$ as defined in sections 6.2 and 6.3 respectively. We also illustrate consistency of $P^P_{\varepsilon_{tol}}$.

9.2. Exact modelling

9.2.1. Data

In the first simulation we consider exact modelling of a signal generated by an AR-system. The signal consists of two components, each being a sum of two sinusoids. To be specific, let $f_1 := 2\pi/100$, $f_2 := 2\pi/120$ and $f_3 := 2\pi/150$. Define $s_k(t) := \sin(f_k \cdot t)$, $k = 1,2,3$, $t \in \mathbb{R}$, and $w_1(t) := s_1(t) + s_2(t)$, $w_2(t) := s_1(t) + s_3(t)$. The data consists of observations of the signals w_1 and w_2 on times $t = 1,\ldots,300$, i.e., $\tilde{w} = (\begin{bmatrix} w_1(t) \\ w_2(t) \end{bmatrix};\ t = 1,\ldots,300) \in (\mathbb{R}^2)^{300}$. The signals are given in figure 11.

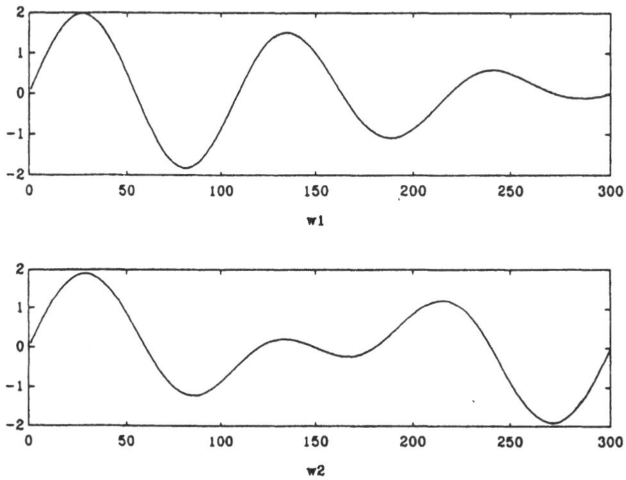

figure 11: data for simulation 9.2.

9.2.2. System

Both w_1 and w_2 are periodic, with period 600 and 300 respectively. Hence $w \in B(R)$ with $R := \begin{bmatrix} \sigma^{600} - 1 & 0 \\ 0 & \sigma^{300} - 1 \end{bmatrix}$. However, there are more powerful models for

w. Observe that for $s(t)=\sin(f.t)$ there holds $s(t+2)+s(t)=2\cos(f).s(t+1)$, hence $s\in B(r)$ with $r(s):=s^2-2\cos(f).s+1=(s-e^{if})(s-e^{-if})$. Defining $p_k(s):=(s-e^{if_k})(s-e^{-if_k})$, $k=1,2,3$, we conclude that $\tilde{w}\in B(R_0)$ with $R_0:=\begin{bmatrix} p_1.p_2 & 0 \\ 0 & p_1.p_3 \end{bmatrix}$.

9.2.3. Model identification

Exact models for the data $\tilde{w}$ are obtained by applying the procedures $\bar{\bar{P}}^D_{\varepsilon_{tol}}$ and $\bar{\bar{P}}^P_{\varepsilon_{tol}}$ with $\varepsilon_{tol}=0$. We denote the resulting models by $B(R^D):=\bar{\bar{P}}^D_0(\tilde{w})$ and $B(R^P):=\bar{\bar{P}}^P_0(\tilde{w})$. These models are identified by using the algorithms of section 7 with $\varepsilon_{tol}=0$. Both models consist of one second order laws and one fourth order law. Let R^D and R^P have elements r^D_{lm} and r^P_{lm} respectively, $l,m=1,2$. The identified laws are given in table 1.

laws:	coefficients of: σ^0	σ^1	σ^2	σ^3	σ^4
r^D_{11}	0.5007	-1.0000	0.5007	0	0
r^D_{12}	-0.2754	0.5502	-0.2754	0	0
r^D_{21}	0.4637	-0.9568	0.5746	-0.1319	0.0507
r^D_{22}	-0.0352	-0.3517	1.0000	-0.8055	0.1920
r^P_{11}	1.2392	-2.4750	1.2392	0	0
r^P_{12}	-0.6815	1.3618	-0.6815	0	0
r^P_{21}	0.6815	-2.7224	4.0818	-2.7223	0.6815
r^P_{22}	1.2392	-4.9490	7.4196	-4.9489	1.2391

table 1: identified AR-laws for simulation 9.2.

9.2.4. Model validation

Two questions arise, namely, whether these AR-laws are equivalent and whether they are equivalent to R_0, i.e., if $B(R^D)=B(R^P)=B(R_0)$.

Direct calculation shows that there exist a constant $\alpha\neq 0$ and unimodular matrices U^D and U^P such that $U^DR^D=U^PR^P=R_I:=\begin{bmatrix} p_2 & \alpha p_3 \\ p_1p_2 & 0 \end{bmatrix}$. So indeed $B(R^D)=B(R^P)$. As $\begin{pmatrix} 0 & 1 \\ p_1 & -1 \end{pmatrix} R_I = \begin{pmatrix} 1 & 0 \\ 0 & \alpha \end{pmatrix} R_0$ it follows that $B(R_I) \subset B(R_0)$, but $B(R_I) \neq B(R_0)$. So the identified laws R^D and R^P are equivalent, but not equivalent to R_0. This is due to the fact that $B(R_0)$ is not the most powerful unfalsified model for $\tilde{w}$. Indeed, a short calculation gives that $p_2 + \alpha p_3 = \alpha' p_1$, where $\alpha := \{\cos(f_1) - \cos(f_2)\}/\{\cos(f_3) - \cos(f_1)\}$ and $\alpha' := \{\cos(f_3) - \cos(f_2)\}/\{\cos(f_3) - \cos(f_1)\}$. Stated otherwise, the space of polynomials $\{s^2 + c.s + 1;\ c \in \mathbb{R}\}$ has dimension two. The most powerful unfalsified model for the generating system is $B(R_0^*)$ with $R_0^* := \begin{pmatrix} p_1 p_2 & 0 \\ 0 & p_1 p_3 \\ p_2 & \alpha p_3 \end{pmatrix}$. It easily follows that $B(R^D) = B(R^P) = B(R_I) = B(R_0^*)$.

The foregoing shows that the identified models correspond to the (most powerful unfalsified) model for the generating system. Hence the generating system is exactly identified. This illustrates the consistency result stated in proposition 8-3.

9.3. Descriptive modelling

9.3.1. Introduction

In the second simulation we model a time series by minimizing the descriptive misfit, given a maximal tolerated complexity, i.e., we use the procedure $P^D_{c_{tol}}$. We will first describe the data and the system generating it, then present the identified model and finally compare this model with the generating system.

9.3.2. Data

The data consists of a two-dimensional time series $\tilde{w} = \binom{w_1}{w_2} \in (\mathbb{R}^2)^{1000}$ and is depicted in figure 12.

9.3.3. System

The data $\tilde{w}$ is generated by the system shown in figure 13. Here s_1 is the noise-free input, n_1 the noise on the input, and $w_1 := s_1 + n_1$ the exactly observed input. The signal s_2 is the output generated by the input w_1. The observed output is $w_2 := s_2 + n_2$.

The signals s_1, s_2 and the noise n_1, n_2 are given in figure 14. For a

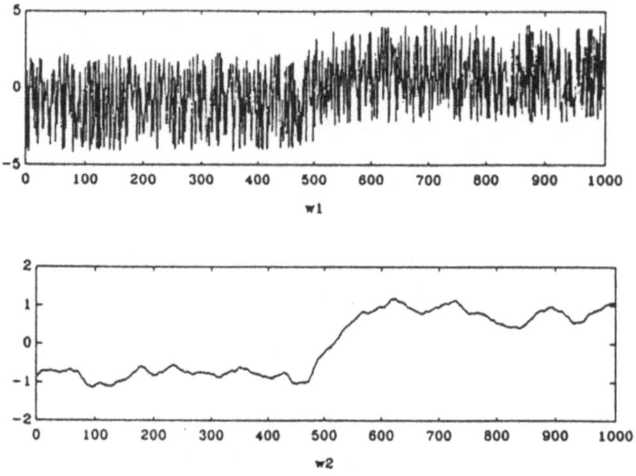

figure 12: data for simulation 9.3.

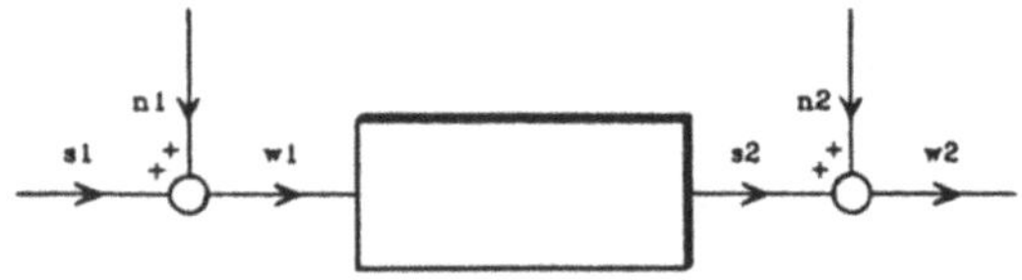

figure 13: generating system for simulation 9.3.

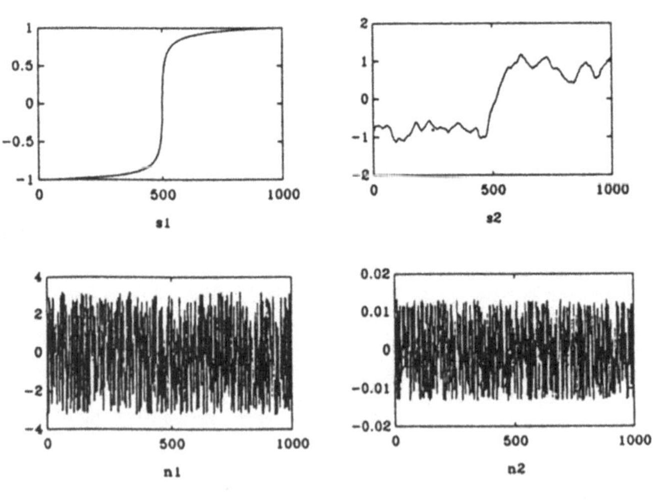

figure 14: signals and noise for simulation 9.3.

signal $s \in \mathbb{R}^T$ and noise $n \in \mathbb{R}^T$ we define the signal to noise ratio in $s+n$ as $\|s\|/\|n\| := \{\sum_{t=1}^{T} s(t)^2 / \sum_{t=1}^{T} n(t)^2\}^{1/2}$. In this simulation the signal to noise ratio for w_1 is ½, for w_2 100.

The system generating s_2 from w_1 is a (symmetric) exponential smoother. For $0<\alpha<1$ we define the exponential smoother e_α as follows. Let l_∞ denote the set of bounded sequences, i.e., $l_\infty := \{w \in \mathbb{R}^{\mathbb{Z}};\ \sup(|w(t)|;\ t \in \mathbb{Z}) < \infty\}$. Then $e_\alpha : l_\infty \to l_\infty$ is defined by $e_\alpha(u) := y$, where $y(t) := \frac{1-\alpha}{1+\alpha} \cdot \sum_{\tau=-\infty}^{\infty} \alpha^{|\tau|} u(t+\tau)$. Note that for u a constant signal, $u(t)=c$ for all $t \in \mathbb{Z}$, the output is $y=u$.

We will embed the graph of e_α $\mathrm{gr}(e_\alpha) := \{(u,y) \in l_\infty^2;\ y = e_\alpha(u)\}$ in an AR-system $B_\alpha \subset (\mathbb{R}^2)^{\mathbb{Z}}$. In order to describe B_α, let $y = e_\alpha(u) = \frac{1-\alpha}{1+\alpha} \cdot (y_- + u + y_+)$, where $y_-(t) := \sum_{\tau=1}^{\infty} \alpha^\tau u(t-\tau)$ and $y_+(\tau) := \sum_{\tau=1}^{\infty} \alpha^\tau u(t+\tau)$. Then $(\sigma-\alpha)y_- = \alpha u$ and $(1-\alpha\sigma)y_+ = \alpha\sigma u$, hence $(\sigma-\alpha)(1-\alpha\sigma)(y_- + u + y_+) = [(1-\alpha\sigma)\alpha + (\sigma-\alpha)(1-\alpha\sigma) + (\sigma-\alpha)\alpha\sigma]u = (1-\alpha^2)\sigma u$. Define $p_\alpha := (s-\alpha)(1-\alpha s)$ and $q_\alpha := \frac{1-\alpha}{1+\alpha} \cdot (1-\alpha^2)s = (1-\alpha)^2 s$, then $\mathrm{gr}(e_\alpha) \subset B_\alpha := B(R_\alpha)$ where $R_\alpha := (-q_\alpha, p_\alpha)$.

In the simulation the signal s_2 is the exponential smoothing of w_1 with $\alpha = 0.95$. Hence the (most powerful unfalsified model of the) generating system is $B(R_g)$ with $R_g = (-q_g, p_g) := (-q_{0.95}, p_{0.95})$. We remark that in identifying the model there is no prior knowledge that w_1 is the input and w_2 the output.

9.3.4. Model identification

Next we analyse the data $\tilde{w}$ by means of $P^D_{c_{tol}}$. We consider models of decreasing complexity, corresponding to requiring one AR-relation of order 5,4,3,2,1 and 0 respectively. For order k the resulting model is indicated by $B_k := B((-q^{(k)}, p^{(k)})) := \{(u,y) \in (\mathbb{R}^2)^{\mathbb{Z}};\ p^{(k)}(\sigma)y = q^{(k)}(\sigma)u\}$, $k=5,4,3,2,1,0$. See table 2. This table also contains the roots of the polynomials $p^{(k)}$, $q^{(k)}$, and the descriptive error $\varepsilon^D_{k,1}(\tilde{w}, B_k)$.

The results in table 2 indicate that little descriptive power is lost by reducing the order from 5 to 2. Moreover, two of the roots of the identified polynomial p turn out to be rather invariant under different orders, while the roots of the identified polynomial q seem to be quite random, although generally one of them is close to 0. It seems reasonable to take c_{tol} such that the corresponding equation structure is $e(c_{tol}) = (0,0,1,0,0,0,\dots)$, i.e., to require one second order relation.

	coefficients of: σ^0	σ^1	σ^2	σ^3	σ^4	σ^5	roots p	roots q	error
order 5: $p^{(5)}$	0.4475	0.0893	-0.5333	-0.5563	0.1161	0.4295	0.9536	0.21	0.0154
$q^{(5)}$	0.0003	-0.0010	-0.0023	-0.0025	-0.0014	-0.0003	1.0548	-0.64±1.07i	
							-1.05	-1.53±0.83i	
							-0.61±0.78i	-18	0.0155
order 4: $p^{(4)}$	0.5482	-0.3488	-0.4063	-0.3417	0.5440		1.0514	0.15	
$q^{(4)}$	0.0003	-0.0014	-0.0016	-0.0017	-0.0001		-0.69±0.73i	-0.56±0.88i	
order 3: $p^{(3)}$	0.5427	-0.6713	-0.2884	0.4144			0.9501	0.037	0.0159
$q^{(3)}$	0.0001	-0.0014	-0.0009	-0.0003			1.0537 -1.31	-1.31±1.55i	
order 2: $p^{(2)}$	0.4061	-0.8168	0.4099				0.9529	5.24	0.0159
$q^{(2)}$	0.0002	-0.0011	0.0002				1.0396	0.15	
order 1: $p^{(1)}$	0.7073	-0.7069					1.0006	1.20	0.0176
$q^{(1)}$	0.0011	-0.0009							
order 0: $p^{(0)}$	0.9806								0.7190
$q^{(0)}$	0.1962								

table 2: identified AR-laws for simulation 9.3.

9.3.5. Model validation

The identified model $B((-q_I, p_I)) := B_2$ is compared with the generating system $B((-q_g, p_g))$ in table 3. This indicates that the AR-law of the identified system is close to the law of the generating system.

	coefficients of: σ^0	σ^1	σ^2	roots	
system: p_g	1	-2.0026	1	0.95	1.0526
q_g	0	-0.0026	0	0	
model: p_I	0.9906	-1.9925	1	0.9529	1.0396
q_I	0.0004	-0.0028	0.0005	0.1537	5.2435

table 3: system and identified model.

We next want to compare the model and the system with respect to their input-output behaviour. So we now will use the prior knowledge that w_1 is an input and w_2 an output. We will compare the impulse responses of the model and the system.

For $B=\{(u,y)\in(\mathbb{R}^2)^{\mathbb{Z}};\ p(\sigma)y=q(\sigma)u\}$ we define the impulse response of B with respect to u as $B^{\delta}:=\{(u,y)\in B;\ u=\delta\}$, where $\delta(0):=1$ and $\delta(t):=0$ for all $t\neq 0$. It can be shown that B^{δ} contains exactly one bounded element if $q\neq 0$, $p\neq 0$ and p has no roots on the unit circle. In this case we call the time series $i\in\mathbb{R}^{\mathbb{Z}}$ such that $(\delta,i)\in B^{\delta}\cap l_{\infty}$ the stable impulse response. The models $B((-q_g,p_g))$ and $B((-q_I,p_I))$ satisfy these conditions. We denote their stable impulse responses by i_g and i_I respectively.Here $i_g(t)=\frac{1-\alpha}{1+\alpha}.\alpha^{|t|}$ and i_I is determined as follows. There exist unique real numbers a_1, a_2, b_1, b_2, d with $|a_1|<1$, $|a_2|>1$ such that $\frac{q_I}{p_I}=\frac{b_1}{s-a_1}+\frac{b_2}{s-a_2}+d$. Define $i_I(0):=d-\frac{b_2}{a_2}$, $i_I(t):=b_1.a_1^{t-1}$ for $t>0$ and $i_I(t):=-b_2a_2^{t-1}$ for $t<0$. It then is a matter of simple calculation to verify that $p_I(\sigma)i_I=q_I(\sigma)\delta$. This corresponds to a causal interpretation of the transferfunction $\frac{b_1}{s-a_1}$ and an anticausal one for $\frac{b_2}{s-a_2}$.

The stable impulse responses i_g of the system and i_I of the identified system are given in figure 15.

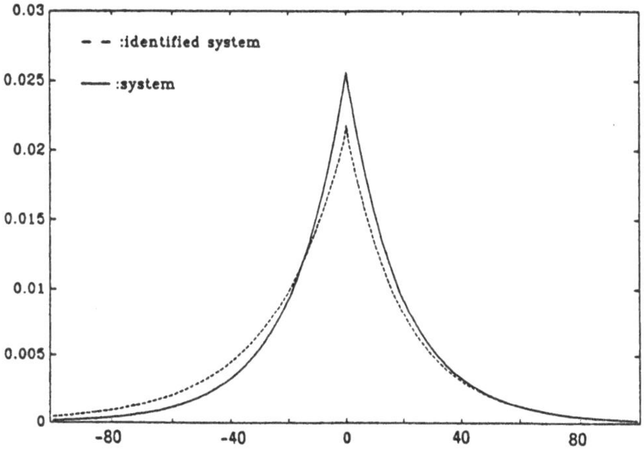

figure 15: impulse responses for simulation 9.3.

9.3.6. Scaling and sampling

We conclude this section with some remarks.

First, the stable impulse response of a system is a highly sensitive function of the AR-coefficients describing the system. For example, in the system $(\sigma-1-\varepsilon)y=u$ with $|\varepsilon|<1$ the stable impulse response is causal if $\varepsilon<0$, anticausal if $\varepsilon > 0$.

Second, the result of the identification algorithm depends on scaling of the variables. In order to illustrate this, consider scaling of the output in the system $B(R_g)$ by a factor $c \neq 0$. Let $B_c := \{(u,y) \in (\mathbb{R}^2)^{\mathbb{Z}};\ p_g(\sigma)y = c.q_g(\sigma)u\}$. Let $\varepsilon := e^D(\tilde{w},(-q_I,p_I))$ denote the descriptive misfit of the identified law $(-q_I,p_I)$ with respect to the data $\tilde{w} = \begin{bmatrix} w_1 \\ w_2 \end{bmatrix}$. Denote the transformed data by $\tilde{w}_c := \begin{bmatrix} w_1 \\ cw_2 \end{bmatrix}$. From definition 5-4 it follows that $e^D(\tilde{w}_c,(-cq_I,p_I)) = \varepsilon.(\|q_I\|^2 + \|p_I\|^2)^{1/2}/(\|q_I\|^2 + c^{-2}.\|p_I\|^2)^{1/2}$. Using the results in table 3, it follows that the descriptive misfit of $(-cq_I,p_I)$ with respect to the scaled data $\tilde{w}_c$ is approximately $c.\varepsilon$. So, e.g., if c is very large then the law $u=0$ has smaller error. In the next section we will illustrate that the predictive procedures prevent these problems of scaling.

Finally, autoregressive modelling is subject to problems of fast sampling. Consider the case that a continuous time system is sampled at a certain sample rate Δ^{-1}. The magnitudes of the AR-coefficients of the sampled system depend on this sample rate. This affects the descriptive quality of the AR-laws, as indicated above. The constant c is related to Δ as $c=\Delta$. It especially seems difficult to identify good approximations of infinite dimensional systems by means of autoregressive modelling in case of high sample rate and small noise. This is only partly due to the smoothness of the resulting signals. It seems contradictory that having a large amount of data, i.e., fast sampling, and good data, i.e., small noise, would be undesirable in identification.

To illustrate this we refer to table 2, where the best AR-law of order 1 is close to $(\sigma-1)w_2=0$ with a small descriptive misfit of 0.0176. If we scale the output appropriately this effect is reduced. For example, $e^D(\tilde{w}_c,(0,\sigma-1)) = c.00176$, while $e^D(\tilde{w}_c,(-c.q_I,p_I)) = 0.0159.(\|q_I\|^2+\|p_I\|^2)^{1/2}/(\|q_I\|^2+c^{-2}\|p_I\|^2)^{1/2}$. So for c sufficiently large the law $(-cq_I,p_I)$ has much better descriptive fit than the law corresponding to smoothness. We remark that decrease in the signal to noise ratio of the output hardly helps in discriminating $(-q_I,p_I)$ from $(0,\sigma-1)$. This is due to the fact that $\|p_I\|/(\|q_I\|^2+\|p_I\|^2)^{1/2} \approx 1$. If $\tilde{w}' = \begin{bmatrix} w_1 \\ w_2' \end{bmatrix}$ with $w_2' = s_2 + c.n_2$, $c>1$, then $e^D(\tilde{w}',(-q_I,p_I)) \approx 0.0159+(c-1).\|n_2\|.\|p_I\|/(\|q_I\|^2+\|p_I\|^2)^{1/2}$ and $e^D(\tilde{w}',(0,\sigma-1)) \approx 0.0176+(c-1).\|n_2\|$, so for c large the errors are nearly the same.

9.4. Predictive modelling

9.4.1. Introduction

In the third simulation we illustrate the difference between descriptive and predictive modelling. We will see that the predictive procedures suffer less from scaling problems. On the other hand, the imposed asymmetry in time, due to the one-step-ahead prediction criterion, sometimes is artificial, in which case the descriptive procedures seem preferable.

We will now first describe the data and the generating system and subsequently analyse the data by means of descriptive and predictive procedures.

9.4.2. Data

The data consists of a three-dimensional time series $\tilde{w} = \text{col}(w_1, w_{21}, w_{22}) \in (\mathbb{R}^3)^{200}$. We will investigate the effect of scaling. In order to illustrate this we will scale w_{22} and identify models for the scaled data $\tilde{w}^{(k)} := \text{col}(\tilde{w}_1^{(k)}, \tilde{w}_2^{(k)}, \tilde{w}_3^{(k)}) := \text{col}(w_1, w_{21}, k.w_{22})$, $k \in \mathbb{R}_+$.

9.4.3. System

The data is generated by the system shown in figure 16.

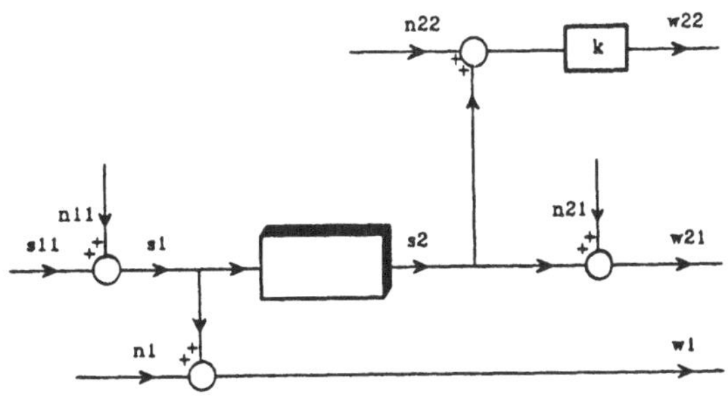

figure 16: generating system for simulation 9.4.

Here s_{11} is the noise-free input, n_{11} noise on the system input, $s_1 := s_{11} + n_{11}$ the input for the system, n_1 noise on the observed input, $w_1 := s_1 + n_1$ the observed input, s_2 the output of the system, n_{21} and n_{22} noise on observed outputs, $w_{21} := s_2 + n_{21}$ and $w_{22} := s_2 + n_{22}$ the observed outputs. The signal to noise ratios are $\|s_{11}\|/\|n_{11}\| = 10$, $\|s_1\|/\|n_1\| = 20$, $\|s_2\|/\|n_{21}\| = 10$ and

$\|s_2\|/\|n_{22}\| = 2$.

The signals, observed data and noise are given in figure 17 for the case $k = 1$ (no scaling on w_{22}).

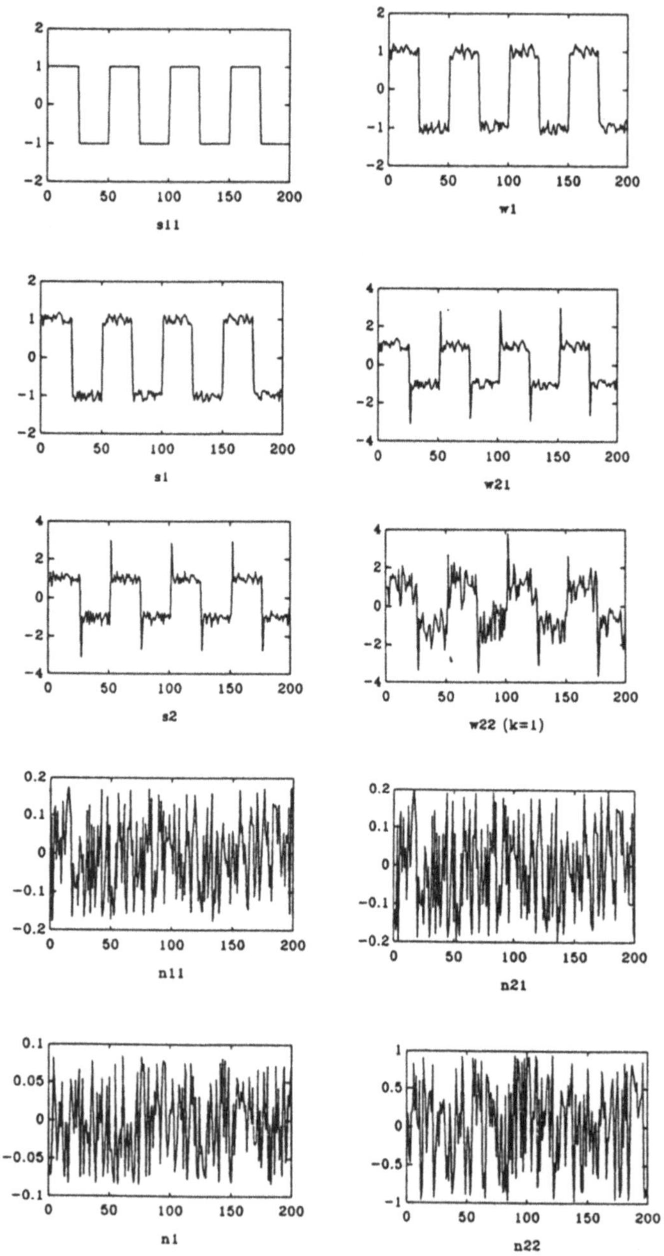

figure 17: data, signals and noise for simulation 9.4.

The system relating s_2 to s_1 is described by $\sigma^2 s_2=(2\sigma-1)s_1$. This corresponds to a simple linear extrapolator $s_2(t):= s_1(t-1) +\{s_1(t-1)-s_1(t-2)\}$.

9.4.4. Model identification and validation

In order to identify a model, we have to reconcile the desires for low complexity and for low misfit. In the simulation we identified the AR-models with best descriptive and predictive fit for orders from 0 up to 4 and for data $\tilde{w}^{(k)}$ corresponding to various scaling constants k. In order to choose a model we compared the increase in fit due to increase in complexity. It turns out that the descriptive misfit decreases only slightly for orders larger than two. Moreover, the results for $k>1$ nearly coincide with those for $k=1$.

The main results of the simulation are summarized in tables 4 and 5. Table 4 contains the best predictive models of orders from 0 up to 4 and for various values of k. Table 5 contains the best descriptive models of orders 0 and 2 and for various k. Specified are the AR-coefficients in $r_1(\sigma)\tilde{w}_1^{(k)}+r_{21}(\sigma)\tilde{w}_2^{(k)}+r_{22}(\sigma)\tilde{w}_3^{(k)}=0$, some of the roots of r_1, r_{21}, r_{22}, and the misfits.

From table 4 it is clear that the model identified by the predictive procedure does not depend on scaling of w_{22}. Moreover, considering the predictive misfits it seems very reasonable to choose a second order model, with predictive misfit 0.12. The model for data $\tilde{w}^{(k)}$ then becomes $r_1^{(k)}(\sigma)w_1^{(k)}+r_{21}^{(k)}(\sigma)w_2^{(k)}+r_{22}^{(k)}(\sigma)w_3^{(k)}=0$, where $r_1^{(k)}(s)=0.08s^2-1.99s+0.96$, $r_{21}^{(k)}(s)=s^2-0.05s+0.01$, $r_{22}^{(k)}\approx k^{-1}(0.01s-0.03)$. So this law is close to the generating system $(-2\sigma+1)s_1+\sigma^2 s_2=0$. The procedure identifies the relation between w_1 and w_{21} as its misfit is due to the noise on w_1 and w_{21}, which is much smaller than the noise on w_{22}. Note finally that, even if $\tilde{w}^{(k)}$ is observed instead of $\tilde{w}=\tilde{w}^{(1)}$, the predictive procedure for all k identifies the same AR-relation for the unscaled variables (w_1,w_{21},w_{22}).

On the other hand, as shown in table 5, the model identified by the descriptive procedures depends strongly on scaling of w_{22}. Roughly speaking, for values of k larger than 0.1 it seems reasonable to choose a model of order 2, which model turns out to be relatively close to the generating system. For values of k smaller than 0.1 it seems reasonable to choose a model of order 0, approximately corresponding to $w_3^{(k)}=k.w_2^{(k)}$.

	order 0			order 1			order 2			order 3			order 4		
	r_1	r_{21}	r_{22}	r_1	r_{21}	r_{22}	r_1	r_{21}	r_{22}	r_1	r_{21}	r_{22}	r_1	r_{21}	r_{22}
k=1															
coeff. σ^0	-0.60	1	-0.44	-1.82	0.48	-0.05	0.96	0.01	-0.03	0.18	0.01	-0.02	-0.18	0.05	-0.00
σ^1				0.40	1	-0.04	-1.99	-0.05	0.01	0.69	-0.02	-0.02	0.30	0.07	-0.02
σ^2							0.08	1	0.00	-1.99	0.09	0.01	0.73	-0.09	-0.02
σ^3										0.08	1	-0.00	-1.99	0.07	0.01
σ^4													0.07	1	-0.00
roots		-		4.62	-0.48	-1.25	0.49	0.02±0.11i	1.65	0.53			0.41±0.21i		
							25.2		-10.3	-0.17			-0.43		
										24.4			26.6		
misfit		0.3250			0.2153			0.1168			0.1149			0.1134	
k=0.1															
coeff. σ^0	-0.60	1	-0.44	-1.82	0.48	-0.46	0.96	0.01	-0.28	0.18	0.01	-0.19	-0.18	0.05	-0.00
σ^1				0.40	1	-0.37	-1.99	-0.05	0.14	0.69	-0.02	-0.21	0.30	0.07	-0.18
σ^2							0.08	1	0.02	-1.99	0.09	0.14	0.73	-0.09	-0.19
σ^3										0.08	1	-0.02	-1.99	0.07	0.13
σ^4													0.07	1	-0.04
roots		-		4.62	-0.48	-1.25	0.49	0.02±0.11i	1.65	0.53			0.41±0.21i		
							25.2		-10.3	-0.17			-0.43		
										24.4			26.6		
misfit		0.3250			0.2153			0.1168			0.1149			0.1134	
k=0.01															
coeff. σ^0	-0.60	1	-0.44	-1.82	0.48	-4.57	0.96	0.01	-2.75	0.18	0.01	-1.89	-0.18	0.05	-0.08
σ^1				0.40	1	-3.65	-1.99	-0.05	1.40	0.69	-0.02	-2.14	0.30	0.07	-1.81
σ^2							0.08	1	0.16	-1.99	0.09	1.40	0.73	-0.09	-1.93
σ^3										0.08	1	-0.17	-1.99	0.07	1.29
σ^4													0.07	1	-0.42
roots		-		4.62	-0.48	-1.25	0.49	0.02±0.11i	1.65	0.53			0.41±0.21i		
							25.2		-10.3	-0.17			-0.43		
										24.4			26.6		
misfit		0.3250			0.2153			0.1168			0.1149			0.1134	

table 4: predictive AR-laws for simulation 9.4.

	order 0	misfit	coeff. order 2: σ^0	σ^1	σ^2	roots	misfit
k=1:							
r_1	1.36	0.3250	1.13	-1.99	0.02	0.57 ; 87.7	0.0561
r_{21}	-2.28		-0.03	-0.12	1	0.24 ; -0.12	
r_{22}	1		-0.03	0.02	-0.00	4.92 ; 1.99	
k=0.2:							
r_1	-0.00	0.1137	1.13	-1.99	0.02	0.57 ; 89.5	0.0559
r_{21}	-0.21		-0.02	-0.13	1	0.20 ; -0.08	
r_{22}	1		-0.19	0.14	-0.02	3.06 ; 2.49	
k=0.14:							
r_1	-0.01	0.0804	1.11	-1.98	0.02	0.57 ; 91.8	0.0555
r_{21}	-0.14		0.01	-0.14	1	0.07±0.09i	
r_{22}	1		-0.43	0.33	-0.08	1.98±1.09i	
k=0.12:							
r_1	-0.01	0.0691	1.08	-1.95	0.02	056; 89.6	0.0547
r_{21}	-0.12		0.06	-0.17	1	0.08±0.23i	
r_{22}	1		-0.80	0.68	-0.24	1.43±1.15i	
k=0.11:							
r_1	-0.01	0.0634	1.02	-1.88	0.02	0.55 ; 76.9	0.0535
r_{21}	-0.11		0.13	-0.22	1	0.11±0.34i	
r_{22}	1		-1.37	1.29	-0.59	1.10±1.06i	
k=0.1:							
r_1	-0.01	0.0577	0.90	-1.72	0.03	0.53 ; 49.4	0.0505
r_{21}	-0.10		0.26	-0.33	1	0.17±0.48i	
r_{22}	1		-2.54	2.71	-1.54	0.88±0.94i	
k=0.09:							
r_1	-0.01	0.0520	0.76	-1.52	0.05	051; 30.3	0.0461
r_{21}	-0.09		0.40	-0.47	1	0.24±0591	
r_{22}	1		-4.06	4.66	-2.96	0.79±0.87i	
k=0.01:							
r_1	-0.00	0.0058	-0.01	0.01	-0.00	0.40 ; 8.64	0.0052
r_{21}	-0.01		-0.01	0.01	-0.02	0.44±0.77i	
r_{22}	1		1.10	-1.39	1	0.70±0.78i	

table 5: descriptive AR-laws for simulation 9.4.

In this way the simulation clearly indicates the effect of scaling of data on the resulting model identified by the descriptive procedures. The model identified by the predictive procedures is invariant under scaling.

9.4.5. Effects of scaling for SISO systems

We conclude this section with a few remarks on the effect of scaling on the identification of single-input single-output (SISO) systems.

In table 6 we give the main results of the simulation experiment consisting of modelling the data $\approx{w}^{(k)} := \text{col}(w_1, k.w_{21})$ for various k by means of the descriptive procedures. From the table of misfits it seems reasonable to accept a second order law , as the second order laws have considerably better fit than lower order laws and nearly as good fit as higher order laws. The table indicates that scaling has little influence on the model for (w_1, w_{21}), as for scaling constant k the identified AR-law $(r_1^{(k)}, r_{21}^{(k)})$ is approximately equal to $(kr_1^{(1)}, r_{21}^{(1)})$.

On the other hand, it turns out that by decreasing the signal to noise ratio for w_{21}, the identified model becomes more sensitive to scaling. Moreover, in section 9.3 we concluded that for the exponential weighting system the identified model is sensitive to scaling. It hence appears that scaling sometimes has influence on the identified model, but that the effect need not always be large. Here we only will give a sketch of an explanation.

For simplicity, consider a second order system $B = \{(w_1, w_2);\ p(\sigma)w_2 = q(\sigma)w_1\}$ with degrees $d((p,q)) = d(p) = 2$. Assume that w_2 is scaled in such a way that $\|p\|^2 = \|q\|^2 = ½$ .Let the data consist of $\tilde{w} = (\tilde{w}_1, \tilde{w}_2)$, $\tilde{w}_1 = w_1 + \varepsilon_1$, $\tilde{w}_2 = w_2 + \varepsilon_2$, where ε_1 and ε_2 are uncorrelated white noise with $\sigma_1 := \|\varepsilon_1\|$ and $\sigma_2 := \|\varepsilon_2\|$. To investigate the effect of scaling, suppose we observe $(c_1\tilde{w}_1, c_2\tilde{w}_2)$, $c_1.c_2 \neq 0$. As the identified models are invariant under a data transformation $(\pm c\tilde{w}_1, \pm c\tilde{w}_2)$, $c \neq 0$, we may consider $\tilde{w}^{(k)} := (\tilde{w}_1, k.\tilde{w}_2)$, with $k := |c_2/c_1|$.

First let $k=1$ and let α denote the descriptive misfit of $(-q, p)$, i.e., $\alpha := \|p\tilde{w}_2 - q\tilde{w}_1\| \approx \frac{1}{2}\sqrt{2}.(\sigma_1^2 + \sigma_2^2)^{½}$. Moreover, let β and γ denote the descriptive misfit of the best first order law for $\tilde{w}_1$ and $\tilde{w}_2$ respectively. For k let e_k^1 denote the descriptive misfit of the best first order law for $\tilde{w}^{(k)}$, and α_k the misfit of $(-kq, p)$, i.e., $\alpha_k := e^D(\tilde{w}^{(k)}, (-kq, p)) = \alpha.k\sqrt{2}/(1+k^2)^{½}$. A relevant indication for the sensitivity to scaling is the influence of k on α_k and

misfit	order 0	1	2	3	4
k=100	0.4812	0.1587	0.0616	0.0564	0.0554
k=10	0.4798	0.1585	0.0616	0.0564	0.0554
k=1	0.3726	0.1370	0.0565	0.0528	0.0520
k=0.1	0.0544	0.0245	0.0134	0.0127	0.0125
k=0.01	0.0055	0.0025	0.0014	0.0013	0.0013

AR-law		coeff. of: σ^0	σ^1	σ^2	roots	
k=100:	r_1	118	-202	3.37	0.59	59.1
	r_{21}	-0.07	-0.12	1	0.33	-0.21
k=10:	r_1	11.8	-20.2	0.34	0.59	59.3
	r_{21}	-0.07	-0.12	1	0.33	-0.21
k=1:	r_1	1.15	-2.00	0.02	0.58	80.0
	r_{21}	-0.06	-0.11	1	0.31	-0.20
k=0.1:	r_1	0.10	-0.19	-0.00	0.52	-111
	r_{21}	-0.03	-0.05	1	0.19	-0.14
k=0.01:	r_1	0.01	-0.02	-0.00	0.51	-98.0
	r_{21}	-0.02	-0.05	1	0.18	-0.13
k=1: predictive:	r_1	0.97	-1.99	0.08	0.50	23.8
	r_{21}	-0.02	-0.04	1	0.17	-0.13

table 6: descriptive misfit and AR-laws for $\widetilde{\widetilde{w}}^{(k)}$.

e_k^1. We assume that for small k $e_k^1 \approx k.\gamma$ and for large k $e_k^1 \approx \beta$. This seems often to be the case. Now if $\alpha\sqrt{2} < \min\{\beta,\gamma\}$ we may expect little sensitivity to scaling, as it seems probable that in this case $e_k^1 > \alpha_k$ for all $k \in \mathbb{R}_+$.

In the case of data $\widetilde{\widetilde{w}}^{(k)} := \text{col}(w_1, kw_{21})$ in this section the underlying system is described by $p(s)=s^2$ and $q(s)=2s-1$. So for $k=1/\sqrt{5}$ we have

$\|kq\| = \|p\|$. Form this we get $\alpha \approx 0.04$, $\beta \approx 0.28$, $\gamma \approx 0.27$. So indeed $\alpha\sqrt{2} < \min\{\beta, \gamma\}$.

On the other hand, for the exponential weighting system of section 9.3 we have $\|p_g\| >> \|q_g\|$. It can be calculated that for $c = 850$ we have $\|cq_g\| \approx \|p_g\|$ and $\alpha \approx 9.5$, $\beta \approx 1.82$, $\gamma \approx 15.3$. So in this case $\beta < \alpha\sqrt{2} < \gamma$. For large values of k we will be unable to identify the generating system. The simulation of section 9.3 corresponds to small k ($k \approx 1/850$).

Finally, if w_1 and w_2 are very smooth we will always have problems in identifying the relationship between w_1 and w_2. In this case $\beta \approx e^D(\tilde{w}_1, \sigma - 1) \approx \sigma_1$ and $\gamma \approx e^D(\tilde{w}_2, \sigma - 1) \approx \sigma_2$, while $\alpha_k \approx (\sigma_1^2 + \sigma_2^2)^{1/2} \cdot k/(1+k^2)^{1/2}$. In this case we may expect $e_k^1 < \alpha_k$ for all k.

9.5. An example illustrating non-optimality

9.5.1. Introduction

In the fourth and final simulation we illustrate the fact that the procedures for modelling, given a maximal tolerated misfit, need not generate models of minimal complexity. This then shows that the procedures $P^D_{\varepsilon_{tol}}$ and $P^P_{\varepsilon_{tol}}$ differ from the (optimal) procedures $P^{*D}_{\varepsilon_{tol}}$ and $P^{*P}_{\varepsilon_{tol}}$ respectively, as indicated in sections 6.2 and 6.3.

We first describe the data and the generating system, then analyse the data by means of the procedures $P^D_{\varepsilon_{tol}}$ and $P^P_{\varepsilon_{tol}}$, and comment on the identified models. We finally illustrate the consistency of $P^P_{\varepsilon_{tol}}$.

9.5.2. Data and system

The data $\tilde{w} = \text{col}(\tilde{w}_1, \tilde{w}_2, \tilde{w}_3) \in (\mathbb{R}^3)^{400}$ is generated by an ARMA-system $M(\sigma^{-1})w = N(\sigma^{-1})n$, where $n = \text{col}(n_1, n_2, n_3)$ consists of three uncorrelated white noise processes with $En_k = 0$, $En_k^2 = 1$, $k = 1, 2, 3$. The matrices M and N are given by $M = \begin{bmatrix} 1 & 0 & 0 \\ 0 & 1 & -1 \\ 0 & 0 & 1-\alpha\sigma^{-1} \end{bmatrix}$ and $N = \begin{bmatrix} 1/2 & 0 & 0 \\ 0 & \beta & 0 \\ 0 & 0 & 1 \end{bmatrix}$ with $\alpha := 1/\sqrt{11}$ and $\beta := \sqrt{1.1}$.

This corresponds to $w_1 = 1/2 . n_1$, $\sigma w_3 = \alpha w_3 + \sigma n_3$, $w_2 = w_3 + \beta n_2$. Figure 18 shows the data $\tilde{w}$, generated by a realization of n.

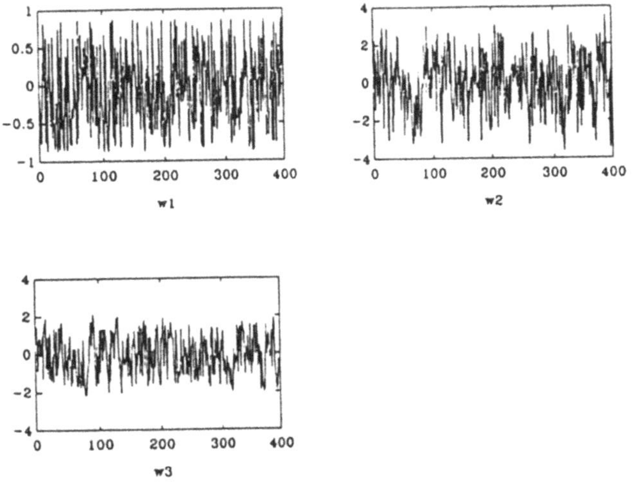

figure 18: data for simulation 9.5.

9.5.3. Model identification

We will identify a model for $\tilde{w}$ by means of descriptive and predictive procedures with (unfavourable) given tolerated misfits.

First we consider $P^D_{\varepsilon_{tol}}$ with $\varepsilon_{tol}=(\ \bar{\varepsilon}_t^{tol}.(1,1,1)\ ;\ t\in\mathbb{Z}_+)$, $\bar{\varepsilon}_0^{tol}:= e_0^D:= 1.6$, $\bar{\varepsilon}_1^{tol}:= e_1^D:= 1.2$, and $\bar{\varepsilon}_t^{tol}:= -1$ for $t>1$. This means that only zero order and first order laws may be used in the identification of a model. The identified model is given in table 7, along with the best (not-allowable) first order law.

Next we consider $P^P_{\varepsilon_{tol}}$ with $\varepsilon_{tol}=(\ \bar{\varepsilon}_t^{tol}.(1,1,1)\ ;\ t\in\mathbb{Z}_+)$, $\bar{\varepsilon}_0^{tol}:= e_0^P:= 1.6$, $\bar{\varepsilon}_1^{tol}:= e_1^P:= 0.95$, and $\bar{\varepsilon}_t^{tol}:= -1$ for $t>1$. The identified model is given in table 7, along with the best (not-allowable) first order law.

9.5.4. Model validation

The identified models are not of minimal complexity, given the maximal tolerated misfit. This is also indicated in table 7. It turns out that both for descriptive and predictive tolerated misfit as given before the model $B^*:= \{w\in(\mathbb{R}^3)^{\mathbb{Z}};\ w_1=0, w_2=0,\ (\sigma-\alpha)w_3=0\}$ satisfies the misfit constraint. This model has complexity $c(B^*)=(1,\frac{1}{2},\frac{1}{3},\frac{1}{4},\dots)$, which is smaller than the complexity of the identified models, which is $(1,1,1,1,\dots)$. It easily follows that $c(B^*)$ is the lowest achievable complexity, given the misfit constraints. However, among these allowable models of lowest complexity there exists none of minimal misfit. For the procedures $\bar{\bar{P}}^D_{\varepsilon_{tol}}$ and $\bar{\bar{P}}^P_{\varepsilon_{tol}}$

	identified model				model B^*			
	w_1	w_2	w_3	misfit	w_1	w_2	w_3	misfit
descr. AR order 0	0.9978	-0.0364	0.0552	0.4992	1	0	0	0.5000
	-0.0661	-0.5347	0.8425	0.6562	0	1	0	1.4938
				1.7197				
order 1: σ^0	-0.0012	-0.8443	-0.5359	1.4470	0	0	$-\alpha$	0.9574
σ^1	0.0012	0.8439	0.5356		0	0	1	
pred. AR order 0	0.9978	-0.0364	0.0552	0.4992	1	0	0	0.5000
	-0.0661	-0.5347	0.8425	0.6562	0	1	0	1.4938
				1.7197				
order 1: σ^0	-0.0004	-0.2937	-0.1865	0.9559	0	0	$-\alpha$	0.9301
σ^1	0.0014	1	0.6348		0	0	1	

table 7: descriptive and predictive AR-laws for simulation 9.5.

there exist models of lowest complexity and minimal misfit, but they seem difficult to compute. Their identification involves the question what is the lowest possible zero order misfit such that there exist first order relations, satisfying the misfit constraint and the orthogonality conditions of the (descriptive or predictive) canonical form.

The procedures $P^D_{\epsilon_{tol}}$ and $P^P_{\epsilon_{tol}}$ first determine as many zero order laws as possible. Requiring three of those laws results in a zero order misfit (1.7197, 0.6562, 0.4992), which is more than tolerated. Hence two zero order laws are accepted. Moreover, the best two laws are chosen. This implies conditions, due to the canonical form, on first order laws. In this simulation there is no allowable first order law satisfying these conditions. The model B^* shows that it is profitable not to take the best two zero order laws in order to get allowable first order laws, i.e., with misfit less than e^D_1 or e^P_1.

9.5.5. Consistency

We finally consider increase of the number of data generated by the ARMA-system. In table 8 we summarize results for the procedure $P^P_{\epsilon_{tol}}$ in

	identified models T=50	T=100	T=400	T=800	$A^P_{\varepsilon_{tol}}$
order 0:					
AR-coeff.					
w_1	0.9999	0.9824	0.9978	0.9961	1
w_2	0.0019	0.1422	-0.0364	-0.0234	0
w_3	0.0161	-0.1210	0.0552	-0.0346	0
misfit	0.5620	0.5161	0.4992	0.4994	0.5000
AR-coeff.					
w_1	-0.0127	0.1797	-0.0661	-0.0547	0
w_2	-0.5286	-0.5440	-0.5347	-0.5246	-0.5257
w_3	0.8488	0.8196	0.8425	0.8471	0.8507
misfit	0.6593	0.6621	0.6562	0.6429	0.6482
AR-coeff.					
w_1	-0.0102				
w_2	0.8489				
w_3	0.5285				
misfit	1.5920	>1,6	>1.6	>1.6	1.6970
order 1:	-				
AR-coeff.					
σ^0: w_1		0.0228	-0.0004	-0.0004	0
w_2		-0.3708	-0.2937	-0.2874	-0.2182
w_3		-0.2511	-0.1865	-0.1772	-0.1348
σ^1: w_1		-0.0614	0.0014	0.0014	0
w_2		1	1	1	1
w_3		0.6771	0.6348	0.6164	0.6180
misfit		0.9296	0.9559	0.9578	0.9759

table 8: consistency of $P^P_{\varepsilon_{tol}}$.

case of $T=50,100,400$ and 800 observations. We also calculated the best first order laws. Observe that for $T=50$ the procedure for this simulation would accept three zero order laws, while for $T=100$ it would accept a first order law. We also give the optimal approximation $A^P_{\varepsilon_{tol}}$, corresponding to the optimal predictive model for ε_{tol} in case the generating system were known. This model can be calculated from covariance matrices, derived from M and N.

The results in table 8 illustrate consistency, as defined in section 8. Note especially that in the limit the best first order law which satisfies the orthogonality conditions of the canonical predictive form has predictive misfit $0.9759 > e_1^P = 0.95$. Hence, almost sure, for a sufficiently large number of observations the procedure $P^P_{\varepsilon_{tol}}$ will only accept two zero order laws.

10. CONCLUSION

In this paper we have described some procedures for approximate modelling of a time series, along with corresponding algorithms. The procedures have been illustrated by means of some numerical simulations.

The procedures determine a deterministic dynamical system which for given data is optimal with respect to a utility of models, depending on the objective of modelling. This utility is expressed in terms of a complexity of models and a measure of fit between data and models. The utility reflects a compromise between the generally conflicting objectives of identifying a simple model and a model which fits the data well. The utility is numerically expressed in terms of canonical parametrizations of dynamical systems. These canonical forms are determined in accordance with the objective of modelling.

The procedures form part of a more general deterministic approach to approximate modelling, as extensively discussed and illustrated in the paper.

The procedures have a clear optimality property as data modelling procedures, in terms of the corresponding utility. A procedure also has an optimal performance as a method of modelling phenomena if it is consistent. This means that nearly optimal models of the phenomenon are identified if the number of observations generated by the phenomenon is sufficiently large. This has been investigated for certain classes of data generating systems and some of the procedures.

We finally mention some topics for future research.

(i) The construction of algorithms for utilities other then $u_{c_{tol}}$ and $u_{\varepsilon_{tol}}$, especially for minimizing the number of unexplained variables (inputs) under a misfit constraint.

(ii) Utilities and algorithms when the purpose of modelling is control.

(iii) Consistency analysis for generating systems of ARMAX type, i.e., with inputs, and the related issue of sufficient excitation.

(iv) Definition of approximate structure of a phenomenon, and corresponding interpretation of stochastic systems, especially of ARMAX type.

(v) Definition of the amount of confidence in identified models, sensitivity with respect to changes in data and tolerated levels of complexity or misfit, and robustness.

REFERENCES

[1] Golub, G.H., and C.F. Van Loan, An analysis of the total least squares problem, *SIAM Journal on Numerical Analysis* 17(6), pp. 883–893, 1980.

[2] Hannan, E.J., *Multiple Time Series*, John Wiley, New York, 1970.

[3] Hannan, E.J., W.T.M. Dunsmuir and M. Deistler, Estimation of vector ARMAX models, *Journal of Multivariate Analysis* 10, pp. 275–295, 1980.

[4] Heij, C., Approximate modelling of deterministic systems, in Curtain, R.F. (ed.), *Modelling, Robustness and Sensitivity Reduction in Control Systems*, pp. 271-283, NATO ASI Series, Springer, Berlin, 1987.

[5] Heij, C., and J.C. Willems, Consistency analysis of approximate modelling procedures, in Byrnes, C.I., C.F. Martin and R.E. Saeks (eds.), *Linear Circuits, Systems and Signal Processing: Theory and Application*, pp. 445-456, North Holland, Amsterdam, 1988.

[6] Jayant, N.S., and P. Noll, *Digital Coding of Waveforms*, Prentice Hall, Englewood Cliffs, New Jersey, 1984.

[7] Kalman, R.E., P.L. Falb and M.A. Arbib, *Topics in Mathematical System Theory*, McGraw–Hill, New York, 1969.

[8] Kendall, M.G., and A. Stuart, *The Advanced Theory of Statistics*, Griffin, London, 1964.

[9] Ljung, L., Convergence analysis of parametric identification methods, *IEEE AC*–23(5), pp. 770–783, 1978.

[10] Ljung, L., *System Identification – Theory for the User*, Prentice Hall, Englewood Cliffs, New Jersey, 1987.

[11] Ljung, L., and P.E. Caines, Asymptotic normality of prediction error estimators for approximate system models, *Stochastics* 3, pp. 29–46, 1979.

[12] Maddala, G.S., *Econometrics*, McGraw–Hill, New York, 1976.

[13] Nieuwenhuis, J.W., and J.C. Willems, Continuity of dynamical systems: a system theoretic approach, *Mathematics of Control, Signals, and Systems* 1(2), pp. 147-165, 1988.

[14] Rissanen, J., Stochastic complexity and modelling, *The Annals of Statistics* 14 (3), pp. 1080–1100, 1986.

[15] Willems, J.C., From time series to linear system. Part I: Finite dimensional linear time invariant systems. Part II: Exact modelling. Part III: Approximate modelling. *Automatica* 22, pp. 561–580, 1986; 22, pp. 675–694, 1986; 23, pp. 87–115, 1987.

[16] Willems, J.C., Models for dynamics, *Dynamics Reported* 2, pp. 171–269, 1989.

IDENTIFICATION - A THEORY OF GUARANTEED ESTIMATES

A.B. KURZHANSKI

Abstract

This paper gives an introduction to the theory of parameter identification and state estimation for systems subjevcted to uncertainties with set-membership bounds on the unknowns.

The situation under discussion may often turn to be more a propos since here the system and the environment are modelled as truly uncertain rather than noisy. The described approach is purely deterministic.

The techniques given here are mainly aimed at problems with nonquadratic constraints with the quadratic case acting as a necessary complimentary tool. Some substantial properties of nonlinear estimation schemes are also indicated.

On the other hand the techniques involved her for the treatment of systems with nonquadratic constraints on the unknowns are proved to have some nontrivial interretalions with those developed in stochastic estimation theory. This may lead to some further estimation schemes that would combine the deterministic and the stochastic models of uncertainty.

The recurrence procedures of this paper are devised into relations that would allow numerical simulations.

Keywords

Identification, state estimation, set-membership constraint, guaranteed estimation, informational domain, stochastic estimation, consistency conditions, uncertain systems, setvalued calculus.

1. INTRODUCTION

A crucial issue in the process of *mathematical modelling* on the basis of *available observations* is the problem of system *parameter identification* under observation noise. The conventional area of applied mathematics within which the problem is usually discussed is *mathematical statistics* [1, 2]. The uncertainties in the system parameters and the observation noise are taken here to be described by stochastic mechanisms. The informational scheme for the identification process usually assumes that there exists an adequate statistical description for the unknowns. Within this framework a fairly complete theory has been developed for linear systems with disturbances modelled by gaussian noise and with quadratic criteria of optimality for the estimates [3, 4]. A large number of investigations is devoted to statistical identification under more general assumptions.

However, the statistical methods are not the only mathematical tools for the treatment of system modelling.

This paper gives an introduction to the theory of *guaranteed identification.* It demonstrates for example that the classical system parameter estimation problem under measurement noise may be posed in a deterministic setting rather than in a traditional probabilistic framework. The adopted model assumes that there is no statistical description for the measurement "noise" or for the disturbances in the system and that the only information on these is restricted to a *set-membership constraint* on their admissible values or realizations. A considerable number of applications in engineering and systems analysis are treated under informational assumptions that justify this approach (see e.g. [5–10]).

The basic techniques that are necessary for the treatment of the problems given here are based on *set-valued calculus* so that the solutions are formulated in the form of set-valued estimators. This approach also assures numerical robustness for the respective approximation schemes. Other results related to the topic of this paper may be found in [11–18].

Let us start with a trivial example. Suppose one is to identify a vector $c \in \mathbf{R}^2$ on the basis of observations $y(k) = c + \xi(k)$, $k = 1 ,\ldots, N ,\ldots$ corrupted by "noise" $\xi(k)$.

Contrary to the conventional approach we will at first assume that there is no statistical data on $\xi(k)$ being available in advance. However we will suppose that a restriction

$$\xi(k) \in Q(k)$$

is given with set $Q(k)$ being known. We will assume that $Q(k)$, $k \geq k_0$ is a convex compact set.

Every single measurement $y(k)$ gives us some information on c, namely it indicates that the following inclusion is true

$$c \in y(k) - Q(k) \tag{1.2}$$

Having had m measurements $y(1) ,\ldots, y(m)$, we observe that inclusion (1.2) should be true for every $k = 1 ,\ldots, m$. Hence, after m observations we will have

$$c \in \bigcap_{k=1}^{m} (y(k) - Q(k)) = C[1 , m]$$

where the set C $[1 , m]$ is the *"guaranteed estimate"* for c after m observations.

It is thus clear that every "new" measurement $y(m + 1)$ introduces an innovation into the estimation process by means of an intersection of the previous estimate $C[1 , m]$ with a "new" set $\{ y(m + 1) - Q(m + 1) \}$, so that

$$C[1 , m + 1] = C[1 , m] \bigcap \{ (y(m + 1) - Q\ (m + 1) \} \tag{1.3}$$

Relation (1.3) is a *recurrence equation* which describes the evolution of the estimate $C[1 , m]$ in m. (Figures 1 and 2 demonstrate set $C[1 , m]$ for $m = 4$ with Q being (1) - a square, (2) - a circle; c^* stands for the unknown value to be estimated). The "accuracy" of the estimate will now depend on the behaviour of the "noise" $\xi(k)$. Let us trace this fact more precisely.

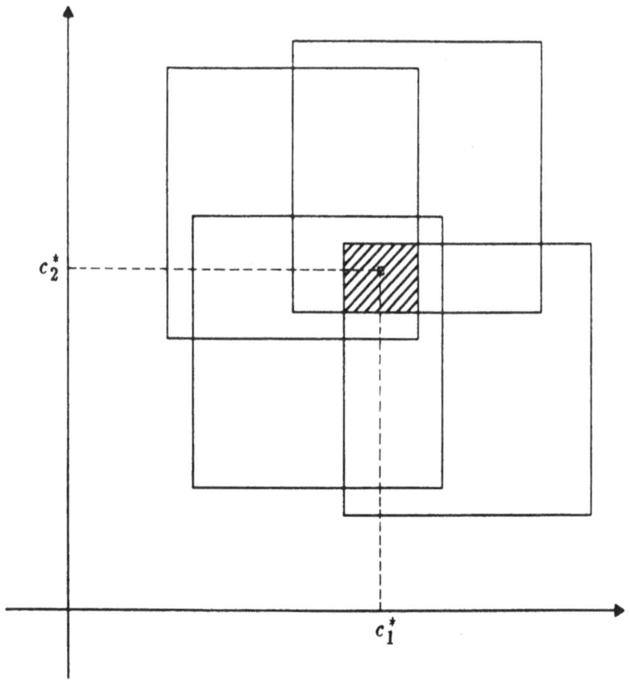

FIGURE 1

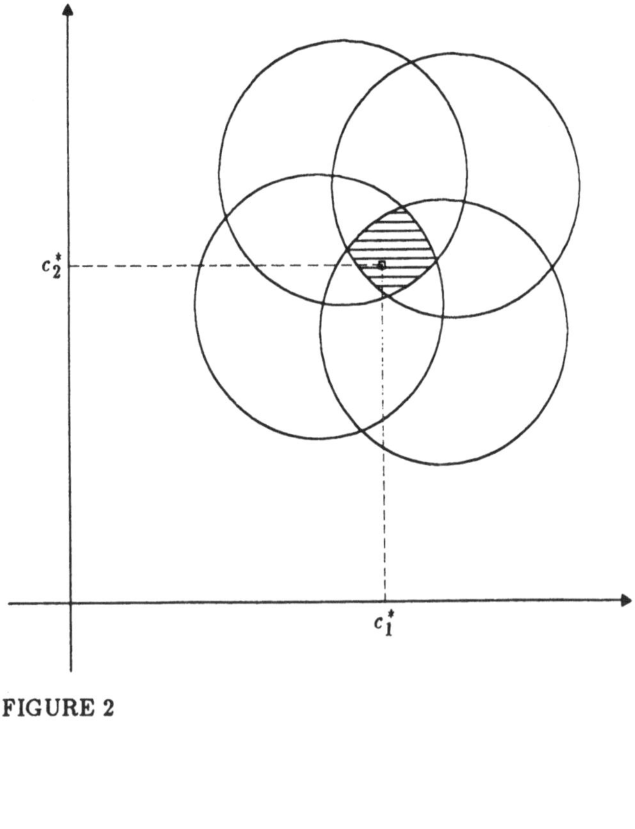

FIGURE 2

Assume c^* , $\xi^*(k)$ are the unknown actual values of c , $\xi(k)$, $k \in [1 ,\ldots, m]$ so that the available measurement is

$$y(k) = c^* + \xi^*(k)$$

Then the estimate

$$C[1\,,\,m] = \bigcap_{k=1}^{m} (c^* - (Q(k) - \xi^*(k))) =$$
$$= c^* + \bigcap_{k=1}^{m} (\xi^*(k) - Q(k))$$

where

$$\bigcap_{k=1}^{m} (\xi^*(k) - Q(k)) = R^*(m)$$

is *the "error set" of the estimation process.* It obviously depends on the behaviour of the "noise" $\xi^*(k)$, $k = 1 ,\ldots, m$.

Let us examine the "worst case" solution (from the point of view of the observer). Suppose

$$\xi^*(k) \equiv 0 \tag{1.4}$$
$$Q(k) \equiv Q\,,\; Q = -Q \quad (k = 1 ,\ldots, m)$$

("noise" constant, and "Q is stationary" and symmetric about the origin). Then, clearly

$$R^*(m) = \bigcap_{k=1}^{m} (-Q(k)) = -Q = Q$$

and the range of the error of estimation is precisely Q. The "guaranteed" error is

$$\max \{ \,||\, q \,||\, \mid q \in Q\}$$

It is obvious here that none of the new measurements do bring any innovation into the estimation process.

In contrast an "adequate" behaviour of $\xi(k)$ may considerably improve the estimation. For example, assume that Q is a square: $Q = \mathbf{S}$,

$$\mathbf{S} = \{q : |\, q_i \,| \leq 1\,,\; i = 1,2\}\,,$$
$$c^* = \begin{pmatrix} c_1^* \\ c_2^* \end{pmatrix} = \begin{Bmatrix} -0.5 \\ 0.25 \end{Bmatrix}$$

is the unknown vector to be identified.

If

$$\xi^*(1) = (1\ ,1)$$
$$\xi^*(2) = (-1\ ,-1)$$

then the error set

$$R^*(2) = \{\xi^*(1) - Q\} \bigcap \{\xi^*(2) - Q\} = \{0\}$$

and the estimation is exact, (Figure 3).

For another example take $Q = S(0)$ to be a unit circle, $m = 3$, $\xi^*(1) = (1{,}0)$, $\xi^*(2) = (0{,}1)$, $\xi^*(3) = (0, -1)$, and the vector c^* $(c_1^* = c_2^* = 2.5)$ is again exactly identified. (Figure 4).

Let us now suppose that *the noise $\xi(k)$ is governed by a random mechanism.* Namely suppose that $\xi(k)$ is a random variable uniformly distributed in $Q = \mathbf{S}$ for any $k = 1\ ,\ldots, \infty$ and that all the vectors $\xi(k)$ are jointly independent.

Taking two points $\xi^{(1)} = (1{,}1)$, $\xi^{(2)} = (-1{,}-1)$, consider two sets

$$Q^{(1)}\ (\epsilon) = Q \bigcap (\xi^{(1)} + \mathbf{S}_\epsilon(0))$$
$$Q^{(2)}\ (\epsilon) = Q \bigcap (\xi^{(2)} + \mathbf{S}_\epsilon(0))$$

where

$$\mathbf{S}_\epsilon\ (0) = \{q : \mid q_i \mid \leq \epsilon\ , i = 1\ , 2\ \}$$

For a random sequence

$$\xi[\cdot] \in \{\xi(k),\ k = 1,\ldots,\infty\}$$

consider the event $A_\epsilon(k)$ that

$$\xi(k) \notin Q^{(1)}(\epsilon) \bigcup Q^{(2)}(\epsilon)$$

for a given k. Denote A_ϵ to be the event that

$$\xi(k) \notin Q^{(1)}(\epsilon) \bigcup Q^{(2)}(\epsilon),\ \forall k$$

Then

$$A_\epsilon = \bigcap_k A_\epsilon(k)$$

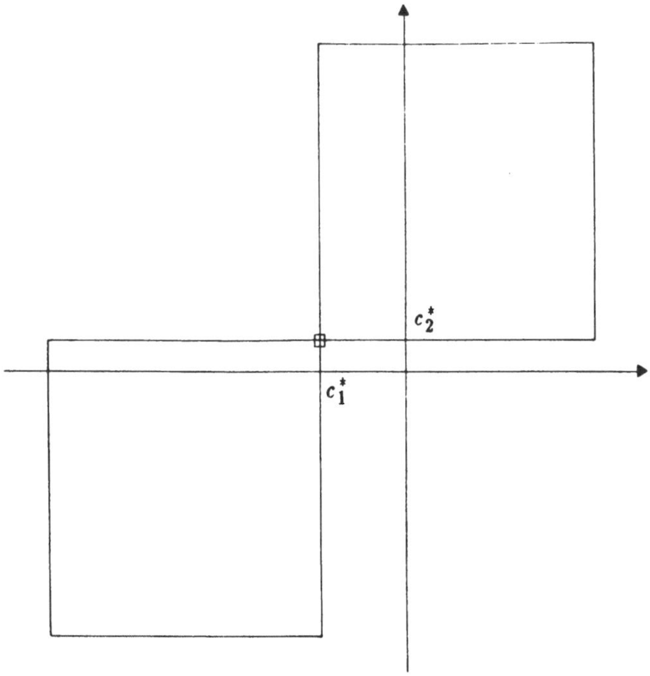

FIGURE 3

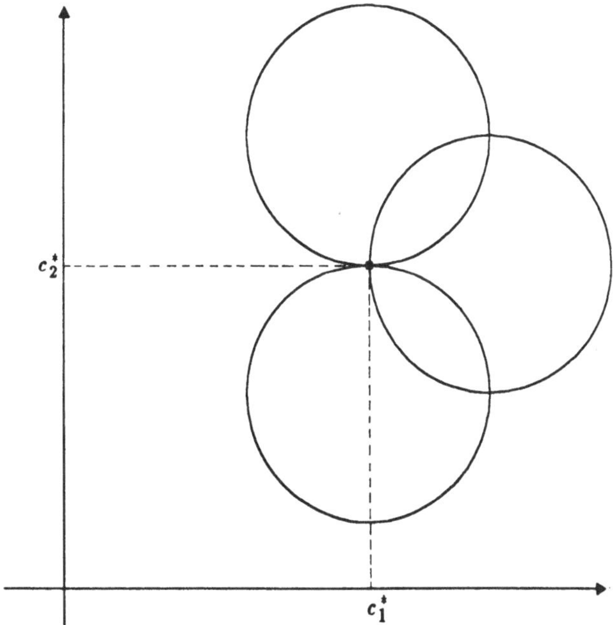

FIGURE 4

and

$$P(A_\epsilon(k)) < p(\epsilon) < 1, \forall k \ .$$

Due to the joint independence of $\xi(k)$, we have

$$P(A_\epsilon) = \prod_{k=1}^{\infty} P\,(A_\epsilon(k)) = 0 \tag{1.5}$$

If we denote

$$A = \bigcup_i \{A_{\epsilon_i}\}, \ \ \epsilon_i > 0\,, \forall\, i = 1\,, \ldots, \infty\,;\ \epsilon_i \longrightarrow 0\,,\ i \longrightarrow \infty$$

and A^c, A^c_ϵ to be the complements of A, A_ϵ, then, obviously, $A^c \subseteq A^c_{\epsilon_i}$ for any $\epsilon_i > 0$ and $A^c = \bigcap_i \{A^c_{\epsilon_i}\}$, so that

$$P(A^c_\epsilon) = 1,\ P(A^c) = \prod_i P\{A^c_{\epsilon_i}\} = 1 \tag{1.6}$$

Hence for any $\epsilon > 0$ the sequence $\xi[\cdot] = \{\xi(k),\ k = 1, \ldots, \infty\}$ will satisfy the inclusions

$$\xi(k') \in Q^{(1)}(\epsilon)$$
$$\xi(k'') \in Q^{(2)}(\epsilon)$$

with probability 1 for some $k = k'$, $k = k''$. (Otherwise, we would have $\xi[\,\cdot\,] \in A_{\epsilon_i}$).

Thus for any $\epsilon_i > 0$, for "almost all" sequences $\xi[\,\cdot\,]$ there exists an $M > 0$ (depending on the sequence) such that for $m > M$ the error set

$$R^*(m) \subseteq (Q^{(1)}(\epsilon) - Q) \bigcap (Q^{(2)}(\epsilon) - Q) = S_\epsilon(0)$$

or otherwise

$$\lim_{m \longrightarrow \infty} h\ (R^*(m)\ ,\ \{0\}) \leq \epsilon_i$$

where

$$h(R^*, \{0\}) = \max\{||z||\ |\ z \in R^*\}$$

and $||z||$ is the Euclidean norm of vector $z \in \mathbf{R}^2$.

It follows that with probability 1 we have

$$h(R^*(m)\ ,\ \{0\}) \underset{m \longrightarrow \infty}{\longrightarrow} 0$$

where $\{0\}$ is a singleton – the null element of $\mathbf{R}^2$.

Therefore, *under the randomness assumptions* of the above *the estimation process is consistent with probability 1*. Under the same assumptions it is clear that *the "worst case"* noise (1.4) ($\xi^*(k) \equiv 0$, $k = 1 ,\ldots, \infty$) *may appear only with probability 0*.

The few elementary facts stated in this introduction develop into a theory of "guaranteed identification" which appears relevant to the treatment of parameter estimation, to dynamic state estimation problems, to the identification of systems with unmodelled dynamics and even to the solution of inverse problems for distributed systems [19]. It may also be propagated to the treatment of some problems for nonlinear systems [20].

The first part of the present paper deals with the simplest identification problem for a linear model describing the respective guaranteed estimates. Here the basic results are those that yield the recurrence relations for the estimates. They also lead to the discussion of the problem of consistency of the identification process.

The second part, written in a more compact form, deals with the "guaranteed" state estimation problem for discrete time linear systems with unknown but bounded inputs. This is followed by an introduction into the basic facts of "guaranteed nonlinear filtering".

The paper mainly deals with nonquadratic constraints on the unknowns. It also deals with nonlinearity and nonstationarity. This is partly done with the aim of reminding the reader that identification and state estimation problems are not merely linear-quadratic and stationary as it may seem from most of the available literature.

A special item discussed in the sequel is the relation between guaranteed and stochastic estimation procedures in the case of non-quadratic constraints on the unknowns.

2. NOTATION

Here we list some conventional notations adopted in this paper:

$\mathbf{R}^n$ will stand for the n-dimensional vector space, while $\mathbf{R}^{m \times n}$ - for the space of $m \times n$ - dimensional matrices, I_n will be the unit matrix of dimension n, $A \otimes B$ - the *Kronecker product* of matrices A , B, so that

$(A \otimes B)$ will be the matrix of the form

$$\begin{bmatrix} a_{11}B \,, & \cdots, & a_{1n}B \\ \cdots & \cdots & \cdots \\ a_{n1}B \,, & \cdots, & a_{nn}B \end{bmatrix}$$

The prime will stand for the transpose and $\bar{A}$ - for an mn - dimensional vector obtained by stacking the matrix $A = \{a^{(1)}, \dots, a^{(n)}\}$, with columns $a^{(i)} \in \mathbf{R}^m$ $(a_j^{(i)} = a_{ij})$, so that $a_{(i-1)h+j} = a_j^{(i)}$, $(i = 1, \dots, n)$, $(j = 1, \dots, m)$, or in other terms

$$\bar{A} = \sum_{i=1}^{n} (e^{(i)} \otimes (A \; e^{(i)}))$$

where $e^{(i)}$ is a unit orth within $\mathbf{R}^n$ $(e_j^{(i)} = \delta_{ij}$, with δ_{ij} the Kronecker delta : $\delta_{ij} = 1$ for $i = j$, $\delta_{ij} = 0$ for $i \neq j)$.

If $\mathbf{C} = \{C\}$ is a set of $(m \times n)$-matrices C, then $\bar{\mathbf{C}}$ will stand for the respective set of mn-vectors $\bar{C} : \bar{\mathbf{C}} = \{\bar{C}\}$.

The few basic operations used in this paper are as follows:

If $<A , B> = tr \; AB'$ is the *inner product of matrices* A , $B \in \mathbf{R}^{m \times n}$ and (p , q) - the *inner product* of vectors p , $q \in \mathbf{R}^n$, then for $x \in \mathbf{R}^n$, $y \in \mathbf{R}^m$ we have

$$\begin{aligned} y \otimes x' &= yx' \in \mathbf{R}^{m \times n} \\ <A , y \otimes x'> &= (A \; x , y) \end{aligned} \tag{2.1}$$

Other matrix equalities used here are

$$(A \otimes B)^{-1} = A^{-1} \otimes B^{-1}$$

$(A , B$ are $n \times n$ dimensional and their determinants $|A| \neq 0$, $|B| \neq 0)$

$$\begin{aligned} (A \otimes B)' &= A' \otimes B' \\ (A \otimes B) \; \bar{K} &= \overline{BKA'} \end{aligned} \tag{2.2}$$

A sequence of integers $i = k, \ldots, s$ will be $[k, s]$. A finite sequence of vectors $\{\xi(i) : i = k, \ldots, s\}$ will be denoted as $\xi\,[k, s]$, while an infinite one $\{\xi(i), i = s, \ldots, \infty\}$ as $\xi\,[s\,;\,\cdot]$ with $\xi\,[1, \cdot] = \xi[\cdot]$. Similar notations will be used for sequences of sets. For example $R[k, s]$ will stand for a sequence of sets $R(i)$, $k \le i \le s$.

Symbols *conv* $\mathbf{R}^n$ and *co* $\mathbf{R}^n$ will denote the varieties of all *convex compact* and *closed convex* subsets of $\mathbf{R}^n$ respectively,

$$\rho(\ell \mid Q) = \sup \{(\ell, q) \mid q \in Q\}$$

will be the *support function* of set $Q \subseteq \mathbf{R}^n$.

With $Q \in conv\ \mathbf{R}^n$ the operation of *sup* in the definition of $\rho(\ell \mid Q)$ may be substituted for *max*. Further on *int* Q will be the set of all *interior points* of Q.

$$S_r(x_0) = \{x : \|\, x - x_0 \,\| \le r\,;\, x, x_0 \in \mathbf{R}^n\}$$

will denote the *Euclidean ball* with center x_0 and radius r, $(\|\, x \,\| = (x, x)^{1/2})$, while $h(P, Q)$ will stand for the *Hausdorff distance* between sets $P, Q \in conv\ \mathbf{R}^n$. Namely

$$h(P, Q) = \min \{\, r : P \subseteq Q + r\,S(0)\,,\ Q \subseteq P + r\,S(0)\,\}\,.$$

The symbol *epi f* stands for the *epigraph*

$$epi\ f = \{z = \{x, y\} : y \ge f(x)\,,\, z \in \mathbf{R}^{n+1}\}$$

of function f - a subset of $\mathbf{R}^{n+1}$ and *co* Q stands for the *convex hull* of set Q with $\overline{co}\ Q$ being the *closure* of *co* Q.

The "time interval" is denoted as $\{1, \ldots, N\} = T_N$

For a given set $P \subseteq \Omega$ the symbol P^c will stand for the *complement* P^c of P.

The basic scheme will be first interpreted through the following "elementary" parameter estimation problem.

1. THE BASIC PROBLEM

Consider a system

$$y(k) = C\,p(k) + \xi(k) \tag{3.1}$$

$$k \in T_s$$

where $y(k)$ is the *available measurement*, $p(k)$ is a *given input*, C is the *matrix parameter to be identified* and $\xi(k)$ is the *unknown disturbance*. We further assume $p \in \mathbf{R}^n$, $y \in \mathbf{R}^m$. Hence $\xi \in \mathbf{R}^m$, $C \in \mathbf{R}^{m \times n}$.

The available additional information on C, $\xi[1, s]$ is given through restrictions on these values which are taken to be specified in advance.

The types of simple restrictions on C, $\xi[1, s]$ to be considered in the sequel are as follows:

$$(\bar{C}-\bar{C}^*)' \mathbb{L}(\bar{C}-\bar{C}^*) + \sum_{k=1}^{s} (\xi(k) - \xi^*(k))' N(k)(\xi(k) - \xi^*(k)) \leq 1 \tag{III.A}$$

where $\mathbb{L} > 0$, $N(k) > 0$ ($\mathbb{L} \in \mathbf{R}^{mn \times mn}$, $N(k) \in \mathbf{R}^{m \times m}$) ξ^*, C^* are given. (This is *the joint quadratic constraint*), or

$$(\bar{C} - \bar{C}^*)' \mathbb{L}(\bar{C} - \bar{C}^*) \leq 1 \tag{III.B}$$

$$\sum_{1}^{s} (\xi[k] - \xi^*[k])' N(k) (\xi[k] - \xi^*[k]) \leq 1$$

which is *the separate quadratic constraint*, or

$$C \in \mathbf{C}_0, \ \xi(k) \in Q(k) \tag{III.C}$$

which is *the geometrical or instantaneous constraint.* Here $\mathbf{C}_0$, $Q(k)$ are assumed to be convex and compact in $\mathbf{R}^{m \times n}$ and $\mathbf{R}^m$ respectively.

The restriction on the pair $\{C, \xi[1, s]\} = \varsigma[1, s]$ (whether given in the form (III.A), (III.B) or (III.C)) will be denoted by a unified relation as

$$\varsigma[1, s] \in \Xi \tag{3.2}$$

where Ξ is a given set in the product space $\mathbf{R}^{m \times n} \times \mathbf{R}^{m \times s}$.

With measurement $y[1, s]$ given, the aim of the solution will be to find the set of all pairs $\varsigma[1, s]$ consistent with (3.1), (3.2) and with given $y[1, s]$. More precisely the solution will be given through the notion of *the informational domain.*

Definition 3.1. The informational domain $\mathbf{C}[s] = \mathbf{C}[1\ ,\ s]$ *consistent with measurement* $y[1\ ,\ s]$ *and restriction (3.2) will be defined as the set of all matrices* C *for each of which there exists a corresponding sequence* $\xi[1\ ,\ s]$ *such that the pair* $\varsigma[1\ ,\ s] = \{C\ ,\ \xi[1\ ,\ s]\}$ *satisfies both restriction (3.2) and equation (3.1) (for the given* $y[1\ ,\ s]$ *).*

Hence the idea of the solution of the estimation problem is to find the set $\mathbf{C}[1\ ,\ s]$ of all the possible values of C each of which (together with an adequate $\xi[1\ ,\ s]$) could generate the given measurement sequence $y[1\ ,\ s]$.

It is obvious that set $\mathbf{C}[s] = \mathbf{C}[1\ ,\ s]$ now contains the unknown actual value $C = C^{\circ}$ which is to be estimated.

With set $\mathbf{C}[s]$ being known, one may also construct a *minmax estimate* $C_*[s]$ of C° - for example through the solution of the problem

$$\max\ \{d(C_0[s]\ ,\ Z)\ |\ Z \in \mathbf{C}[s]\} = \tag{3.3}$$

$$= \min_{C} \left\{ \max\{d(C\ ,\ Z)\ |\ Z \in \mathbf{C}[s]\},\ C \in \mathbf{C}[s] \right\} = \epsilon(s)\ ,$$

where $d(\cdot\ ,\cdot)$ is some metric in the space $\mathbf{R}^{m \times n}$.

The element $C_0[s]$ is known as the *Chebyshev center* for set $\mathbf{C}[s]$.

Once $C_0[s]$ is specified, the estimation error $d(C_0[s]\ ,\ C^{\circ}) \le \epsilon(s)$ *is guaranteed* by the procedure.

However, for many purposes, especially under a nonquadratic constraint (III.C), it may be convenient to describe *the whole set* $\mathbf{C}[s]$ rather than the minmax estimate $C_*[s]$.

If s varies and even $s \to \infty$ it makes sense to consider the *evolution* of $\mathbf{C}[s]$ and its *asymptotic behaviour* in which case the estimation process may turn to be *consistent*, i.e.

$$\lim_{s \to \infty} \mathbf{C}[s] = \{C^{\circ}\} \tag{3.4}$$

The convergence here is understood in the sense that

$$\lim_{s \to \infty} h\ (\mathbf{C}[s]\ ,\ C^{\circ}) = 0$$

where $h(\mathbf{C}', \mathbf{C}'')$ is the *Hausdorff metric* (see Introduction), and C° is a singleton in $\mathbf{R}^{m \times n}$.

In some particular cases the equality (3.4) may be achieved in a finite number s_0 of stages s when for example

$$\mathbf{C}[s] = C^{\circ}, \; s^{\circ} > 1,$$

The main discussion will be further concerned with the nonquadratic geometrical constraint (III.C). However it is more natural to start with the simplest "quadratic" restriction (III.A). In this case, as we shall see, the set $\mathbf{C}[s]$ turns to be an ellipsoid and the respective equations for $\mathbf{C}[s]$ arrive in explicit form.

4. THE JOINT QUADRATIC CONSTRAINT. RECURRENCE EQUATIONS

As equation (3.1) yields

$$\xi(k) = y(k) - Cp(k)$$

the set $\mathbf{C}[s]$ consists of all matrices C that satisfy (III.A), i.e.

$$(\bar{C} - \bar{C}^*)' \,\mathbb{L}\, (\bar{C} - \bar{C}^*) + \tag{4.1}$$
$$\sum_{k=1}^{s} (y(k) - Cp(k) - \xi^*(k))' N(k)(y(k) - Cp(k) - \xi^*(k)) \leq 1$$

In view of the equality (2.2) which here turns into

$$I_m \, C \, p = (p' \otimes I_m) \, \bar{C}$$

we may rewrite (4.1) as

$$(\bar{C} - \bar{C}^*)' \,\mathbb{P}[s](\bar{C} - \bar{C}^*) - 2(\mathbb{D}[s] \,, \bar{C} - \bar{C}^*) + \gamma^2[s] \leq 1$$

where

$$\mathbb{P}[s] = \mathbb{L} + \sum_{k=1}^{s} P(k)$$
$$P(k) = (p(k) \otimes I_m) \, N(k)(p'(k) \otimes I_m)$$
$$\mathbb{D}[s] = \sum_{k=1}^{s} D(k)$$

$$D'(k) = y^{*\prime}(k)\ N(k)\ (p'(k) \otimes I_m)$$

$$\gamma^2(s) = \sum_{k=1}^{s} y^{*\prime}(k)\ N(k)\ y^*(k) \tag{4.2}$$

$$y^*(k) = y(k) - C^* p(k) - \xi^*(k) \tag{4.3}$$

Hence the result is given by

Theorem 4.1. The set $\mathbf{C}[s]$ *is an ellipsoid defined by the inequality*

$$((\bar{C} - \bar{C}^* - \mathbb{P}^{-1}[s]\,\mathbb{D}[s])', \mathbb{P}[s]\,(\bar{C} - \bar{C}^* - \mathbb{P}^{-1}[s]\,\mathbb{D}[s])) \leq 1 - h^2[s] \tag{4.4}$$

with center

$$\bar{C}_o[s] = \mathbb{P}^{-1}[s]\ \mathbb{D}[s] + \bar{C}^*$$

Here

$$h^2[s] = \gamma^2(s) - (\mathbb{D}[s]\ ,\ \mathbb{P}^{-1}[s]\ \mathbb{D}[s]) \tag{4.5}$$

$$\mathbb{P}[s] = \mathbb{P}[s-1] + P(s)\ ,\ \mathbb{D}[s] = \mathbb{D}[s-1] + D(s) \tag{4.6}$$

$$\gamma^2(s) = \gamma^2(s-1) + y^{*\prime}(s)\ N(s)\ y^*(s)\ ,\ \gamma(0) = 0 \tag{4.7}$$

$$\mathbb{P}[0] = \mathbb{L}\ ,\ \mathbb{D}(0) = 0$$

$$\mathbb{P}^{-1}[s] = \mathbb{P}^{-1}[s-1] - \mathbb{P}^{-1}[s-1]\,G(s-1)\,K^{-1}(s-1)\,G'(s-1)\,\mathbb{P}[s-1] \tag{4.8}$$

$$G(s-1) = p(s-1) \otimes I_m$$

$$K(s-1) = N^{-1}(s-1) + G'(s-1)\ \mathbb{P}[s-1]\ G(s-1)$$

Relations (4.4) - (4.8) are *evolutionary equations* that describe the dynamics of the set $\mathbf{C}[s]$ (which is an *ellipsoid*) and its *center* $C_0[s]$ which *coincides* precisely *with the minmax estimate* $C_*[s]$ for $\mathbf{C}[s]$ (assuming $d(C\ ,\ Z)$ of (3.3) is taken to be the Euclidean metric).

Remark 6.1 A standard problem of *statistical estimation* is to find the *conditional distribution* of the values of a matrix C after s measurements due to equation (3.1) where $\xi(k)$, $k \in [1, \infty)$ are non correlated gaussian variables with given mean values $E\xi(k) = \xi^*(k)$ and covariance matrices

$$E\xi(k)\xi'(k) = N^{-1}(k)\ .$$

The initial gaussian distribution for the vector $\bar{C}$ is taken to be given with $E\bar{C} = \bar{C}^*$, $E\bar{C}\bar{C}' = \mathbb{L}^{-1}$.

A standard application of the least-square method or of some other conventional (e.g. bayesian or maximal likelihood) techniques yields an estimate

$$\bar{C}_*[s] = \mathbb{P}^{-1}[s]\mathbb{D}[s] + \bar{C}^*$$

with $\mathbb{P}[s]$, $\mathbb{D}[s]$ governed by equations (4.6), (4.8) [4]. The estimate is therefore similar to that of theorem 4.1: $\bar{C}_*[s]$ coincides with $\bar{C}_0[s]$. Here, however, the analogy ends – equations (4.5), (4.7) are specific only for the guaranteed estimates. The estimation errors for the stochastic and for the guaranteed deterministic solutions are defined through different notions and are therefore calculated through different procedures.

The next step is to specify the "worst case" and "best case" disturbances for the estimation process. From the definition (4.3) of $y^*(k)$ it is clear that if the actual values $\varsigma^\circ[1\ ,s] = \{\xi^\circ\ [1\ ,\ s]\ ,\ C^\circ\}$ for $\varsigma[1\ ,\ s] = \{\xi[1\ ,\ s]\ ,\ C\}$ are taken to be

$$\varsigma^\circ[1\ ,\ s] = \varsigma^*[1\ ,\ s]\ ,\ C^\circ = C^* \tag{4.9}$$

then

$$y^*[1\ ,\ s] \equiv 0\ ,\ \mathbb{D}[s] \equiv 0$$

and therefore

$$h^2[s] = 0 \tag{4.10}$$

The ellipsoid $C[1\ ,\ s]$ is then the "largest" possible in the sense that it includes all the ellipsoids derived through other measurements than the "worst" one

$$y_w(k) = C^*\ p(k) + \xi^*(k)\ ,\ k \in [1\ ,\ s]$$

(Note that whatever are the admissible values of $y[1\ ,\ s]$, all the respective ellipsoids $C[s]$ have *one and the same center* $C_0[s]$ and matrix $\mathbb{P}[s]$. They differ only through $h[s]$ in the right hand part of (4.4)).

The "smallest" possible ellipsoid is the one that turns to be a singleton. It is derived through the "best possible" measurement $y^{(b)}[1\ ,\ s]$. The latter is defined by the pair

$$\{C^{(b)}\ ,\ \xi^{(b)}[1\ ,\ s]\}$$

where $C^{(b)} = C^*$ and $\xi^{(b)}[1\,,\,s]$ satisfies conditions

$$\sum_{k=1}^{s} (\xi^{(b)}(k) - \xi^*(k))' N(k)(p'(k) \otimes I_m) = 0 \tag{4.11}$$

$$\sum_{k=1}^{s} (\xi^{(b)}(k) - \xi^*(k))' N(k)(\xi^{(b)}(k) - \xi^*(k)) = 1 \tag{4.12}$$

With $C^{(b)} = C^*$ and with (4.11), (4.12) fulfillled we have

$$\begin{aligned} y(k) &= C^* p(k) + \xi^{(b)}(k) \\ y^*(k) &= \xi^{(b)}(k) - \xi^*(k) \end{aligned} \tag{4.13}$$

which yield $\mathbf{D}(k) \equiv 0$, $k \in [1\,,\,s]$ and further on, due to (4.5), (4.12), (4.11)

$$h^2[s] = \gamma^2[s] = 1$$

Hence from (4.4) it follows that $\mathbf{C}(s)$ is a singleton

$$\mathbf{C}(s) = C_o[s]$$

It is worth to observe that the set $\Xi_b(\cdot)$ of disturbances $\xi^{(b)}[1\,,\,s]$ which satisfy (4.11), (4.12) is nonvoid. Indeed, to fulfill (4.12) it suffices that $s > m$, det $N \neq 0$ and

$$(\eta_i[1\,,\,s], p_j[1\,,\,s]) = 0$$

for any $i,j \in [1\,,\,m]$. Here

$$\eta'(k) = (\xi^{(b)}(k) - \xi^*(k))'N(k)$$

Relation (4.11) defines a linear subspace $\mathbf{L}_\eta^{(k)}$ generated by vectors $\eta(k)$ and therefore also a linear subspace $\mathbf{L}_\xi$ generated by respective "vectors"

$$\bar{\xi}[1\,,\,s] = \xi^{(b)}[1\,,\,s] - \xi^*[1\,,\,s]$$

due (4.14). The required values

$$\bar{\xi}^{(b)}[1\,,\,s] = \xi^{(b)}[1\,,\,s] - \xi^*[1\,,\,s]$$

are then determined through the relation

$$\bar{\xi}^{(b)}[1\,,\,s] \in \mathbf{L}_\xi \bigcap \sigma_N(1)$$

where $\sigma_N(1)$ is the sphere

$$\sum_{k=1}^{s} \bar{\xi}'(k)\, N(k)\, \bar{\xi}(k) = 1$$

The last results may be given in the form of

Lemma 4.1. (a) The "worst case" guaranteed estimate given by the "largest" ellipsoid $\mathbf{C}[s]$ *is generated by the measurement*

$$y_W[1\,,\,s] = C^* p[1\,,\,s] + \xi^*[1\,,\,s]$$

(b) *The "best case" guaranteed estimate given by a singleton* $\mathbf{C}[s] = C_0$ *is generated by the measurement*

$$y^{(b)}[1\,,\,s] = C^*\, p[1\,,\,s] + \xi^{(b)}[1\,,s]$$

where $\xi^{(b)}[1\,,\,s]$ *is any sequence* $\xi[1\,,\,s]$ *that satisfies (4.11), (4.12).*

Case (b) indicates that *exact identifiability is possible even in the presence of disturbances.*

The terms used in the relations of the above are also relevant for exact identifiability in the absence of disturbances.

5. Exact Identifiability in the Absence of Disturbances

The equation

$$y(k) = Cp(k) \tag{5.1}$$

may be rewritten as

$$y(k) = (p'(k) \otimes I_m)\bar{C}$$

which yields

$$(p(k) \otimes I_m)\; N(k)\; y(k) = (p(k) \otimes I_m)\; N(k)\; (p'(k) \otimes I_m)\bar{C}$$

for $k \in [1\,,\,s]$. This leads to equation

$$\mathbb{D}[s] = \mathbb{P}(s)\;\bar{C} \tag{5.2}$$

Hence for resolving (5.2) it suffices for the matrix $\mathbb{P}(s)$ to be invertible.

The matrix $\mathbb{P}[s]$ may be rewritten as

$$\mathbb{P}[s] = \sum_{k=1}^{s} N(k) \otimes p(k)\; p'(k) = \sum_{k=1}^{s} (p(k)\; p'(k) \otimes N(k))$$

The invertibility of $\mathbb{P}[s]$ with $N(k) = I_m$ is then ensured if $W[s] = \sum_{k=1}^{s} p(k)p'(k)$ is nonsingular.

Lemma 5.1 For the exact identifiability of matrix C in the absence of disturbances it is sufficient that

$$det\ \mathbb{P}[s] \neq 0$$

where $\mathbb{P}[s]$ is an $m^2 \times m^2$ matrix.

With $N(k) = I_m$ it is sufficient that

$$det\ W[s] \neq 0$$

where $W[s]$ is $m \times m$ dimensional.

In traditional statistics $W[s]$ is known as the *informational matrix*. We shall now proceed with the treatment of other types of constraints.

6. SEPARATE QUADRATIC CONSTRAINTS

Let us treat constraints (III.B) by substituting them with an equivalent system of joint constraints.

$$\alpha\,(\bar{C} - \bar{C}^*)'\,\mathbb{L}(\bar{C} - \bar{C}^*) + \tag{6.1}$$
$$+ (1 - \alpha) \sum_{k=1}^{s} (\xi[k] - \xi^*[k])'\,N(k)(\xi[k] - \xi^*[k]) \leq 1$$

which should be true for any $\alpha \in (0\ ,\ 1]$.

For any given $\alpha \in (0\ ,\ 1]$, the respective domain $\mathbf{C}_\alpha[s]$ will be an ellipsoid of type (4.4) with $\mathbb{L}$ substituted for $\mathbb{L}_\alpha = \alpha\mathbb{L}$ and $N(k)$ for $N_\alpha = (1 - \alpha)N(k)$. The actual domain $\mathbf{C}[s]$ for constraint (III.B) should therefore satisfy the equality

$$\mathbf{C}[s] = \{\bigcap \mathbf{C}_\alpha\,[s]\ |\ 0 < \alpha \leq 1\} \tag{6.2}$$

The latter formula shows that the calculations for $\mathbf{C}[s]$ may be *decoupled* into those for a series of ellipsoids governed by formulae of type (4.4)-(4.8) in which the matrices $\mathbb{L}\ ,\ N(s)$ are substituted for $\mathbb{L}_\alpha\ ,\ N_\alpha(s)$ respectively, each with a specific value of

$\alpha \in (0\ ,\ 1]$.

Thus each array of relations (4.4)-(4.8), $\mathbb{L} = \mathbb{L}_\alpha\ ,\ N[1\ ,\ s] = N_\alpha[1\ ,\ s]$, produces an ellipsoid $\mathbf{C}_\alpha[s]$ that includes $\mathbf{C}[s]$. An approximation $\mathbf{C}^{(r)}\ [s]$ to $\mathbf{C}[s]$ from above may be reached through an intersection of any finite number of ellipsoids

$$\mathbf{C}^{(r)}\ [s] = \bigcap_{j=1}^{r} \mathbf{C}_{\alpha_j}[s] \tag{6.3}$$

where α_j runs through a fixed number of r preassigned values $\alpha_j \in (0\ ,\ 1];\ j = 1\ ,\dots,\ r$.

By intersecting over *all the values of* $\alpha \in (0\ ,\ 1]$ we will reach the *exact solution* (6.2). These facts may be summarized in

Lemma 6.1 The set $\mathbf{C}[s]$ *for constraint (6.1) may be presented as an intersection (6.2) of ellipsoids* $\mathbf{C}_\alpha[s]$ *each of which is given by relations (4.4)-(4.8) with* $\mathbb{L}\ ,\ N\ [1\ ,\ s]$ *substituted for* $\mathbb{L}_\alpha\ ,\ N_\alpha[1\ ,\ s]$.

Restricting the intersection to a finite number r *of ellipsoids* $\mathbf{C}_{\alpha_j}[s]$ *as in (6.3), one arrives at an approximation of* $\mathbf{C}[s]$ *from above:*

$$\mathbf{C}[s] \subseteq \mathbf{C}^{(r)}[s]\ .$$

It is not difficult to observe that for obtaining the exact solution $\mathbf{C}[s]$ it suffices to have only a denumberable sequence of values $\alpha_j\ ,\ j = 1\ ,\dots,\ \infty$.

The relations given here are trivial. However they indicate that the calculation of $\mathbf{C}[s]$ may be done by *independent parallel calculations* for each of the ellipsoids $\mathbf{C}_\alpha\ [s]$.

This suggestion may be further useful for the more complicated and less obvious problems of the sequel.

Another option is to approximate $\mathbf{C}[s]$ by a *polyhedron*. This may require the knowledge of the projections of set $\mathbf{C}[s]$ on some preassigned directions $\ell^{(i)} \in \mathbf{R}^n$.

Since $\mathbf{C}[s]$ is obviously a convex compact set, it may also be described by its *support function*, [21]

$$\rho(\ell \mid \mathbf{C}[s]) = \max \{ (\ell\ ,\ \bar{C}) \mid \bar{C} \in \bar{\mathbf{C}}[s] \},\ \ell \in \mathbf{R}^{mn}\ ,$$

Denote

$$f(\ell) = inf \{\rho(\ell \mid \bar{C}_\alpha[s]) \mid \alpha \in (0, 1]\}$$

The function $f(\ell)$, being positively homogeneous, may turn to be *nonconvex*.

We may convexify it by introducing $(co\ f)(\ell)$ - a closed convex function such that

$$\overline{co}\ (epi\ f) = epi\ (co\ f).$$

The support function may now be calculated as follows.

Theorem 6.1 Assume $f(0) = 0$. *Then* $\rho(\ell \mid \bar{C}[s]) = (co\ f)\ (\ell)$.

The function $f(\ell)$ defines a convex compact set $\mathbf{C}[s]$ as one that consists of all those $\bar{C} \in \mathbf{R}^{mn}$ that satisfy

$$(\ell, \bar{C}) \leq f(\ell), \forall\ \ell \in \mathbf{R}^{mn} \tag{6.4}$$

or in other words

$$\mathbf{C}[s] = \{ C : (\ell, \bar{C}) \leq \rho(\ell \mid \bar{C}_\alpha [s]), \forall\ \alpha \in (0, 1], \ell \in \mathbf{R}^n\}$$

However (6.4) is equivalent to

$$(\ell, C) \leq (co\ f)\ (\ell), \forall\ \ell \in \mathbf{R}^{mn}$$

according to the definition of *co f*. Being closed, convex and positively homogeneous, *co f* turns to be the support function for $\mathbf{C}[s]$.

This result shows that provided $\mathbf{C}[s]$ is nonvoid, $(f(0) = 0)$, the function $\rho(\ell \mid \bar{C}[s])$ may be estimated through a direct minimization of $\rho(\ell \mid \bar{C}_\alpha[s])$ over α - rather than through the procedure of calculating the "infimal convolution" of the supports $\rho(\ell \mid \bar{C}_\alpha[s])$ as required by conventional theorems of convex analysis.

The knowledge of $\rho(\ell \mid \bar{C}[s])$ allows to construct some *approximations* from above for $\mathbf{C}[s]$. Taking, for example r directions $\ell^{(i)} \in \mathbf{R}^{mn}$, $(i = 1, \ldots r)$ we may solve optimization problems in $\alpha \in (0, 1] : \rho_i[s] = inf \{ \rho(\ell^{(i)} \mid \bar{C}_\alpha[s]) \mid \alpha \in (0, 1] \}$

Denoting

$$L_i[s] = \{C : (\ell^{(i)}, \bar{C}) \leq \rho_i[s]\}$$

we may observe

$$\mathbf{C}[s] \subseteq \{ \bigcap L_i[s] \mid 1 \leq i \leq r \} = \mathbf{L}_r[s]$$

Where $\mathbf{L}_r[s]$ is *an mn-dimensional polyhedron with r faces.*

7. GEOMETRICAL CONSTRAINTS

Returning to equation (3.1) assume that the restrictions on $\xi(k)$ and C that are given in advance are taken to be *geometrical* (i.e. of type III (C)). Namely

$$\xi(k) \in Q(k) \, , \, k \in [1 \, , \, s] \tag{7.1}$$

$$C \in \mathbf{C}_0 \tag{7.2}$$

where $Q(k)$, $\mathbf{C}_0$ are convex compact sets in $\mathbf{R}^m$ and $\mathbf{R}^{m \times n}$ respectively. The *informational set* $\mathbf{C}[s]$ will now consist of all those matrices C that satisfy (7.2) and also generate the measured value $y[1 \, , \, s]$ together with some disturbance $\xi[1 \, , \, s]$ that satisfies (7.1).

Using standard techniques of convex analysis and matrix algebra we come to the following sequence of operations.

The system equations (3.1), (7.1) may be transformed into

$$y(k) \in (p'(k) \otimes I_m)\bar{C} + Q(k) \, ;$$

since $I_m \, C \, p = (p' \otimes I_m) \, \bar{C}$ according to (2.2).

The set $\mathbf{C}[s]$ will then consist of all matrices C such that for every $k \in [1 \, , \, s]$ we have

$$\psi'(k)(p'(k) \otimes I_m) \, \bar{C} \leq (\psi(k) \, , \, y(k)) + \tag{7.3}$$
$$+\rho(\psi(k) \mid - \, Q(k)) \, ,$$

together with

$$(\bar{\Lambda} \, , \, \bar{C}) \leq \rho(\bar{\Lambda} \mid \mathbf{C}_0) \tag{7.4}$$

for any $\psi(k) \in \mathbf{R}^m$, $\bar{\Lambda} \in \mathbf{R}^{mn}$.

(Recall that symbol $\rho(\psi \mid Q)$ stands for the value of the support function

$$\rho(\psi \mid Q) = sup \, \{(\psi \, , \, q) \mid q \in Q\}$$

of the set Q at point ψ.)

This leads to the inequality

$$\sum_{k=1}^{s} \psi'(k)(p'(k) \otimes I_m)\bar{C} + (\bar{\wedge}, \bar{C}) \leq$$
$$\leq \sum_{k=1}^{s} \{(\psi(k), y(k)) + \rho(\psi(k) \mid -Q(k))\} + \rho(\bar{\wedge} \mid \bar{\mathbf{C}}_o)$$

for any $\psi(k) \in \mathbf{R}^m$, $\bar{\wedge} \in \mathbf{R}^{mn}$

Therefore, with $\bar{\wedge} \in \mathbf{R}^{mn}$ given we have*

$$(\bar{\wedge}, \bar{C}) \leq \rho(\bar{\wedge}' - \sum_{k=1}^{s} \psi'(k)(p'(k) \otimes I_m) \mid \bar{\mathbf{C}}_o) + \tag{7.5}$$
$$+ \sum_{k=1}^{s} ((\psi(k), y(k)) + \rho(\psi(k) \mid -Q(k)))$$

For an element $C \in \mathbf{C}[s]$ it is necessary and sufficient that relation (7.5) is true for any $\psi(k) \in \mathbf{R}^m$, $k \in [1, s]$.

Hence we come to

Lemma 7.1. The informational domain $\mathbf{C}[s]$ *consistent with measurement* $y[1, s]$ *and with restrictions (7.1), (7.2) is defined by the following support function.*

$$\rho(\wedge \mid \mathbf{C}[s]) = f(\wedge) \tag{7.6}$$

where

$$f(\wedge) = inf\,\{\rho(\bar{\wedge}' - \sum_{k=1}^{s} \psi'(k)(p'(k) \otimes I_m) \mid \mathbf{C}_o) +$$
$$+ \sum_{k=1}^{s} \psi'(k)\, y(k) + \rho(\psi(k) \mid -Q(k)) \mid \psi(k) \in \mathbf{R}^m,\ k = [1, s]\,\}$$

The proof of Lemma 7.1 follows from (7.5) and from the fact that $f(\wedge)$ is a convex, positively homogeneous function, [21].

A special case arrives when there is no information on C at all and therefore $\mathbf{C}_o = \mathbf{R}^{m \times n}$. Following the previous schemes we come to

---

* When using the symbol $\rho(p \mid Q)$ for the support function of set Q at point p we will not distinguish a vector-column p from a vector-row p'.

Lemma 7.2. Under restrictions (7.1), $\mathbf{C}_0 = \mathbf{R}^{m \times n}$, *the set* $\mathbf{C}[s]$ *is given by the support function.*

$$\rho(\wedge \mid \mathbf{C}[s]) = \tag{7.7}$$
$$= inf\left\{\sum_{k=1}^{s} \{\rho(-\psi(k) \mid Q(k)) + \psi'(k)\, y(k)\}\right\}$$

over all vectors $\psi(k)$ that satisfy

$$\sum_{k=1}^{s} \psi'(k)\,(p'(k) \otimes I_m) = \bar{\wedge}' \tag{7.8}$$

A question may however arise which is whether in the last case the set $\mathbf{C}[s]$ will be bounded.

Lemma 7.3. Suppose $\mathbf{C}_0 = \mathbf{R}^{m \times n}$ *and the matrix* $\{p(1), \ldots, p(s)\} = P(s)$ *for* $s \geq n$ *is nonsingular. Then the set* $\mathbf{C}[s]$ *is bounded.*

Taking equation (7.8) it is possible to solve it in the form

$$\psi(k) = (p'(k) \otimes I_m)\,(I_m \otimes W(s))^{-1}\,\bar{\wedge} \tag{7.9}$$

where as before

$$W[s] = \sum_{k=1}^{s} (I_m \otimes p(k))(p'(k) \otimes I_m)$$

Indeed (7.8) may be transposed into

$$\sum_{k=1}^{s} (I_m \otimes p(k))\psi(k) = \bar{\wedge} \tag{7.10}$$

and the solution may be sought for in the form

$$\psi(k) = (p'(k) \otimes I_m)\ell \tag{7.11}$$

In view of (7.8) this yields equation

$$(I_m \otimes W[s])\ell = \bar{\wedge} \tag{7.12}$$

where the matrix $W[s]$ is invertible (the latter condition is ensured by the linear independence of vectors $p(k)$, $k = 1 \ldots s$, $s \geq n$). Equations (7.10)-(7.11) produce the solution (7.9).

Substituting $\psi(k)$ of (7.9) into (7.7) it is possible to observe that the support function $\rho(\wedge \mid \mathbf{C}[s])$ is equibounded in $\bar{\wedge}$ over all $\bar{\wedge} \in \mathbf{S}_1^{mn}(0)$ where $\mathbf{S}_1^{mn}(0)$ is a unit ball in $\mathbf{R}^{mn}$. This proves the boundedness of $\mathbf{C}[s]$.

Remark 7.1 Assuming that $\xi[s]$ is bounded by a quadratic constraint (III.B) with $\mathbb{L} = 0$ (so that there is no initial bound on C), and that $P(s)$ is nonsingular, the set $\mathbf{C}[s]$ again remains bounded.

The result of Lemma 7.3 therefore remains true when the geometrical constraint on $\xi[k]$ is substituted by a quadratic constraint on $\xi[\cdot]$. It is not difficult to observe that the result still remains true when $\xi[\cdot]$ is bounded in the metric of space ℓ_p:

$$\sum_{i=1}^{s} ((\xi[k] - \xi^*[k])' N(k)(\xi[k] - \xi^*[k]))^{p/2} \le 1$$

with $1 \le p \le \infty$,

8. RECURRENCE EQUATIONS FOR GEOMETRICAL CONSTRAINTS

One could already observe that equations (4.4)-(4.8) of theorem 4.1 are given in a recurrent form so that they would describe the evolution of the set $\mathbf{C}[s]$ that estimates the unknown matrix C. The next step will be to derive recurrence evolution equations for the case of geometrical constraints.

Starting with relation (7.5), substitute

$$\psi'(k) = \bar{\wedge}' M(k)$$

where $M(k) \in \mathbf{R}^{mn \times m}$, $1 \le k \le s$.

Then (7.5) will be transformed into the following inequality

$$(\bar{\wedge}, \bar{C}) \le \rho(\bar{\wedge}' \mid (I_{mn} - \sum_{k=1}^{s} M(k)(p'(k) \otimes I_m))\bar{\mathbf{C}}_o) + \tag{8.1}$$
$$+ \sum_{k=1}^{s} (\bar{\wedge}', M(k)\, y(k)) + \rho(\bar{\wedge} \mid M(k)(-Q(k)))$$

Denote the sequence of matrices $M(k) \in \mathbf{R}^{mn \times m}$, $k \in [1,\ldots,s]$ as $M[1\ ,\ s]$.

Lemma 8.1 In order that $C \in \mathbf{C}[s]$ it is necessary and sufficient that (8.1) would hold for any $\bar{\lambda} \in \mathbf{R}^{mn}$, and any sequence $M[1\ ,\ s] \in \mathbf{M}[1\ ,\ s]$.

The proof is obvious from (7.5), (8.1) and Lemma 7.1. Hence in view of the properties of support functions for convex sets we come to the following assertion.

Lemma 8.2 In order that the inclusion

$$C \in \mathbf{C}[s]$$

would be true it is necessary and sufficient that

$$\bar{C} \in C(s\ ,\bar{\mathbf{C}}_o\ ,\ M[1\ ,\ s])$$

for any sequence $M[1\ ,\ s] \in \mathbf{M}[1\ ,\ s]$ where

$$C(s\ ,\bar{\mathbf{C}}_o\ ,\ M[1\ ,\ s]) = (I_{mn} - \sum_{k=1}^{s} M(k)\ (p'(k) \otimes I_m))\ \bar{\mathbf{C}}_o + \\ + \sum_{k=1}^{s} M(k)\ (y(k) - Q(k))$$

From Lemma 8.2 it now follows

Lemma 8.3. The set $\mathbf{C}[s]$ may be defined through the equality

$$\bar{\mathbf{C}}[s] = \bigcap \{\ C\,(s\ ,\bar{\mathbf{C}}_o\ ,\ M[1\ ,\ s])\ \mid\ M[1\ ,s] \in \mathbf{M}[1\ ,\ s]\ \}$$

In a similar way, assuming the process starts from set $C[s]$ at instant s, we have

$$\bar{\mathbf{C}}[s+1] \subseteq (I_n - M(s+1)\,(p'(s+1) \otimes I_m))\ \bar{\mathbf{C}}[s] + \\ + M(s+1)(y(s+1) - Q(s+1)) = C(s+1\ ,\ \bar{\mathbf{C}}[s]\ ,\ M(s+1)) \tag{8.2}$$

for any $M(s+1) \in \mathbf{R}^{mn \times n}$ and further on

$$\bar{\mathbf{C}}[s+1] = \bigcap \{\ C\,(s+1\ ,\ \bar{\mathbf{C}}[s]\ ,\ M)\ \mid\ M \in \mathbf{R}^{mn \times n}\} \tag{8.3}$$

This allows us to formulate

Theorem 8.1 The set $\mathbf{C}[s]$ satisfies the recurrence inclusion

$$\bar{\mathbf{C}}\,[s+1] \subseteq C(s+1\ ,\ \bar{\mathbf{C}}[s]\ ,\ M),\ \mathbf{C}[0] = \mathbf{C}_0 \tag{8.4}$$

- whatever is the matrix $M \in \mathbf{R}^{mn \times n}$ - and also the recurrence equation (8.3).

The relations of the above allow to construct numerical schemes for approximating the solutions to the guaranteed identification problem.

Particularly, (8.4) may be decoupled into a variety of systems

$$\bar{C}_M\,[s+1] \subseteq C(s+1\,,\bar{C}_M[s]\,,M(s))\,,\mathbf{C}[0]=\mathbf{C}_0 \tag{8.5}$$

each of which depends upon a sequence $M[1\,,s]$ of "decoupling parameters". It therefore makes sense to consider

$$\mathbf{C}_U\,[s] = \{\bigcap \mathbf{C}_M[s] \mid M[1\,,s]\} \tag{8.6}$$

Obviously $\mathbf{C}\,[s] \subseteq \mathbf{C}_U\,[s]$

From the linearity of the right-hand side of (8.2) and the convexity of sets $\mathbf{C}_0$, $Q(s)$ it follows that actually $\mathbf{C}[s] = \mathbf{C}_U[s]$.

Lemma 8.4 The set $\mathbf{C}[s] = \mathbf{C}_U[s]$ *may be calculated through an intersection (8.6) of solutions* $\mathbf{C}_M[s]$ *to a variety of independent inclusions (8.5) parametrized by sequences* $M[1\,,s]$.

This fact indicates that $\mathbf{C}[s]$ may be reached by *parallel computations* due to equations (8.5). The solution to each of these equations may further be substituted by approximative set-valued solutions with ellipsoidal or polyhedral values. The precise techniques for these approximations however lie beyond the scope of this paper.

An important question to be studied is whether the estimation procedures given here may be consistent. It will be shown in the sequel that there exist certain classes of identification problems for which the answer to this question is affirmative.

9. GEOMETRICAL CONSTRAINTS. CONSISTENCY CONDITIONS

We will discuss this problem assuming $\mathbf{C}_0 = \mathbf{R}^{m \times n}$. Then the support function $\rho(\wedge) \mid \mathbf{C}[s])$ for set $\mathbf{C}[s]$ is given by (7.7), (7.8).

The measurement $y(k)$ may be presented as

$$y(k) = (p'(k) \otimes I_m)\, \bar{C}^* + \xi^*(k) \;, \; (k = 1, \ldots, s) \tag{9.1}$$

where $\bar{C}^*$ is the actual vector to be identified, $\xi^*(k)$ is the unknown actual value of the disturbance.

Substituting (9.1) into (7.7), (7.8) we come to

$$\begin{aligned} &\rho(\wedge \mid \mathbf{C}[s]) = \\ &= inf\left\{ \sum_{k=1}^{s} \rho(\psi(k) \mid \xi^*(k) - Q(k)) + \right. \\ &\left. + \sum_{k=1}^{s} \psi'(k)(p'(k) \otimes I_m)\, \bar{C}^* \right\}, \end{aligned}$$

over all vectors $\psi(k)$ that satisfy

$$\psi[1\,,\,s] \in \Psi[s\,,\wedge] \tag{9.2}$$

where

$$\Psi[s\,,\,\wedge] = \{\psi[1\,,\,s] : \sum_{k=1}^{s} \psi'(k)(p'(k) \otimes I_m) = \bar{\wedge}'\}$$

This is equivalent to

$$\rho(\wedge \mid \mathbf{C}[s]) = (\bar{\wedge}, \bar{C}^*) + \rho(\wedge \mid R^*[s]) \;,$$

where

$$\begin{aligned} &\rho(\wedge \mid R^*[s]) = \\ &= inf\left\{ \sum_{k=1}^{s} \rho(\psi(k) \mid \xi^*(k) - Q(k)) \mid \psi[1\,,\,s] \in \Psi[s\,,\,\wedge] \right\} = \varphi(\wedge) \end{aligned} \tag{9.3}$$

In other terms

$$\bar{\mathbf{C}}[s] \subseteq \bar{C}^* + R^*[s]$$

where $R^*[s]$ is the *error set* for the estimation process. The support function for $R^*[s]$ is given by (9.3).

Since $\xi^*(k) \in Q(k)$ we have

$$\rho(\bar{\Lambda} \mid R^*[s]) \geq 0 \,, \forall \, \bar{\Lambda} \in \mathbf{R}^{m \times n}$$

Hence every sequence $\psi^0 \, [1 \,,\, s] \in \Psi \, (s \,,\, \Lambda)$ that yields

$$\sum_{k=1}^{s} \rho(\psi(k) \mid \xi^*(k) - Q(k)) = 0$$

will be a minimizing element for problem (9.3).

The estimation process will be consistent within the interval $[1 \,,\, s]$ if

$$R^*[s] = \{0\}$$

or, in other terms, if

$$\rho(\Lambda \mid R^*[s]) = 0 \,, \forall \, \Lambda \in \mathbf{R}^{m \times n} \tag{9.4}$$

Lemma 9.1 In order that $\rho(\bar{\Lambda} \mid R^*[s]) = 0 \,, \forall \, \bar{\Lambda} \in \mathbf{R}^{m \times n}$ *it is necessary and sufficient that there would exist mn + 1 vectors* $\bar{\Lambda}^{(i)} \in \mathbf{R}^{mn}$, $i = 1 \,, \ldots, \, mn$, *such that*

$$\sum_{i=1}^{mn+1} \alpha_i \, \bar{\Lambda}^{(i)} \neq 0 \,, \{\forall \, \alpha : (\alpha \,,\, \alpha) \neq 0 \,, \alpha_i \geq 0 \,, \forall \, i \in [1 \,, \ldots, \, mn + 1]\} \tag{9.5}$$

$$(\alpha = \alpha_1 \,, \ldots, \alpha_{mn \, + \, 1})$$

and

$$\rho(\Lambda^{(i)} \mid R^*[s]) = 0 \,, \forall \, i \in [1 \,, \ldots, \, mn + 1]$$

Vectors $\bar{\Lambda}^{(i)}$ that satisfy (9.5) are said to form a *simplicial basis* in $\mathbf{R}^{mn}$.

Every vector $\bar{\Lambda} \in \mathbf{R}^{mn}$ may then be presented as

$$\bar{\Lambda} = \sum_{i=1}^{mn+1} \alpha_i \, \bar{\Lambda}^{(i)} \,, \, \alpha_i \geq 0$$

Hence for any $\bar{\Lambda} \in \mathbf{R}^{mn}$ we have

$$\rho(\Lambda \mid R^*[s]) = \rho\left(\sum_{i=1}^{mn+1} \alpha_i \, \bar{\Lambda}^{(i)} \mid R^*[s]\right) \leq$$

$$\leq \sum_{i=1}^{mn+1} \alpha_i \rho(\bar{\Lambda}^{(i)} \mid R^*[s]) = 0$$

In view of (9.4) this yields $R^*[s] = \{0\}$.

We will now indicate some particular classes of problems when the inputs and the disturbances are such that they ensure the conditions of Lemma 9.1 to be fulfilled.

Condition 9.A

(i) *The disturbances $\xi^*(k)$ are such that they satisfy the equalities*

$$(\xi^*(k)\ ,\ \psi^*(k)) = \rho(\psi^*(k)\ |\ Q(k)) \tag{9.6}$$

for a certain r-periodic function $\psi^*(k)$ $(r \geq m)$ that yields

Rank $\{\psi^*(1)\ ,\dots,\ \psi^*(r)\} = m.$

(ii) *The input function $p(k)$ is q-periodic with $q \geq n+1$*

Among the vectors $p(k)$, $(k = 1\ ,\dots,\ q)$ one may select a simplicial basis in $\mathbf{R}^n$, i.e. for any $x \in \mathbf{R}^n$ there exists an array of numbers $\alpha_k \geq 0$ such that

$$x = \sum_{k=1}^{q} \alpha_k\, p(k)$$

(iii) *Numbers r and q are relative prime.*

Lemma 9.2 Under Condition 9.A the error set $R^*[s] = 0$ for $s \geq rq$.

We will prove that $R^*[s_0] = 0$ for $s_0 = rq$. The condition $R^*[s] = 0$ for $s \geq s_0$ will then be obvious.

Due to (9.3), the objective is to prove that under Condition 9.A there exists for every $\wedge \in \mathbf{R}^{m \times n}$ a set of vectors $\psi^0(k)$, $k = 1\ ,\dots,\ s^0$, such that

$$\sum_{k=1}^{s_0} \rho(\psi^0(k)\ |\ \xi^*(k) - Q(k)) = 0\ , \tag{9.7}$$

$$\psi^0\,[1\ ,\ s_0] \in \Psi[s_0\ ,\ \wedge]\ .$$

Condition 9.A implies that there exists such a one-to-one correspondence $k = k(i\ ,\ j)$ between pairs of integers $\{i\ ,\ j\}$ $(i \in [1\ ,\dots,\ r]\ ,\ j \in [1\ ,\dots,\ q])$ and integers $k \in [1\ ,\dots,\ s_0]$ that

$$p(k) = p(i)\ ,\ \psi(k) = \psi(j) \tag{9.8}$$

Indeed, if k^* is given, then it is possible to find a pair i^* , j^*, so that

$$k^* = i^* + \gamma r \ , \ k^* = j^* + \sigma q \ ,$$

where γ , σ are integers. Then we assume $p(k^*) = p(i^*)$, $\psi(k^*) = \psi(j^*)$.

The latter representation is unique in the sense that pair i^* , j^* may correspond to no other number k^{**} than k^*.

(If, on the contrary, there would exist a $k^{**} \geq k^*$ such that

$$k^{**} = i^* + \gamma_0 r \ , \ k^{**} = j^* + \sigma_0 q \ ,$$

then we would have

$$k^{**} - k^* = (\gamma_0 - \gamma) r$$
$$k^{**} - k^* = (\sigma_0 - \sigma) q$$

and $k^{**} - k^*$ would be divided by $s_0 = rq$ without a remainder. Since $k^{**} - k^* < s_0$, it follows that $k^{**} = k^*$).

As the number of pairs $\{i \ , j\}$ is so and as each pair $\{i \ , j\}$ corresponds to a unique integer $k \in [1 \ , s_0]$, the function $k = k(i \ , j)$ is a one-to-one correspondence.

Thus if $\wedge \in \mathbf{R}^{m \times n}$ and sequence $\psi^* \ [1 \ , s]$ satisfies Condition 9.A (i), then there exists a sequence $x[1 \ , s_0]$, $(x(k) \in \mathbf{R}^n)$, such that

$$\sum_{i=1}^{r} \psi^*(i) \ x'(i) = \wedge$$

Due to Condition 9.A (ii)

$$x(i) = \sum_{j=1}^{q} \alpha_{ij} \ p(j)$$

for some values $\alpha_{ij} \geq 0$.

Therefore

$$\sum_{i=1}^{r} \ \sum_{j=1}^{q} \alpha_{ij} \ \psi^*(i) \ p'(j) = \wedge \tag{9.9}$$

Assigning to every pair $\{i \ , j\}$ the value $k = k(i \ , j)$ we may renumerate the values α_{ij} with one index, substituting ij for $k = k(i \ , j)$. Having in mind (9.8), we may rewrite (9.9) as

$$\sum_{k=1}^{s_0} \alpha_k \, \psi^*(k) \, p'(k) = \wedge \tag{9.10}$$

The transition from (9.9) to (9.10) is unique. Hence, for each $\wedge \in \mathbf{R}^{m \times n}$ there exists a sequence $\alpha[1 \, , s_0]$ of nonnegative elements $\alpha_k \geq 0$ such that

$$\sum_{k=1}^{s_0} \alpha_k \, \psi^{*\prime}(k) \, (p'(k) \otimes I_m) = \bar{\wedge}' \tag{9.11}$$

Substituting $\psi^0(k) = \alpha_k \psi^*(k)$ and taking into account equalities (9.6) we observe that (9.7) is fulfilled. Namely

$$\sum_{k=1}^{s_0} \rho(\alpha_k \, \psi^*(k) \mid \xi^*(k) - Q(k)) = 0$$

while (9.11) yields $\psi^0 \, [1 \, , s] \in \Psi[s_0 \, , \wedge]$. Lemma 9.2 is thus proved.

A second class of problems that yield consistency is described by

Condition 9.B.

(i) *function $p(k)$ is periodic with period $q \leq n$. The matrix $W[q] = \sum_{k=1}^{q} p(k) \, p'(k)$ is nonsingular,*

(ii) *the disturbances $\xi(k)$ are such that if $\{\bar{\wedge}^{(i)}\}$, $i = 1 \, , \ldots, \, mn + 1$ is a given simplicial basis in $\mathbf{R}^{mn}$ and vectors $\psi^{(i)}(k) \in \mathbf{R}^m$ are those that yield*

$$\sum_{k=1}^{q} \psi^{(i)\prime}(k)(p'(k) \otimes I_m) = \bar{\wedge}^{(i)} \tag{9.12}$$

then the sequence $\xi(j)$, $j = 1 \, , \ldots , q(mn + 1)$ does satisfy conditions

$$(\xi(k + i) \, , \psi^{(i)}(k)) = \rho(\psi^{(i)}(k) \mid Q(k)) \tag{9.13}$$
$$(k = 1 \, , \ldots , q \; ; \; i = 1 \, , \ldots , \, m \, (n + 1))$$

Lemma 9.3 Under Condition 9.B the set $R[s] = \{0\}$ for $s \geq q(mn + 1)$

The proof of this Lemma follows from Lemma 7.1 and from direct substitution of (9.12), (9.13) into (9.3) (since the required set of vectors $\psi^{(i)}(k)$ does always exist due to condition $\mid W(q) \mid \neq 0$)

A simple particular case when Lemma 9.3 works is when C is a vector ($C \in \mathbf{R}^n$) and when the restriction on $\xi(k)$ is $\mid \xi(k) \mid \leq \mu$.

Then $\wedge^{(i)} \in \mathbf{R}^n$ and (9.12) turns into

$$\sum_{k=1}^{q} \psi^{(i)}(k)\ p'\ (k) = \wedge^{(i)}$$

where $\psi^{(i)}(k)$ are scalars.

Relations (9.13) now yield

$$\xi(k+i) = \mu\ sign\ \psi^{(i)}(k) \tag{9.14}$$

Therefore the "best" disturbance $\xi(j) = \pm\mu$ now depends only upon the signs of $\psi^{(i)}(k)$, $j = i + k$. Here *the order of pluses and minuses is predetermined* by relation (9.14). However a natural question does arise. This is whether the consistency condition would still hold (at least asymptotically, with $h\,(R\,[s]\,,\{0\}) \longrightarrow 0\,,\, s \longrightarrow \infty)$) if $\xi(j)$ would attain its values at random.

The answer to the last question is given below.

Condition 9.C

(i) *function* $p(k)$, $k = 1\,,\ldots,\,\infty$, *is periodic with period* $q \leq n$; *the matrix* $W(q)$ *is non-singular.*

(ii) *the sequence* $\xi(i)$ *is formed of jointly independent random variables with identical nondegenerate probabilistic densities, concentrated on the set*

$$Q(k) \equiv Q\ ,\ Q \in comp\ \mathbf{R}^m\ ,\ int\ Q \neq \emptyset$$

Condition (ii) means in particular that for every convex compact subset $Q_\epsilon \subseteq Q$, ($Q_\epsilon \in comp\ \mathbf{R}^m$) of measure $\epsilon > 0$ the probability

$$P\{\xi(k) \in Q_\epsilon\} = \delta > 0\,,\, \forall\, k \in [1\,,\infty]$$

At the same time it will not be necessary for values of the distribution densities of the variables $\xi(i)$ to be known.

Lemma 9.4 Under Condition 9.C the relation

$$h(R^*[s]\,,\{0\}) \longrightarrow 0\,,\ s\longrightarrow\infty$$

holds with probability 1.

We will prove that for every $\epsilon > 0$ with probability 1 for a sequence $\xi[\cdot]$ there exists a number $N > 0$ such that for $s \geq N$ one has

$$h(R^*[s]\,,\{0\}) \leq \epsilon \tag{9.15}$$

Since $W(q)$ is nonsingular, there exists for a given $\Lambda \in \mathbf{R}^{m\times n}$ a sequence $\psi^0\,[1\,,q]$ such that

$$\sum_{k=1}^{q} \psi^0(k)\,p'(k) = \Lambda$$

Let $\xi^0(k) \in Q$ denote a respective sequence of elements that satisfy the relations

$$(\xi^0(k)\,,\psi^0(k)) = \rho(\psi^0(k)\mid Q) \tag{9.16}$$

It is clear that elements $\xi^0(k)$ belong to the *boundary* ∂Q of set Q. Without loss of generality we may assume that all the vectors $\xi^0(k)$ are chosen among the *extremal points* of Q.

(A point $\xi^0 \in Q$ is said to be *extremal* for Q if it cannot be presented in the form

$$\xi^0 = \alpha\,\xi^{(1)} + (1-\alpha)\,\xi^{(2)}\,,\ 0<\alpha<1\,,$$

for any pair of elements $\xi^{(1)}\,,\xi^{(2)} \in Q$.)

Hence each $\xi^0(k)$ of (9.16) is either already extremal - if (9.16) gives a unique solution, - or could be chosen among the extremal points for set $\Xi_\Lambda = \{\xi : (\xi\,,\psi^0(k)) = \rho(\psi^0(k)\mid Q)$ which yields extremality of $\xi^0(k)$ relative to Q).

Consider a sequence of Euclidean balls $S_\delta\,(\xi^0(k))$ with centers at $\xi^0(k)$ and radii $\delta > 0$. Denote

$$Q_\delta(k) = Q \bigcap S_\delta(\xi^0(k))$$

Then with *int* $Q \neq \emptyset$ the measure $\mu(Q_\delta(k)) > 0$ for any $\delta > 0$.

Let us consider q infinite sequences

$$\xi(qj + k) , \quad (j = 0 ,\ldots, \infty;\ k = 1 ,\ldots, q) \tag{9.17}$$

generated by the "noise" variable $\xi(i)$.

Denote $A_\delta(k)$ to be the event that

$$\xi\ (qj + k) \notin Q_\delta(k) ,\ (j = 1 ,\ldots, \infty)$$

and

$$A(k) = \bigcup \{A_{\delta_i}(k) \mid \delta_i > 0 ,\ \delta_i \longrightarrow 0 ,\ i \longrightarrow \infty \}$$

Then obviously $P(\xi\ [\cdot] \in A_{\delta_i}(k)) = 0$ for any $\delta_i > 0$ (due to the joint independence of the variables $\xi(i)$) and due to a Lemma by Borel and Cantelli [22] we have (for any $k = 1 ,\ldots, q$)

$$P(\xi\ [\cdot] \in A^c(k)) = 1$$

Hence with probability 1 for a sequence $\xi[\cdot]$ there exists a number $j(k)$ such that

$$\xi(qj(k) + k) \in Q_\delta(k) \tag{9.18}$$

Denoting $\bigcap_{k=1}^{q} \bar{A}(k) = B$, we observe

$$P(\xi\ [\cdot] \in B) = P(\xi[\cdot] \in \bigcap_{k=1}^{q} A^c(k)) = \tag{9.20}$$

$$= \prod_{k=1}^{q} P(\xi\ [\cdot] \in A^c\ (k)) = 1$$

due to the joint independence of the random variables $\xi(i)$.

Hence each sequence $\xi^*[\cdot]$ may be decoupled into q nonintersecting subsequences (9.17) each of which, with probability 1, satisfies for any $\delta > 0$ the inclusion (9.18) for some $i = qj(k) + k$ (due to (9.20)).

Therefore, with $\delta > 0$ given, we may select

$$\psi^*(i) = \psi^0(k)$$
$$\text{for } i = qj(k) + k\ ,\ k = 1 ,\ldots, q\ ,$$

$$\psi^*(i) = 0\,,\, i \neq qj(k) + k\,, \qquad (9.21)$$
$$N = qj(q) + q$$

Substituting $\psi^*(i)\,,\, \xi^*(i)$ into (9.3) and using the periodicity of $p(i)$ $(p(qj+k) = p(k)\,,\, j = 1\,,\ldots,\,\infty\,;\, k = 1\,,\ldots,\, q)$

we have

$$\rho(\wedge \mid R[N]) = \sum_{i=1}^{N} \rho(-\psi^*(i) \mid \xi^*(i) - Q) =$$

$$= \sum_{k=1}^{q} \rho(\psi^*(qj(k)+k) \mid \xi^*(qj(k)+k) - Q) \qquad (9.22)$$

with

$$\sum_{i=1}^{N} \psi^*(i)\, p'(i) = \sum_{k=1}^{q} \psi^*(qj(k)+k)\, p'(qj(k)+k) = \wedge$$

$$\xi^*(qj(k)+k) \in Q_\delta(k)$$

In view of (9.16), (9.21), (9.22) and the definition of $Q_\delta(k)$ one may observe

$$\rho(\wedge \mid R[N]) =$$
$$= \sum_{k=1}^{q} (\rho(\psi^0(k) \mid \xi^0(k) - Q) +$$
$$+ \rho(\psi^0(k) \mid \xi^*(qj(k)+k) - \xi^0(k))) \leq$$
$$\leq \delta \sum_{k=1}^{q} ||\, \psi^0(k)\, ||$$

Therefore, with $\wedge$, σ given, one may select $\psi^0\,[1\,,\,q]$, δ, so that

$$\delta \sum_{k=1}^{q} ||\, \psi^0\,(k)\, || \leq \sigma$$

Summarizing the discussion of the above we observe that for every $\wedge \in \mathbf{R}^{m \times n}$, $\sigma > 0$, there exists a number $N(\wedge\,,\sigma)$ that ensures $\rho(\wedge \mid R[s]) \leq \sigma\,,\, s \geq N,\, N = N(\wedge\,,\sigma)$.

If $\bar{\wedge}_0^{(i)} = e^{(i)}$ is an orthonomal basis in $\mathbf{R}^{mn}$ $(e_j^{(i)} = \delta_{ij}\,;\, j = 1\,,\ldots,mn)$ and

$$N_0(\sigma) = \max\{N(\bar{\wedge}_0^{(i)}\,,\sigma)\,,\, N(-\bar{\wedge}_0^{(i)}\,,\sigma)\,\}\,,\, (i = 1\,,\ldots,\, mn)\,,$$

then

$$\rho(\pm e^{(i)} \mid R[s]) \leq \sigma \text{ , } (\forall\, i = 1 \text{ ,..., } mn) \text{ , } s \geq N_0\,(\sigma)$$

and

$$h\{R[s] \text{ , } \{0\}\} \leq \sqrt{mn}\ \sigma$$

Taking $\epsilon = \sqrt{mn}\ \sigma$, $N = N_0\,\sigma$ we arrive at the relation (9.15). Lemma 9.4 is now proved.

The examples given in Cases A and C indicate two important classes of disturbances $\xi(k)$ of which one consists of *periodic functions* and the other of *a sequence of equidistributed independent random variables.* In both cases one may ensure consistency of the identification process. However this requires some additional assumptions on the inputs $p(k)$. Basically this means that function $p(k)$ should be periodic and its informational matrix should be nondegerate as indicated in the precise formulations, (see also [23, 24]).

10. IDENTIFICATION OF THE COEFFICIENTS OF A LINEAR AUTONOMOUS DYNAMIC SYSTEM

Consider a dynamic process governed by a linear system

$$x(k+1) = Ax(k) + B\ u(k) + \xi(k) \qquad (10.1)$$
$$k \in [0 \text{ , } s]$$

The *input* $u(k)$ and the *output* $y = x(k)$ are taken here to be *given*, the constant *coefficients A, B are to be identified* and the *input noise* $\xi(k)$ is taken to be *unknown but bounded* by a geometrical constraint

$$\xi(k) \in Q(k) \text{ , } k \in [0 \text{ , } s] \qquad (10.2)$$

Here as usual $x \in \mathbf{R}^n$, $u \in \mathbf{R}^p$, $v \in \mathbf{R}^q$, $A \in \mathbf{R}^{n \times n}$, $B \in \mathbf{R}^{n \times p}$ and there is some additional information on A,B. Namely it is assumed that

$$A \in \mathbf{A} \text{ , } B \in \mathbf{B} \text{ , } \qquad (10.3)$$

where A , B are convex and compact sets in the matrix space of respective dimensions.

We will derive a recurrence equation for the related informational domains. These are given by the following definition.

Definition 10.1 The informational domain $A[s] \times B[s] = H[s]$ consistent with system (10.1), restrictions (10.2), (10.3) and measurement $x(k)$, $k \in [0 , s]$ is the set of all matrix pairs $\{A , B\}$ for each of which there exists a sequence $\xi[0 , s] \in Q[0 , s]$ such that relations (10.1)-(10.3) would be fulfilled.

Since the input $u[0 , s]$ is taken to be given, the domain $H[s]$ will obviously depend upon $u[0 , s]$:

$$H[s] = H[s , u[0 , s]) = H(s , \cdot)$$

In order to solve the estimation problem we introduce a matrix C and a vector $p(k)$.

$$C = [A, B], p(k) = \begin{pmatrix} x(k) \\ u(k) \end{pmatrix}$$

Then taking

$$y(k) = x(k + 1) ,$$

we come to the standard measurement equation of § 3:

$$y(k) = Cp(k) + \xi(k)$$

Applying the recurrence equation of (8.2) we come to the relations that describe the dynamics of set $H(s , u[0 , s]) = H[s]$.

The consistency theorems of § 9 may be applied if there is some additional information on A , B and on the known inputs $u[0 , s]$ that would ensure that the conditions of these theorems would be fulfilled.

Another formal scheme for obtaining a recurrence equation for $H[s]$ may be presented as follows. Introducing a vector

$$z = \begin{pmatrix} \bar{A} \\ \bar{B} \end{pmatrix}$$

and an $n \times n(n + m)$- matrix

$$\mathrm{G}(k) = (x'(k) \otimes I_n , u'(k) \otimes I_n)$$

we arrive at the system

$$z(k+1) = z(k) , \tag{10.4}$$

$$y(k) = \mathrm{G}(k)\, z(k) + \xi(k) , 0 \le k \le s , \tag{10.5}$$

where the aim is to identify the *informational domain* $Z(s) = H[s]$ of the states of system (10.4) consistent with measurement $y[0 , s]$ and constraints (10.2), (10.3).

Following formally the results of § 13 (formula (13.6) for the one-stage process) and rewriting them in terms of the notations of this paragraph we come to the recurrence relation

$$Z(k+1) \subseteq \bigcap_M \{(I - M'\, \mathrm{G}(k))\, Z(k) + \tag{10.6}$$

$$+ M(y(k) - Q(k))\} , Z(0) = \begin{pmatrix} \bar{A} \\ \bar{B} \end{pmatrix}$$

$$Z \in \mathbf{R}^{n(n+m)} , M \in \mathbf{R}^{n(n+m) \times n}$$

which at each stage is true for any matrix $M \in \mathbf{M}^{n(n+m) \times n}$. According to the conventional scheme we arrive at

Lemma 10.1 The set-valued estimate for the vector C of coefficients for system (10.1) *is given by the solution* $Z(s) = H(s)$ *for equation* (10.6).

It is now natural to consider in greater detail the issue of *state estimation* for linear systems with unknown but bounded measurement noise and input disturbances. We will start with the first case.

11. THE OBSERVATION PROBLEM

Consider a recurrence equation

$$x(k+1) = A(k)\, x(k) , x(k_0) = x^0 , \tag{11.1}$$

$$x \in \mathbf{R}^n , A(k) \in \mathbf{R}^{n \times n} , k \ge k_0 ,$$

together with a measurement equation

$$y(k) = g'(k)\, x(k) + \xi(k) , k \ge k_0 + 1$$

with vector $g(k) \in \mathbf{R}^n$ and "noise" $\xi(k)$ restricted by a geometrical constraint.

$$\xi(k) \in Q(k)\ ,\ Q(k) \in comp\ \mathbf{R}^m$$

The objective is to estimate the initial vector x^o by processing a given measurement $y[1\ ,\ s]$, taking $A(k)\ ,\ g(k)\ ,\ Q(k)$ to be given in advance. We will further call this *the observation problem* (in the presence of unknown but bounded "noise" with set-membership bounds on the unknowns).

Observing that $x(s) = S(s)\ x^o$, where $S(s)$ is the solution to the matrix equation

$$S(k+1) = A(k)\ S(k)\ ,\ S(k_o) = I_n$$

we may denote

$$p'(k) = g'(k)\ S(k) \tag{11.2}$$

transforming our problem to the conventional form of § 3 with

$$y(k) = p'(k)\ x^o + \xi(k)$$

and with x^o replacing the unknown C.

The condition for the identifiability of x^o in the absence of "noise" now turns to be again $|\ W(s)\ | \neq 0$ with

$$W(s) = \sum_{k=k_o}^{s} S'(k)\ g(k)\ g'(k)\ S(k) \tag{11.3}$$

The latter relation is known as the *observability condition* [3, 4] for system (11.1) with measurement

$$y(k) = g'(k)\ x(k) \tag{11.4}$$

Condition $|W(s)| \neq 0$ is obviously ensured if vectors $p(k) = S'(k)\ g(k)$, $(k = 1\ ,\dots,\ k)$ are linearly independent.

The general solution will now consist in constructing the informational domains $\mathbf{X}^o[s]$ for the vector x^o. They are the direct substitutes for $\mathbf{C}[s]$.

Following (8.2), (11.2) we will have a system of recurrence relations

$$\begin{aligned} X_M^o\ (k+1) \subseteq (I_n - M(k+1)\ g'(k+1)\ S(k+1))X_M^o(k) + \\ +M(k+1)(y(k+1) - Q(k+1))\ ,\ X(k_o) = X^o \end{aligned} \tag{11.5}$$

$$S(k+1) = A(k)\, S(k)\ ,\ S(k_o) = I_n$$

which are true for any sequence $M[k_o + 1\ ,\ s]$.

The results of the previous paragraph then leads us to

Lemma 11.1 The solution x^o to the observation problem may be estimated from above by

$$\mathbf{X}^o[s] = \{ \bigcap X^o_M(s) \mid M[k_o + 1\ ,\ s]\} \tag{11.6}$$

Namely

$$x^o \in \mathbf{X}^o[s]\ ,\ \forall\ s > k_o + 1$$

The solution will be consistent with

$$h\,\{ X^o_M\,(k)\ ,\ x^o\} \longrightarrow 0\ ,\ k \longrightarrow \infty \tag{11.7}$$

if for example the problem falls under one of the conditions 9A - 9C of the previous paragraph.

Particularly, for an autonomous system (11.1), this will be ensured if

(a) the function $p(k) = g' S(k)$ is n-periodic,

(b) the vectors

$$g'\ ,\ g'A\ ,\ldots,\ g'A^{n-1}$$

are linearly independent (the system (11.1), (11.4) is completely observable).

(c) the noise is uniformly distributed in the interval $Q(k) \equiv Q = -\ Q$.

Lemma 11.2 Under conditions (a) - (c) *the solution* $\mathbf{X}^o[s]$, (11.5), (11.6) *to the observation problem is consistent in the sense of* (11.7).

A simple example, when the conditions of Lemma 11.1 are satisfied, is given by a system (11.1) in $\mathbf{R}^3$

$$g' = (1\ ,0\ ,0)\ ,\ A = \begin{pmatrix} 0 & 1 & 0 \\ 0 & 0 & 1 \\ 1 & 0 & 0 \end{pmatrix}\ ,\ \mid \xi(k) \mid \ \leq 1$$

Here

(a) $p(k) = p(3i + j)$ is periodic with period 3, $j = 1, 2, 3$; $i = 0, \ldots, \infty$; $3i + j = k$

(b) $p(j) = e^{(j)} = \delta_{kj}$, $k = 1, 2, 3$ so that $p(1)$, $p(2)$, $p(3)$ are linearly independent,

(c) $\xi(k)$ is taken to be equidistributed in the interval [-1,1].

The solution to this problem may be given by a polyhedral approximation so that, assuming $X^o[k]$ given, we will seek for an approximation of $X^o[k]$ by a polytope $X^o[k+1]$ through the formula

$$\rho(\ell \mid \mathbf{X}^o[k+1]) = inf\,\{H(\ell, m, \mathbf{X}^o[k]) \mid m\}$$

$$H(\ell, m, \mathbf{X}^o[k]) = \{\rho(\ell \mid (I_n - m' p(k+1))\, \mathbf{X}^o[k]) + (\ell, m)\, y(k+1) +$$

$$+ \rho(-\ell \mid m'\, Q(k+1))\}, \ \ell \in \mathbf{R}^3, m \in \mathbf{R}^3,$$

taking for each step a set of orthonormal vectors $\{e^{(i)}\}$ with a set of vectors $\{-e^{(i)}\}$, and assuming $\ell = e^{(i)}$, $\ell = -\, e^{(i)}$, $(i = 1, \ldots, 3)$

Therefore, in order to define $\mathbf{X}^o[k+1]$ with $\mathbf{X}^o[k]$ given, we will have to solve 6 independent unconstrained minimization problems, in 3 variables each, so that the vertices of $\mathbf{X}^o[k+1]$ would be given by 3 coordinates each, selected from the variety of numbers

$$\rho(+e^{(i)} \mid \mathbf{X}[k+1]),\ -\rho\,(-e^{(i)} \mid \mathbf{X}[k+1]),\ (i = 1, 2, 3).$$

A simpler algorithm involves only one optimization problem (in three variables, the coordinates of m) so that one should minimize in m the function

$$V_A\,(m, k+1) = \prod_{i=1}^{3} \left[H(e^{(i)}, m, \mathbf{X}^o[k]) + H(-e^{(i)}, m, \mathbf{X}^o\,[k])\right]$$

which for a given m, is equal to the volume of a polyhedron $\mathbf{X}(m, k+1) \supseteq \mathbf{X}[k+1]$ The last inclusion is true for any $m \in \mathbf{R}^3$ and one should therefore seek for the optimal m. The projections of $X[k]$ on the axes $\{x_1, x_2\}$, $\{x_1, x_3\}$ are shown in Figure 5.

A separate issue is the construction of an ellipsoidal approximation for $\mathbf{X}[k+1]$.

A more complicated problem is to estimate the state of a linear system with unknown input on the basis of measurement corrupted by noise. We will therefore deal with the problem of guaranteed state estimation for a linear system subjected to unknown but bounded disturbances with nonquadratic restrictions on the unknowns.*

* The treatment of quadratic constraints is known well enough and may be found in references [15, 16]

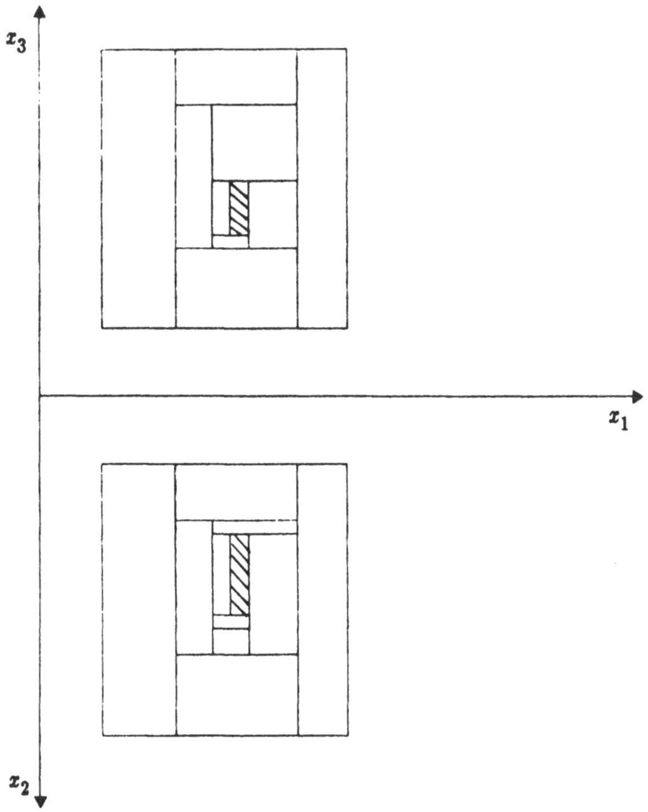

FIGURE 5

12. UNCERTAIN DYNAMIC SYSTEMS

An uncertain dynamic system is understood here to be a discrete-time multistage process, described by an n-dimensional equation

$$x(k+1) = A(k)\, x\,(k) + B(k)\, v\,(k) \tag{12.1}$$

where $A(k)$, $B(k)$, $k = 0 ,\ldots, s$ are given matrices. The input $v(k)$, and the initial stage x^0 are vectors of finite-dimensional spaces $\mathbf{R}^p$ and $\mathbf{R}^n$ respectively. They are assumed to be unknown being restricted in advance by instantaneous "geometric" constraints

$$x(0) = x^0 \in X^0 \ , \ v(k) \in P(k) \ , \ k = 0 ,\ldots, s \ , \tag{12.2}$$

where X^0, $P(k)$ are given convex and compact sets. It is further assumed that direct measurements of the state $x(k)$ are impossible, the available information on the process dynamics being generated by the equation

$$y(k) = G(k)\, x\,(k) + \xi(k)\ ;\ k = 1\ ,...,s \tag{12.3}$$

with measurement vector $y(k) \in \mathbf{R}^m$ and matrix $G(k)$ given. The disturbances $\xi(k)$ are unknown and restricted as before by an inclusion

$$\xi(k) \in Q(k) \tag{12.4}$$

with convex compact set $Q(k) \in \mathbf{R}^m$ given in advance.

We will use the symbol $x(k\ ,\ v\,[0\ ,\ k-1]\ ,\ x^0)$ to denote the end of the trajectory $x(j)$ for system (12.1) formed for $[0\ ,\ k]$ with $v[0\ ,\ k-1]\ ,\ x^0$ given.*

Let us assume that after s stages of system operation there appeared a measurement sequence $y[1\ ,\ s]$, generated due to relations (12.1)-(12.4).

The knowledge of $y[1\ ,\ s]$ will allow us to consider the following construction.

Definition 12.1 An informational domain $X[s] = X(s\ ,0\ ,X^0)$ *will be defined as the set that consists of the ends* $x(s\ ,\ v[0\ ,\ s-1]\ ,\ x^0)$ *of all those trajectories* $x(j)$ *formed for the interval* $j \in [0\ ,\ s]$ *that could generate the measured sequence* $y[1\ ,\ s]$ *under constraints* (12.2)-(12.4).

More generally, with $y[k+1\ ,\ \ell]\ ,\ (k+1 \leq \ell)$ *and* $F \in co\ \mathbf{R}^n$ *given,* $X(\ell\ ,\ k\ ,\ F)$ *will be the set of the ends* $x(\ell\ ,\ v[k\ ,\ \ell-1]\ ,\ x^*)$ *of the trajectories* $x(j)$ *of system* (12.1) *that start at stage* k *from state* $x(k) = x^*$ *and are consistent with realization* $y[k+1\ ,\ \ell]$ *due to equation* (12.3) *with constraints*

$$x^* \in F\ ,\ v(i) \in P(i)\ ,\ k \leq i \leq \ell - 1\ ,$$
$$\xi(j) \in Q(j)\ ,\ k+1 \leq j \leq \ell\ ,$$

The dynamics of the total system (12.1)-(12.3) will now be determined by the evolution of sets $X[s]$. It is clear that set $X[s]$ includes the unknown actual state of system

* In order to simplify some further notations of this paragraph we will generally start the process at stage $k_0 = 0$ instead of arbitrary $k_0 = k^*$, although the basic system is *nonstationary*.

(12.1).

In particular $X[s] = X(s, 0, X^0)$.

From the definitions of the above it is possible to verify the following assertions.

Lemma 12.1 Assume F, $P(k)$, $Q(k)$ to be convex compact sets in spaces $\mathbf{R}^n$, $\mathbf{R}^p$, $\mathbf{R}^m$ respectively. Then each of the sets $X(s, \ell, F)$ will be convex and compact.

Lemma 12.2 Whatever is the set $F \subseteq \mathbf{R}^n$, the following equality is true ($s \geq \ell \geq k$)

$$X(s, k, F) = X(s, \ell, X(\ell, k, F)) \tag{12.5}$$

Condition (12.5) indicates that the transformation $X(s, k, F)$ possesses a *semi-group property generating a generalized dynamic system* in the space of convex compact subsets of $\mathbf{R}^n$. The generalized system will then absorb all the informational and dynamic features of the total process. Here each $X[s]$ contain all the prehistory of the process and the process evolution for $r > s$ depends only upon $X[s]$ but not upon the previous $X[i]$, $i < s$.

The general description of $X[s]$ requires a rather cumbersome procedure which does not follow directly from § § 7,8. Our objective is to obtain a description of sets $X[s]$ which are the *set-valued state estimators* for the system (12.1)-(12.4). The situation therefore justifies the consideration of approximation techniques based on solving some auxiliary deterministic or even stochastic estimation problems. In order to explain the procedures, we will start with an elementary one-stage solution.

13. GUARANTEED STATE ESTIMATION. THE ONE-STAGE PROBLEM

Consider the system

$$z = Ax + Bv, \quad y = Gz + \xi \tag{13.1}$$

where

$$x, z \in \mathbf{R}^n, \quad v \in \mathbf{R}^p, \quad \xi \in \mathbf{R}^m,$$

and the matrices A , B , G are given. Knowing the constraints

$$x \in X \, , v \in P \, , \xi \in Q \, , \tag{13.2}$$

where

$$X \in comp \, \mathbf{R}^n \, , P \in comp \mathbf{R}^p \, , Q \in comp \, \mathbf{R}^q$$

and knowing the value y, one has to determine the set Z of vectors z consistent with equations (13.1) and inclusions (13.2).

Denote

$$Z_s = AX + BP$$
$$Z_y = \{z : y - Gz \in Q\}$$

Then obviously

$$Z = Z_s \bigcap Z_y \tag{13.3}$$

Standard considerations yield a relation for the support function

$$\rho(\ell \mid Z) = \max \{(\ell \, , z) \mid z \in Z\}$$

Applying the convolution formula of convex analysis [21]

$$\rho(\ell \mid Z) = inf \, \{\rho(\ell^* \mid Z_s) + \rho \, (\ell^{**} \mid Z_y) \mid \ell^* + \ell^{**} = \ell\}$$

Lemma 13.1 The support function $\rho(\ell \mid Z) = \psi(\ell)$ where

$$\psi(\ell) = inf \, \{\Phi(\ell \, , p) \mid p \in \mathbf{R}^m\} \tag{13.4}$$
$$\Phi(\ell \, , p) = \rho(A' \, \ell - A' \, G' \, p \mid X) + \rho(B' \, \ell - B' \, G' \, p \mid P)$$
$$+ \rho(-p \mid Q) + (p \, , y) \, ,$$

The set Z may be given in another form. Indeed whatever the vectors $\ell \, , p \, , \ell \neq 0$ are, it is possible to represent (13.4) $p = M\ell = p[\ell \, , M]$ where matrix $M \in \mathbf{R}^{m \times n}$. Relation (13.4) will then turn into

$$\psi(\ell) = inf \, \{\Phi(\ell \, , p[\ell \, , M]) \mid M \in \mathbf{R}^{m \times n}\} \tag{13.5}$$

Problem (13.5) will be referred to as the *dual problem* for (13.3). The latter relation yields the inclusion

$$Z \subseteq (I_n - M' \, G) \, (AX + BP) + M'(y - Q) = R(M) \tag{13.6}$$

which will be true for any matrix M.

Equality (13.5) thus leads to *set-valued duality relations* in the form of (13.6) and further on in the form of

Lemma 13.2 The following equality is true

$$Z = \{\bigcap R(M) \mid M\} \tag{13.7}$$

over all matrices $M \in \mathbf{R}^{m \times n}$. *Here set* Z *is a "guaranteed" estimate for* z *which may be calculated due to* (13.5).

The necessity of solving (13.5) gives rise to the question of whether it is possible to calculate $\rho(\ell \mid Z)$ in some other way, for example, by the variation of the relations for some kind of *stochastic estimation problem*. A second question is whether there exist any general relations between the solutions to the guaranteed and to the stochastic filtering problems.

In fact it is possible to obtain an inclusion that would combine the properties of both (13.6) and of conventional relations for the linear-quadratic Gaussian estimation problem.

14. RELATION BETWEEN GUARANTEED AND STOCHASTIC ESTIMATION. THE ONE-STAGE PROBLEM

Having fixed a certain triplet $h = \{x , v , \xi\}$ that satisfies (13.2) (the set of all such triplets will be further denoted as H), consider the system

$$w = A(x + q) + Bv \; , \; y = Gw + \xi + \eta \; , \tag{14.1}$$

where q , η are independent Gaussian stochastic vectors with zero means

$$Eq = 0 \; , \; E\eta = 0 \; ,$$

and with covariance matrices

$$Eqq' = L \qquad E\eta\eta' = N$$

where L , N are positive definite. Assume that after one random event the vector y has appeared due to system (14.1). The conditional expectation $E(w \mid y)$ may then be determined for example by means of a Bayesian procedure or by a least-square method. We have

$$E(w \mid y) = Ax + A\mathbf{P}A'\ G'\ N^{-1}(y - GAx - GBv - \xi) + Bv\ , \qquad (14.2)$$
$$\mathbf{P}^{-1} = L^{-1} + A'\ G'\ N^{-1}\ GA$$

or in accordance with a conventional matrix transformation [25].

$$\mathbf{P} = L - LA'\ G'K^{-1}\ GAL\ , \qquad (14.3)$$
$$K = N + GALA'\ G'\ ,$$

an equivalent condition

$$\bar{w}_y = E(w \mid y) = Ax + ALA'\ G'\ K^{-1}\ (y - GAx - (GBv + \xi)) + Bv \qquad (14.4)$$

We observe that the conditional variance

$$E((w - \bar{w}_y)(w - \bar{w}_y)' \mid y) = A\mathbf{P}A' \qquad (14.5)$$

does not depend upon h and is determined only by pair

$$\wedge = \{L\ ,\ N\}$$

where $L > 0$, $N > 0$. (In the latter case further we will write $\wedge > 0$.)

Therefore we may consider the set of all conditional mean values

$$W(\wedge) = \{\bigcup \bar{w}_y \mid h \in H\}$$

that correspond to all possible $h \in H$. Here

$$W(\wedge) = (I_n - ALA'\ G'\ K^{-1}\ G)\ (AX + BP) + ALA'\ G'\ K^{-1}\ (y - Q) \qquad (14.6)$$

Having denoted

$$\Psi(\wedge) = \ K^{-1}\ GALA'$$

we come to

Lemma 14.1 The set $W(\wedge)$ is convex and compact: $W(\wedge) \in$ comp $\mathbf{R}^n$. The following equality is true

$$\rho(\ell \mid W(\wedge)) = \Phi(\ell\ ,\ p(\ell\ ,\ \wedge)) \qquad (14.7)$$

where

$$p(\ell\,,\wedge) = \Psi(\wedge)\,\ell$$

We may now observe that function $\Phi(\ell\,, p(\ell\,,\wedge))$ differs from $\Phi(\ell\,, p[\ell\,, M])$ used in (13.5) by a mere substitution of $p(\ell\,,\wedge)$ by $p[\ell\,, M]$. Comparing (14.7) and (13.5), we conclude

Lemma 14.2 Whatever is the pair $\wedge > 0$, *the inclusion*

$$Z \subseteq W(\wedge) \tag{14.8}$$

is true.

We will see that by varying $\wedge$ in (14.8) it is possible to achieve an exact description of set Z.

In order to prove this conjecture some standard assumptions are required.

Assumption 14.1 The matrix GA is of rank m.

We shall also make use of the following relation:

Lemma 14.3 Under assumption 14.1 take $\wedge = \wedge(1\,,\alpha) = \{I_n\,,\alpha I_m\}$. *Then* $\Psi(\wedge(1\,,\alpha))\,G' \longrightarrow I_m$ *with* $\alpha \longrightarrow 0$.

The given relation follows from equality $\Psi(\wedge(1\,,\alpha))\,G' = (\alpha I_m + D)^{-1}D$ where matrix $D = GALAG'$ is nonsingular, $L = I_n$.

Theorem 14.1 The inclusion $z \in Z$ *is true if and only if for any* $\ell \in \mathbf{R}^n$, $\wedge > 0$ *we have*

$$(\ell\,, z) \le \rho(\ell \mid W(\wedge)) = f(\ell\,,\wedge) \tag{14.9}$$

Inequality (14.9) follows immediately from the inclusion $z \in Z$ due to Lemma 14.2. Therefore it suffices to show that (14.9) yields $z \in Z$. Suppose that for a certain z^* the relation (14.9) is fulfilled, however $z^* \bar{\in} Z = Z_s \bigcap Z_y$. First assume that $z^* \bar{\in} Z_y$. Then there exists an $\epsilon > 0$ and a vector p^* such that

$$(-p^*\,, y) + (G'\,p^*\,, z^*) > \rho(-p^* \mid Q) + \epsilon \tag{14.10}$$

Now we will show that it is possible to select a pair of values ℓ^*, $\wedge^*$ that depend upon p^* and are such that

$$(\ell^*, z^*) > \rho(\ell^* \mid W(\wedge^*)) = f(\ell^*, \wedge^*) \tag{14.11}$$

Indeed, taking $\ell^* = G' p^*$, $\wedge(1, \alpha) = \{I_n, \alpha I_m\}$ we have

$$f(\ell^*, \wedge(1, \alpha)) = \Phi(\ell, \wedge(1, \alpha)) \pm ((p^*, y) + \rho(-p^* \mid Q)) \tag{14.12}$$

From Lemma 14.3 and condition

$$p(\ell^*, \wedge(1, \alpha)) = K^{-1}(\alpha)\ GAI_n\ A'\ G'\ p^*\ ,\ K(\alpha) = \alpha I_m + GAA'G'$$

it follows that

$$p(\ell^*, \wedge(1, \alpha)) \longrightarrow p^*\ ,\ \alpha \longrightarrow 0 \tag{14.13}$$

But then from condition (14.13), from Lemma 14.2 and from the properties of function $f(\ell, \wedge)$ it also follows that for any $\epsilon > 0$ there exists an $\alpha_0(\epsilon)$ such that for $\alpha \leq \alpha_0(\epsilon)$ the inequality

$$|\ f(\ell^*, \wedge(1, \alpha)) - ((p^*, y) + \rho(-p^* \mid Q))\ | \leq \epsilon/2 \tag{14.14}$$

is true.

Comparing (14.10), (14.12), (14.14) we observe that for $\alpha \leq \alpha_0(\epsilon)$.

$$(\ell^*, z^*) = (G' p^*, z^*) \geq f(\ell^*, \wedge(1, \alpha)) + \epsilon/2\ .$$

Therefore, with $\wedge^* = \wedge(1, \alpha^*)$, $\alpha < \alpha_0(\epsilon)$ the pair $\{\ell^*, \wedge^*\}$ yields the inequality (14.11).

Now assume $z^* \bar{\in} Z_s$. Then there exists a vector ℓ^0 for which

$$(\ell^0, z^*) \geq \varsigma(\ell^0) + \sigma\ ,\ \sigma > 0\ .$$

where

$$\varsigma(\ell) = \rho(A'\ \ell \mid X) + \rho(B'\ \ell \mid P)$$

Taking $\ell = \ell^0$, $\wedge = \wedge(1, \alpha)$ we find:

$$\Psi(\wedge(1, \alpha)) \longrightarrow 0\ ,\ \alpha \longrightarrow \infty\ .$$

But then for any $\sigma \longrightarrow 0$ there exists a number $\alpha^0(\sigma)$ such that

$$|\ f(\ell^0, \wedge(1, \alpha)) - \varsigma(\ell^0)\ | \leq \sigma/2$$

provided $\alpha > \alpha^0(\sigma)$. Hence, for $\alpha > \alpha^0(\sigma)$ we have

$$(\ell^0, z^*) \geq f(\ell^0, \alpha) + \sigma/2$$

contrary to (14.9). The theorem is thus proved.

From the given proof it follows that Theorem 14.1 remains true if we restrict ourselves to the one parametrical class

$$\wedge^{(1)} = \{\wedge(1, \alpha)\}, \wedge(1, \alpha) = \{I_n, \alpha I_m\}$$

Therefore, the theorem yields:

Corollary 14.1 Under the conditions of Theorem 14.1 the inclusion $z \in Z$ is true if and only if for any $\ell \in \mathbf{R}^n$ we have

$$(\ell, z) \leq f_1(\ell), \tag{14.15}$$

where

$$f_1(\ell) = inf\,\{f(\ell, \wedge(1, \alpha)) \mid \alpha > 0\}$$

Being positively homogeneous, the function $f_1(\ell)$ may, however, turn out to be nonconvex, its lower convex bound being the *second conjugate* $f_1^{**}(\ell)$, [21]. Here $g^*(q) = sup\,\{(\ell, q) - g(\ell)\}$ is the conjugate and $g^{**}(\ell) = (g^*)^*(\ell)$

The convexification of $f_1(\ell)$ in (14.15) will not violate this inequality. In other words, (14.15) will yield

Corollary 14.2 Under the conditions of Theorem 14.1, we have

$$\rho(\ell \mid Z) = f_1^{**}(\ell) \leq f_1(\ell) \tag{14.16}$$

However, if we move on to a broader class $\wedge^{(2)} = \{L, N\}$ where $L > 0$ and $N > 0$ depend together on at least m independent parameters it is possible to achieve a direct equality immediately, i.e.

$$\rho(\ell \mid Z) = f_2(\ell) \tag{14.17}$$

where

$$f_2(\ell) = inf\,\{f(\ell, \wedge) \mid \wedge \subseteq \wedge^{(2)}\} = f_1^{**}(\ell), \tag{14.18}$$

Problem (14.18) will be called *the stochastically dual* for (13.5). The following assertion is true.

Theorem 14.2 Under assumption 14.1 relations (14.17), (14.18) *are true, where the infimum is taken over all* $L > 0$, $N > 0$.

The proof of Theorem 14.2 is rather long and will be omitted in this text. It may be found in paper [26].

The stochastic dual problem (14.18) may therefore replace (13.6).

On the other hand we may again turn to set-valued duality, now in terms of a stochastic problem. Due to Corollary 14.1 the set of inequalities (14.15) will lead us to

Lemma 14.3 The following equality is true

$$Z = \{\bigcap W(\Lambda) \mid \Lambda \in \Lambda^{(1)}\} \tag{14.19}$$

The relations of this paragraph indicate that set Z may be described by deterministic relations (13.7) as well as by approximations (14.19) generated due to the stochastic estimation problems of the above.

The results of this paragraph allow to devise solutions to multistage problems.

15. A MULTI-STAGE SYSTEM

Returning to system (12.1)-(12.4) let us seek for $X[s] = X(s , k_0 , X^0)$. We further introduce notations

$$Y(k) = \{x : y(k) - G(k)\, x \in Q(k)\}$$

and $X^*(s , j , F)$ is the solution $X(s)$ to the equation

$$X(k+1) = A(k)\, X(k) + B(k)\, P(k) \,,\, j \leq k < s - 1 \tag{15.1}$$

with $X(j) = F$. Then it is possible to verify the following recurrent equation similar to (13.3).

Lemma 15.1 Assume $y[k_0 + 1 , k]$ *to be the realization for the measurement vector* y *of system (12.3), (12.1). Then the following condition is true.*

$$X[k] = X(k , k_0 , X^0) = X^* (k , k - 1 \mid X[k-1]) \bigcap Y(k) \tag{15.2}$$

Formula (15.2) indicates that the innovation introduced by the k-th measurement $Y(k)$ appears in the form of an intersection. Therefore $X^*(k, k-1 \mid X[k-1])$ is the estimate for the state of the system on stage k *before* the arrival of the k-th measurement while $X[k]$ is the estimate obtained *after* its arrival.

Relations (15.2) may be interpreted as a *recurrence equation*. One may rewrite them in a somewhat different way, namely through (13.6) and (13.7). Applying (13.7) for each stage k we come to

Lemma 15.2 The set $X[k]$ satisfies the following recurrence equation

$$X[k+1] = \{\bigcap (I_n - M'G(k))(A(k)X[k] + B(k)\mathbf{P}(k)) + \\ + M'(y(k) - Q(k)) \mid M\}$$
$$X[k_0] = X^0$$

A nonlinear version of this scheme is given further in § § 18–20. However, the topic of this paragraph is another procedure. It is the scheme of *stochastic filtering approximation* which follows from the results of § 14, (Theorem 14.1). Together with (12.1, (12.3) consider the system (involving almost sure equalities)

$$w(k+1) = A(k)\, w(k) + B(k)\, v(k) + C(k)\, u(k) \tag{15.3}$$
$$k = k_0, 1, \ldots, s-1 \,;\; w(k_0) = x^0 + w^0 ,$$
$$z(k) = G(k)\, w(k) + \xi(k) + \eta(k) ,\; u(k) \in \mathbf{R}^q , \tag{15.4}$$

where the inputs x^0, $v(k)$, $\xi(k)$ are deterministic, subjected to "instantaneous" constraints

$$x^0 \in X^0 ,\; v(k) \in \mathbf{P}(k) ,\; \xi(k) \in Q(k) \;,$$

while w^0, $u(k)$, $\eta(k)$ are independent stochastic Gaussian vectors with

$$\bar{w}^0 = Ew^0 = 0 ,\; \bar{u}(k) = Eu(k) = 0 ,$$
$$\bar{\eta}(k) = E\eta(k) = 0 ,\; Ew^0\, w^{0\prime} = P^0 , \tag{15.5}$$
$$Eu(k)\, u'(k) = L(k) ,\; E\eta(k)\, \eta'(k) = N(k) ,$$

where L, N are positive definite.

Suppose that after $k - k_0$ stages for system (15.3), (15.4) measurement $z[k_0, k] \in \mathbf{R}^{m(k-k_0)}$ has been realized. Having fixed the triplet

$$\xi[0\,,\,k] = \{x^0\,,\,v[k_0\,,\,k-1]\,,\,\xi[k_0\,,\,k]\}$$

and having denoted $\omega(k) = \{\,v(k-1)\,,\,\xi(k)\,\}$, $D(k) = \{\,P(k-1)\,,\,Q(k)\,\}$ we may find a recursion for the conditional mean value

$$\bar{w}(k+1) = E\{w(k+1) \mid \omega(k)\,,\,\bar{w}(k)\,,\,z(k+1)\}$$

Define

$$\begin{aligned} W[k+1\,,\,F] &= W(k+1\,,\,L(k)\,,\,N(k+1)\,,\,F) \\ &= \bigcup \{\,\bar{w}[k+1] \mid \omega(k) \in D(k)\,,\,\bar{w}(k) \in F\,\} \end{aligned}$$

From Theorems 14.1, 14.2 and Lemma 14.3 we come to the following propositions

Theorem 15.1 Suppose Assumption 14.1 holds for $A = A(k)$, $G = G(k+1)$, $k \in [k_0\,,\,s]$ and the sequence of observations $y[k_o\,,\,s]$, $z[k_o\,,\,s]$ for system (12.1), (12.3) *and* (15.3), (15.4) *coincide: $y[k_o\,,\,s] = z[k_o\,,\,s]$. Then the following relation is true*

$$\begin{aligned} X[s] &= \{\bigcap W(s\,,\,L\,,\,N\,,\,X[s-1]) \mid \wedge \in \wedge^{(1)}\}\,,\, s > k_0\,, \\ X[k_0] &= X^0\,,\,\wedge = \{L\,,\,N\}\,,\,P^0 = 0\,, \end{aligned} \tag{15.6}$$

moreover, with $P^0 = 0$ and

$$f_i(\ell\,,\,s) = inf\,\{\rho\,(\ell \mid W(s\,,\,L\,,\,N\,,\,X[s-1])\}$$

over all $(L\,,\,N) = \wedge \subset \wedge^{(i)}$, $i = 1\,,\,2$, we have

$$\rho(\ell \mid X[s]) = f_1^{**}\,(\ell\,,\,s),\ \rho(\ell | X[s]) = f_2(\ell\,,\,s)\,,$$

where the second conjugate is taken in the variable ℓ.

Theorem 15.2 Under the condition of Theorem 15.1 for each positive definite matrix pair $\{\,L(k-1)\,,\,N(k)\,\} = \wedge(k)$, the following inclusions are valid

$$\begin{aligned} X[k+1] &\subseteq W(k+1\,,\,L(k)\,,\,N(k+1)\,,\,X[k]) \\ &= R(k+1\,,\,\wedge(k+1)\,,\,X[k])\,,\,k \geq 0\,, \end{aligned} \tag{15.7}$$

where

$$\begin{gathered} R(k+1\,,\,\wedge(k+1)\,,\,X[k]) = (I_n - H(k+1)\;G(k+1))\;(A(k)\;X[k] + B(k)P(k) + \\ +\;H(k+1)\;(y(k+1) - Q(k+1))\,, \\ X[0] = X^0\,, \\ H(k+1) = C(k)\;L(k)\;C'(k)\;G'(k+1)\;K^{-1}\,(k+1)\,, \\ K(k+1) = N(k+1) + G(k+1)\;C(k)\;L(k)\;C'(k)\;G'(k+1)\,, \end{gathered}$$

The recurrence relations (15.7) thus allow a complete description of $X[s]$ through equation (15.6). Solving the system

$$W(k+1) = R(k+1\,,\wedge(k+1)\,,\,W(k))\,,$$
$$W(0) = X^0$$

we find

$$X[k+1] \subseteq W(k+1)$$

where

$$\rho(\ell \mid X[k+1]) = inf\,\{\rho(\ell \mid W(k+1)) \mid \wedge(j+1)\,;\,j = k_0\,,\ldots,\,k\,;\,P^0 = 0\}$$

with each pair $\wedge(j+1) = \{\,L(j)\,,\,N(j+1)\,\}$ belonging to the class $\wedge^{(2)}$. The total number of parameters over which the minimum is sought for does not exceed km.

The procedure given above is similar to the one given in (14.2). It is justified if the sets $X[k]$ are to be known for each $k > 0$. Note that in any way with *arbitrary* $L(j)\,,\,N(j+1)\,,\,j = 0\,,\ldots,\,k-1$, the set $W(k)$ always *includes* $X[k]$.

Let us now assume that the desired estimate is to be found for only a fixed stage $s > k_0$. Taking $z[k_0\,,\,s]$ to be known and triplet $\xi[k_0\,,\,s]$ for system (15.3), (15.4) to be fixed, we may find the conditional mean values

$$\bar{w}(k) = E\{\,w(k) \mid z[k_0+1\,,\,k]\,,\,\xi[k_0\,,\,k]\}$$

and the conditional covariance

$$\mathbf{P}(k) = E\{\,w(k) - \bar{w}(k))\,(w(k) - \bar{w}(k))' \mid z[k_o+1\,,\,k]\,,\,\xi[k_0\,,\,k]\}$$

where

$$Ew(k_0) = x^0\,,\,\mathbf{P}(k_0) = P^0$$

Denoting

$$\bar{w}[k\,,\,j\,,\,F] = E\{\,w(k) \mid z[j+1\,,\,k]\,,\,v[j\,,\,k-1]\,,\,\xi[j+1\,,\,k]\,,\,\bar{w}(j)\}$$
$$\bar{W}[k\,,\,j\,,\,F] = \bigcup E\{\,w(k) \mid z[j+1\,,\,k]\,,\,v[j\,,\,k-1] \in P[j\,,\,k-1]\,,$$
$$\xi[j+1\,,\,k] \in Q[j+1\,,\,k]\,,\,\bar{w}(j) \in F\}$$
$$\bar{W}[k\,,\,k_0\,,\,X^0] = \bar{W}(k)\,,$$

and having in view the Markovian property for the process (15.3), (15.4) it is possible to conclude the following:

Lemma 15.3 The equality

$$\bar{W}(k) = \bar{W}[k\,,\,j\,,\,\bar{W}(j)] \tag{15.8}$$

holds for any $j\,,\,k\,,\,j \le k$.

The corresponding formulae that generalize (14.2), (14.3) have the form

$$\begin{gathered}\bar{W}(k+1) = (E - S(k+1)\;G(k+1))\;(A(k)\;\bar{W}(k) + B(k)P)\\ + S(k+1)\;(z(k+1) - Q)\;,\\ S(k+1) = D(k+1)\;G'(k+1)\;K^{-1}\;(k+1)\;,\\ P(k+1) = D(k) - D(k)\;G'(k+1)\;K^{-1}(k+1)\;G(k+1)D(k)\;,\\ D(k) = A(k)\;P(k)\;A'(k) + C(k)\;L(k)\;C'(k)\\ K(k+1) = N(k+1) + G(k+1)\;D(k)\;G'(k+1)\\ P(k_{0)} = L\;,\end{gathered} \tag{15.9}$$

If we again suppose $z[k_0\,,\,s] = y[k_0\,,\,s]$, then due to the inclusions

$$\bar{W}(k+1) \supseteq \bar{W}[k+1\,,\,k\,,\,X[k]]\;,\;k > k_0$$

that follow from Lemma 14.2 and to the monotonicity property

$$\bar{W}[k+1\,,\,k\,,\,F_1] \subseteq \bar{W}[k+1\,,\,k\,,\,F_2]\;,\;F_1 \subseteq F_2\;,$$

that follows from (15.9) we obtain in view of (15.8)

$$X[k] \subseteq \bar{W}(k)\;,\;\text{for } k > 1 \tag{15.10}$$

Consider the following condition:

Assumption 15.1 The system (12.1), (12.3), $v\,[0\,,\,s-1] = 0\,,\,\xi[1\,,\,s] = 0$ *is completely observable on* $[k_o\,,\,s]$.

The given property is defined for example in [4].

In the latter case the following proposition is true:

Theorem 15.3 Under the conditions of Theorem 15.1 and assumption 15.1 assume $y[k_0\,,\,s] = z[k_0\,,\,s]$. *Then the equality*

$$X[s] = \{\bigcap \bar{W}(s) \mid P^0\,,\,N(k{+}1)\,,\,L(k)\,,\,k = k_0\,,\ldots,\,s-1\} \tag{15.11}$$

is true for any $P^0 > 0$ *and any diagonal* $N(k) > 0$, $L(k) > 0$. *Moreover for the given class of matrices we have*

$$\rho(\ell \mid X[s]) = f^{**}(\ell, s) \,, f^{**}(\ell, s) = \inf\{\rho(\ell \mid \bar{W}(s)) \mid P^0, L > 0, N > 0, k \in [k_0, s]\} \tag{15.12}$$

Therefore, the precise estimate is again attained here through a minimization procedure.

Remark 15.1 The relations (15.9), (15.10) may therefore be treated as follows

(a) In the case of a *set-membership* description of uncertainty as in (12.2), (12.4) with $u(k) \equiv 0$, $\eta(k) \equiv 0$, equations (15.9), (15.10) contain *complete information* on $X[k+1]$ as stated in Theorem 15.3.

(b) In the case of *both* set-membership and stochastic uncertainty, as in (15.3)-(15.5), equation (15.9) describes *the evolution of the set of the mean values of the estimates.*

(c) In the case of pure *stochastic* uncertainty with sets X^0, $P(k)$, $Q(k)$ consisting of one element $(x^0$, $p(k)$, $q(k))$ each, the relation (15.9) turns out to be an *equality* which coincides with the conventional equations of *Kalman's filtering theory.*

Remark 15.2 Following the scheme of Theorem 14.1 it is possible to demonstrate that relation (15.11) holds for P^0 , $N(k)$, $L(k)$ selected as follows:

$$P^0 = \beta \, I_n \,, \; N(k) = \alpha(k) \, I_m \,, \; L(k) = \beta(k) \, I_n$$

where

$$\beta > 0 \quad , \quad \alpha(k) > 0 \quad , \quad \beta(k) \geq 0 \quad , \quad k \in [k_0 \,, s]$$

Example

Consider a two-dimensional system

$$x(k+1) = \begin{pmatrix} 1, & \epsilon \\ -\epsilon\omega^2, & 1 \end{pmatrix} x(k) \tag{15.13}$$

with a scalar observation

$$y = x_1 + \xi, \quad \xi \in Q = \{\xi : |\xi| \leq \nu\} \; . \tag{15.14}$$

The initial state $x^0 \in X^0$ where $X^0 = x^*(0) + \mathrm{S}$, x^* is given and $\mathrm{S} = \{x : |x_i| \leq 1;\ i = 1, 2\}$ is a square.

The aim is to estimate the state $x(k)$ at each stage k. Making use of formula (13.6) at each stage k, we will estimate $X[k+1] = X(k+1, k, X[k])$ by a rectangle $\boldsymbol{X}[k]$ oriented along the axes $\{x_1, x_2\}$. Here the calculations are as follows.

If X is a rectangle such that $X = x^* + \boldsymbol{X}$ where

$$\boldsymbol{X} = \{x : |x_1| \leq \mu_1, |x_2| \leq \mu_2\} \ ,$$

then

$$\rho(l|X) = (l, x^*) + \mu_1|l_1| + \mu_2|l_2| \tag{15.15}$$

Thus we may calculate some values of the function $\rho(l|X(k+1, k, X(k)))$ with $X(k)$ given. Using formula (13.6) for our example we have

$$F(M) = (I_n - M'G)A = \begin{pmatrix} 1 - m_1, & 0 \\ - \quad m_2, & 1 \end{pmatrix} A = \begin{pmatrix} 1 - m_1, & \epsilon(1 - m_1) \\ - m_2 - \epsilon\omega^2, & - \epsilon m_2 + 1 \end{pmatrix}$$

$$M = (m_1, m_2)$$

Therefore

$$\rho(l|X(k+1, k, X[k])) = \tag{15.16}$$
$$= \inf\{\rho(l'F(M)|X[k]) + \rho(l'M'|y(k) - Q)\} \ ,$$

Starting with rectangle X^0 and calculating $\rho(l|X[1])$ for

$$l = \begin{pmatrix} 1 \\ 0 \end{pmatrix}, \begin{pmatrix} 0 \\ 1 \end{pmatrix}, \begin{pmatrix} -1 \\ 0 \end{pmatrix}, \begin{pmatrix} 0 \\ -1 \end{pmatrix}$$

due to formulae (15.15), (15.16), we define a rectangle $\boldsymbol{X}[1] \supseteq X[1]$ – the "smallest" rectangle that includes $X[1]$ and is oriented along the axes $\{x_1, x_2\}$. Further on, taking $\boldsymbol{X}[1]$ instead of $X[1]$, and repeating the procedure, we come to a rectangle $\boldsymbol{X}[2]$ etc. Thus, after k stages, we will find a rectangle

$$\boldsymbol{X}[k] \supseteq X(k, 0, X^0) = X[k]$$

which is an upper estimate for $X[k]$.

The respective calculations were done for a system described by relations (15.13), (15.14) with $y(k)$ being an actual realization of the system generated by an initial vector $x^* \in X^0$ unknown to the observer and by an unknown "noise" $\xi(k)$ that attains either of the values $+\mu$ or $-\mu$ due to a random mechanism.

The results of the simulations for several starting sets X^0 are given in Figures 6–8 with $\epsilon = 0.2$, $\omega^2 = 1.2$, $\nu = 0.5$. In Figure 9 we have the same problem with an additional "horizontal" input disturbance

$$\begin{pmatrix} 0 \\ 1 \end{pmatrix} v(k)$$

added to the right hand part of (15.13), assuming $v(k)$ being unknown, random and uniformly distributed in the interval $-0.25 \leq v(k) \leq 0.25$. The calculations are the same as before except that due to (13.6) we have to substitute $\rho(l'F(M)|X(k))$ by

$$\rho(l'F(M)|X(k)) + \rho(l'(I_2 - M'G)|BP)$$

where

$$BP = \{p : p_1 = 0, |p_2| \leq 0.25\}$$

The ideas of the above allow to approach *nonlinear systems*. Some of the basic facts related to guaranteed nonlinear filtering are given in the sequel.

16. NONLINEAR UNCERTAIN SYSTEMS

Consider a multistage process described by an n-dimensional *recurrence inclusion*

$$x(k+1) \in F(k, x(k)), \qquad k \geq k_0 \geq 0 \tag{16.1}$$

where $k \in [k_0, \infty)$, $x(k) \in \mathbb{R}^n$, $F(k, x(k))$ is a given multivalued map from $[k_0, \infty) \times \mathbb{R}^n$ into comp $\mathbb{R}^n$.

As before suppose the initial state $x(k_0) = x^0$ of the system is confined to a preassigned set:

$$x^0 \in X^0 , \tag{16.2}$$

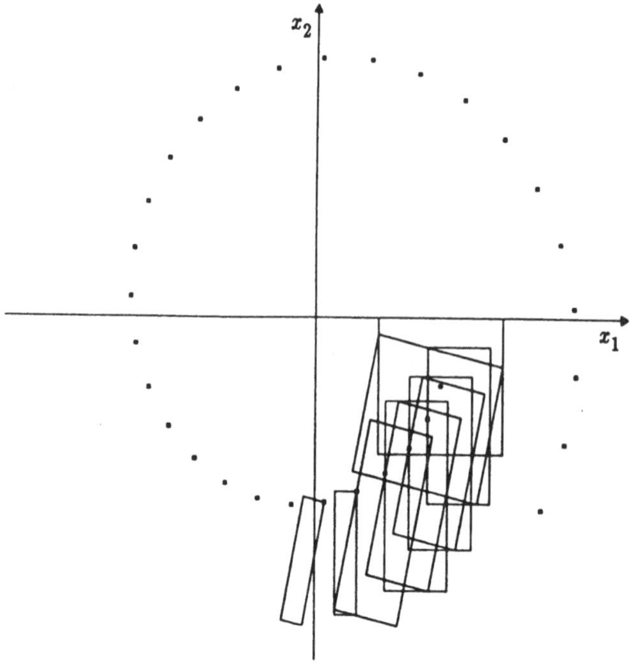

FIGURE 6

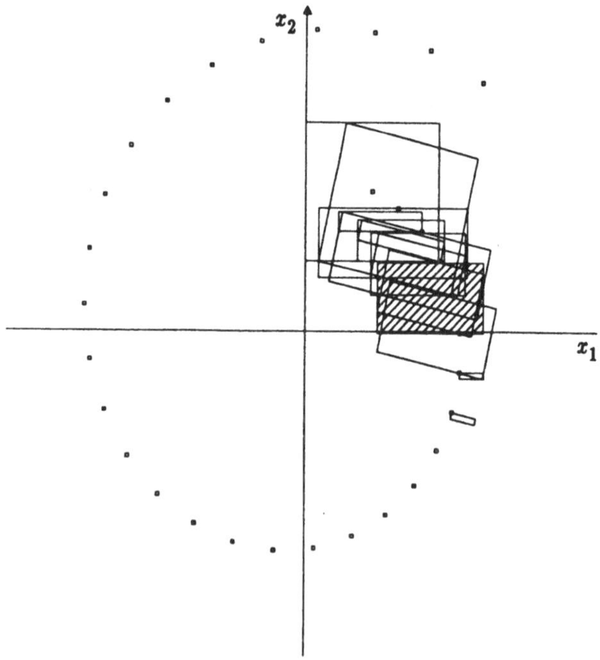

FIGURE 7

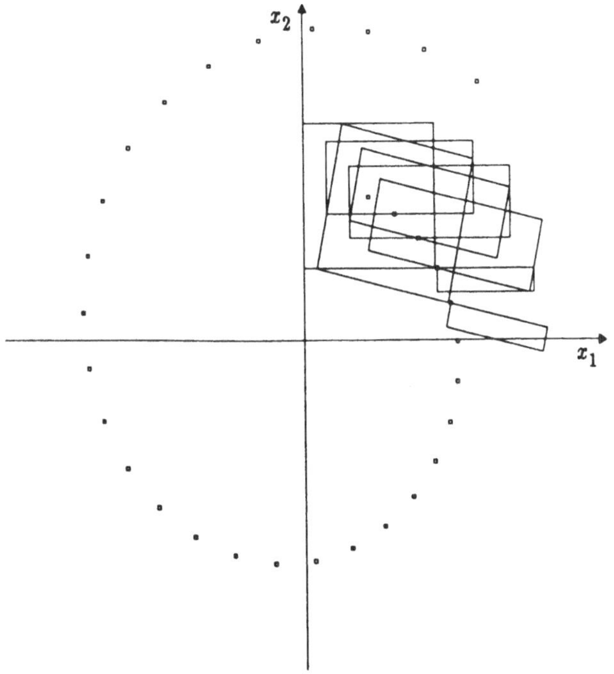

FIGURE 8

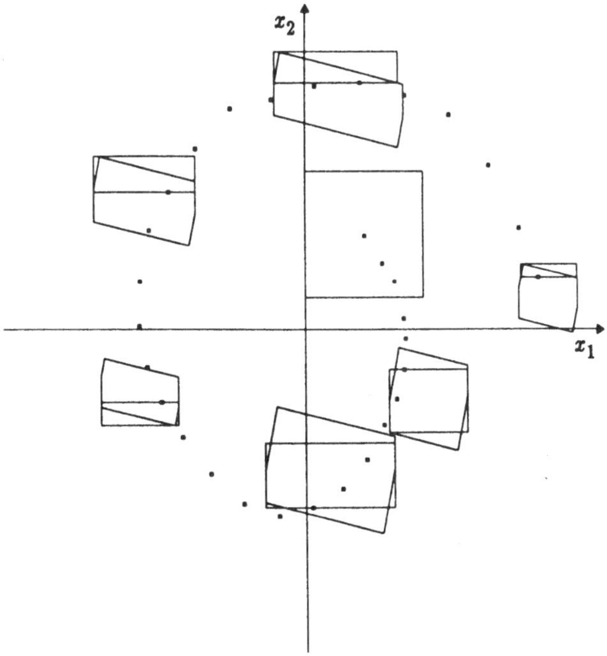

FIGURE 9

Let $Q(k)$ be a multivalued map from $[k_0, \infty)$ into comp $\mathbb{R}^m$ and $G(k)$ - a single-valued map from $[k_0, \infty]$ into the set of $m \times n$-matrices. The pair $G(k)$, $Q(k)$, *introduces a state constraint*

$$G(k)x(k) \in Q(k)\,, \quad k \geq k_0+1 \ , \tag{16.3}$$

on the solutions of system (16.1).

The subset of $\mathbb{R}^n$ that consists of all the points of $\mathbb{R}^n$ through which at stage $s \in [k_0, \infty)$ there passes at least one of the trajectories $x(k, k_0, x^0)$, that satisfy constraint (16.3) for $k \in [k_0, \tau]$, will be denoted as $X(s \mid \tau, k_0, x^0)$.

If set $Q(k)$ of (16.3) is of a specific type

$$Q(k) = y(k) - \tilde{Q}(k)$$

where $y(k)$ and $\tilde{Q}(k)$ are given, then (16.3) transforms into

$$y(k) \in G(k)x(k) + \tilde{Q}(k) \tag{16.4}$$

which could be interpreted as an *equation of observations* for the uncertain system (16.1) given above. Sets $X(s \mid \tau, k_0, X^0)$ therefore give us *guaranteed estimates* of the unknown states of system (16.1) on the basis of an observation of vector $y(k)$, $k \in [k_0, \tau]$ due to equation (16.4).

For various relations between s and τ this reflects the following situations

(a) for $s = \tau$ - the problem of *"guaranteed filtering"*

(b) for $s > \tau$ - the problem of *"guaranteed prediction"*

(c) for $s < \tau$ - the problem of *"guaranteed refinement"*

The aim of this paper will first be to study the informational sets $X(\tau \mid \tau, k_0, X^0) = X(\tau, k_0, X^0)$ similar to those of the above and their evolution in "time" τ.

The sets $X(k, k^0, x^0)$ may also be interpreted as *attainability domains* for system (16.1) under the state space constraint (16.3). The objective is therefore to describe the evolution of these domains. A further objective will be to describe the more complicated sets $X(s \mid \tau, k_0, x^0)$ and their evolution

17. A GENERALIZED NONLINEAR DYNAMIC SYSTEM

From the definition of sets $X(s \mid \tau, k^0, x^0)$ it follows that the following properties are true.

Lemma 17.1. *Whatever are the instants t,s,k, $(t \geq s \geq k \geq 0)$ and the set $\mathbf{F} \in \operatorname{comp} \mathbb{R}^n$, the following relation is true*

$$X(t,k,\mathbf{F}) = X(t,s,X(s,k,\mathbf{F}))\,. \tag{17.1}$$

Lemma 17.2. *Whatever are the instants $s,t,\tau,k,l(t \geq s \geq l;\ \tau \geq l \geq k;\ t \geq \tau)$ and the set $\mathbf{F} \in \operatorname{comp} \mathbb{R}^n$ the following relation is true*

$$X(s \mid t,k,\mathbf{F}) = X(s \mid t,l,X(l \mid \tau,k,\mathbf{F}))\,. \tag{17.2}$$

Relation (17.1) shows that sets $X(k,\tau,X)$ again satisfy a *semigroup property* which allows to define a *generalized dynamic system* in the space $2^{\mathbb{R}^n}$ of all subsets of $\mathbb{R}^n$. On the other hand, (17.2) is a more general relation which is true when the respective intervals of observation may overlap.

In general the sets $X(s \mid t,k,\mathbf{F})$ need not be either convex or connected. However, it is obvious that the following is true

Lemma 17.3. *Assume that the map F is linear in x:*

$$F(k\,,\,x) = A(k)x + P$$

where $P \in \operatorname{conv} \mathbb{R}^n$. Then for any set $\mathbf{F} \in \operatorname{conv} \mathbb{R}^n$ each of the sets $X(s \mid t,k,\mathbf{F}) \in \operatorname{conv} \mathbb{R}^n (t \geq s \geq k \geq 0)$.

Therefore the next step will be to describe the evolution of the set $X[k] = X(k\,,\,k_o\,,\,X^o)$. This will be later given in the form of a decoupling procedure. However it is convenient to commence with a description of the one-stage problem.

18. THE ONE-STAGE PROBLEM

Consider the system

$$z \in F(x), \quad Gz \in Q, \quad x \in X,$$

where $z \in \mathbb{R}^n$, $X \in \operatorname{comp} \mathbb{R}^n$, $Q \in \operatorname{conv} \mathbb{R}^m$, $F(\kappa)$ is a multivalued map from $\mathbb{R}^n$ into $\operatorname{conv} \mathbb{R}^n$, G is a linear (single-valued) map from $\mathbb{R}^n$ into $\mathbb{R}^m$.

It is clear that the sets $F(X) = \{\bigcup F(x) \mid x \in X\}$ need not be convex.

Let Z, Z^* respectively denote the sets of all solutions to the following systems:

(a) $z \in F(X), \quad Gz \in Q,$

(b) $z^* \in \operatorname{co} F(X), \quad Gz^* \in Q,$

It is obvious that the following statement is true

Lemma 18.1. *The sets Z, $\operatorname{co} Z$, Z^* satisfy the following inclusions*

$$Z \subseteq \operatorname{co} Z \subseteq Z^* \tag{18.1}$$

Denote

$$\Phi(l,p,q) = (l - G'p, q) + \rho(-p \mid Q)$$

Then the function $\Phi(l,p,q)$ may be used to describe the sets $\operatorname{co} Z, Z^*$. The techniques of nonlinear analysis yield

Lemma 18.2. *The following equalities are true*

$$\rho(l \mid Z) = \rho(l \mid \operatorname{co} Z) = \sup_q \inf_p \Phi(l,p,q) \quad , \quad q \in F(X),\ p \in \mathbb{R}^m \tag{18.2}$$

$$\rho(l \mid Z^*) = \inf_p \sup_q \Phi(l,p,q) \quad , \quad q \in F(X),\ p \in \mathbb{R}^m \tag{18.3}$$

The sets $\operatorname{co} Z$, Z^* are convex due to their definition. However it is not difficult to give an example of a nonlinear map $F(x)$ for which Z is nonconvex and the functions $\rho(l \mid \operatorname{co} Z)$, $\rho(l \mid Z^*)$ do not coincide, so that the inclusions $Z \subset \operatorname{co} Z$, $\operatorname{co} Z \subset Z^*$ are strict.

Indeed, assume $X = \{0\}$, $x \in \mathbf{R}^2$

$$F(0) = \{x : 6x_1 + x_2 \leq 3 \ , \ x_1 + 6x_2 \leq 3 \ , \ x_1 \geq 0 \ , \ x_2 \geq 0\}$$
$$G = (0 \ , 1) \ , \ Q = (0 \ , 2) \ .$$

Then

$$Y = \{x : Gx \in Q\} = \{x : 0 \leq x_2 \leq 2\}$$

The set $F(0)$ is a nonconvex polyhedron O K D L in Figure 10a while set Y is a stripe. Here, obviously, set Z which is the intersection of $F(0)$ and Y, turns to be a nonconvex polyhedron O A B D L, while sets co Z , Z^* are convex polyhedrons O A B L and O A C L respectively (see Figures 10b and 10c). The corresponding points have the coordinates

A = (0 , 2), B = (1/2 , 2), C = (1 , 2), D = (3/7 , 3/7), K = (0 , 3), L = (3 , 0),
O = (0 , 0).
Clearly $Z \subset \text{co } Z \subset Z^*$.

This example may also serve to illustrate the existence of a "duality gap", [21] between (18.2) and (18.3).

For a linear-convex map $F(x) = Ax + P$ $(P \in \text{conv}\,\mathbb{R}^n)$ there is no distinction between Z, $\text{co}\,Z$, and Z^*:

Lemma 18.3 *Assume $F(x) = Ax + P$ where $P \in \text{conv}\,\mathbb{R}^n$, A is a linear map from $\mathbb{R}^n$ into $\mathbb{R}^n$. Then $Z = \text{co}\,Z = Z^*$.*

The description of Z, $\text{co}\,Z$, Z^* may however be given in a "decoupled" form which, allows to present all of these sets as the intersections of some parametrized varieties of convex multivalued maps of relatively simple structure.

19. THE ONE STAGE PROBLEM - A DECOUPLING PROCEDURE.

Whatever are the vectors $l, p (l \neq 0)$ it is possible to present $p = M'l$ where M belongs to the space $\mathbb{M}^{m \times n}$ of real matrices of dimension $m \times n$. Then, obviously,

$$\rho(l \mid Z) = \sup_q \inf_M \Phi(l\,,\, M'l\,,\, q) = \rho(l \mid \text{co}\,Z)\;,\; q \in F(X),\; M \in \mathbb{M}^{n \times m}, \tag{19.1}$$

$$\rho(l \mid Z^*) = \inf_M \sup_q \Phi(l\,,\, M'l\,,\, q) \qquad q \in F(X),\; M \in \mathbb{M}^{n \times m}$$

or

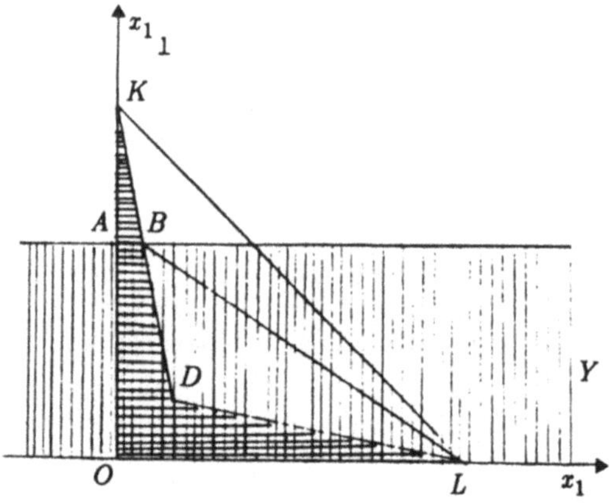

FIGURE 10a

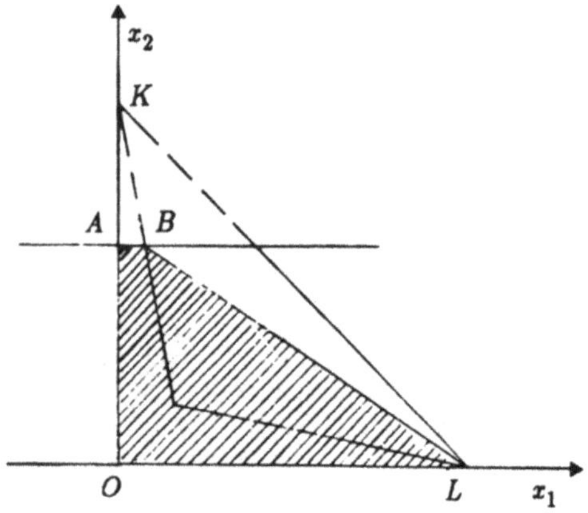

FIGURE 10b

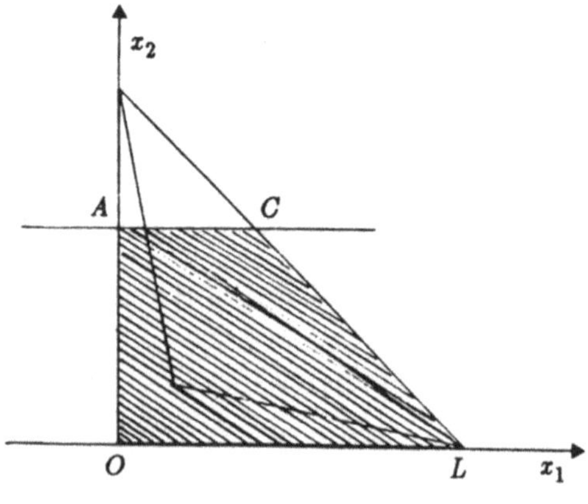

FIGURE 10c

$$\rho(l \mid Z^*) = \inf\{\Phi(l\,,\, M'l) \mid M \in \mathbf{M}^{n \times m}\}, \tag{19.2}$$

where

$$\Phi(l\,,\, M'l) = \bigcup\{\Phi(l\,,\, Ml\,,\, q) \mid q \in \operatorname{co} \boldsymbol{F}(x)\} =$$
$$= \rho((E - G'\, M')l \mid \operatorname{co} \boldsymbol{F}(X)) + \rho(-M'l \mid Q)\ .$$

From (19.1) it follows

$$Z \subseteq \bigcup_{q \in \boldsymbol{F}(X)} \bigcap_{M} R(M,q) \subseteq \bigcap_{M} \bigcup_{q \in \boldsymbol{F}(X)} R(M,q),\ M \in \mathbf{M}^{n \times m} \tag{19.3}$$

where

$$R(M,q) = (E_n - MG)q - MQ\,.$$

Similarly (19.2) yields

$$Z^* \subseteq \bigcap_{M} \bigcup_{q \in \operatorname{co} \boldsymbol{F}(X)} \{(E_n - MG)q - MQ\}\,. \tag{19.4}$$

Moreover a stronger assertion holds.

Theorem 19.1. *The following relations are true*

$$Z = Z(X) = \bigcup_{q \in \boldsymbol{F}(X)} \bigcap_{M} R(M,q) \tag{19.5}$$

$$Z* = Z*(X) = \bigcap_{M} R(M, \operatorname{co} F(X)) \tag{19.6}$$

where $M \in \mathbf{M}^{m \times n}$.

Obviously for $\boldsymbol{F}(x) = AX + P, (X, P \in \operatorname{co} \mathbb{R}^n)$ we have $\boldsymbol{F}(X) = \operatorname{co} F(X)$ and $Z = Z^* = \operatorname{co} Z$.

This *first* scheme of relations may serve to be a basis for constructing multistage procedures. Another procedure could be derived from the following *second* scheme. Consider the system

$$z \in \boldsymbol{F}(x) \tag{19.7}$$

$$Gx \in Q, \tag{19.8}$$

for which we are to determine the set of all vectors z consistent with inclusions (19.7), (19.8). Namely, we are to determine the restriction $\boldsymbol{F}_Y(x)$ of $\boldsymbol{F}(x)$ to set Y. Here we have

$$\boldsymbol{F}_Y(x) = \begin{cases} \boldsymbol{F}(x) & \text{if } x \in Y \\ \phi & \text{if } x \in Y \end{cases}$$

where as before $Y = \{ x : Gx \in Q \}$.

Lemma 19.1 *Assume $F(x) \in \mathrm{comp}\, \mathbb{R}^n$ for any x and $Q \in \mathrm{conv}\, \mathbb{R}^m$. Then*

$$F_Y(x) = \bigcap_L (F(x) - LGx + LQ)$$

over all $n \times m$ matrices L, $(L \in \mathbf{M}^{n \times m})$.

Denote the null vectors and matrices as $\{0\}_m \in \mathbf{R}^m$, $\{0\}_{m,n} \in \mathbf{R}^{m \times n}$, and the $(n \times m)$ matrix L_{mn} as

$$L_{mn} = \begin{bmatrix} I_m \\ \{0\}_{n-m,\,m} \end{bmatrix}$$

Suppose $x \in Y$. Then $\{0\}_m \in Q - Gx$ and for any $(n \times m)$ -matrix L we have $\{0\}_n \in L\,(Q - Gx)$. Then it follows that for $x \in Y$.

$$F(x) \subseteq \bigcap_L (F(x) + L(Q - Gx)) \subseteq F(x)$$

On the other hand, suppose $x \bar{\in} Y$.

Let us demonstrate that in this case

$$\bigcap_L \{ F(x) + L(Q - Gx) \} = \phi .$$

Denote $A = F(x)$, $B = Q - Gx$. For any $\lambda > 0$ we then have

$$\bigcap_L (A + LB) \subseteq (A + \lambda L_{mn} B) \cap (A - \lambda L_{mn} B)$$

Since $\{0\}_m \notin B$ we have $\{0\}_n \notin L_{mn}B$. Therefore there exists a vector $l \in \mathbb{R}^n$, $l \neq 0$ and a number $\gamma > 0$ such that

$$(l,x) \geq \gamma > 0 \quad \text{for any} \quad x \in L_m B ,$$

Denote

$$\mathbb{L} = \{ x : (l,x) \geq \gamma \} .$$

Then $\mathbb{L} \supseteq L_m B$ and

$$(A + \lambda L_{mn} B) \cap (A - \lambda L_{mn} B) \subseteq (A + \lambda \mathbb{L}) \cap (A - \lambda \mathbb{L})$$

Set A being bounded there exists a $\lambda > 0$ such that

$$(A+\lambda \mathbb{L}) \cap (A-\lambda \mathbb{L}) = \phi .$$

Hence

$$\bigcap_L (A+LB) = \phi$$

and the Lemma is proved.

If in addition to (19.7), (19.8) we have

$$x \in X \tag{19.9}$$

then the set Z_0 consistent with (19.7)-(19.9) may be presented as

$$Z_0(X) = \bigcup_{x \in X} \bigcap_L (F(x) - LGx + LQ) \tag{19.10}$$

Therefore each of the sets $Z(x)$, $Z^o(x)$ $(x \in X)$ may be respectively decoupled into the calculation of either set-valued functions $R(M\ ,\ q)$ or

$$R_0\ (L\ ,\ x) = F(x)\ -\ LGx\ +\ LQ$$

according to (19.5), (19.10). It may be observed that each of these are also applicable when $Z(X)$, $Z_0(X)$ are *disconnected.*

In the linear-convex case

$$F(x) = Ax + P\ ,\ P \in \text{conv}\ \mathbf{R}^n\ ,$$

we have

$$Z(X) = \bigcap_M \{(E\ -\ MG)(AX\ +\ P)\ +\ MQ\}$$

$$Z_0(x) = \bigcap_L \{(A\ -\ LG)X + P\ +\ LQ\}$$

20. SOLUTION TO THE PROBLEM OF NONLINEAR "GUARANTEED" FILTERING

Returning to system (16.1)-(16.3) we will look for the sequence of sets $\boldsymbol{X}[s] = \boldsymbol{X}(s,k_0,X^0)$ together with two other sequences of sets. These are

$$\boldsymbol{X}^*[s] = \boldsymbol{X}^*(s,k_0,X^0)$$

- the solution set for system

$$x(k+1) \in \operatorname{co} F(k, X^*[k]), \quad X^*[k_0] = X^0 \tag{20.1}$$

$$G(k+1)\, x(k+1) \in Q(k+1), \; k \geq k_0 \tag{20.2}$$

and $X_*[s] = X_*(s, k_0, X^0)$ which is obtained due to the following relations:

$$X_*[s] = \operatorname{co} Z[s] \tag{20.3}$$

where $Z[k+1]$ is the solution set for the system

$$z(k+1) \in F(k, X_*[k]), \quad Z[k^0] = X^0, \tag{20.4}$$

$$G(k+1) z(k+1) \in Q(k+1), \quad k \geq k_0. \tag{20.5}$$

The sets $X_*[\tau]$, $X^*[\tau]$ are obviously convex. They satisfy the inclusions

$$X[\tau] \subseteq X_*[\tau] \subseteq X^*[\tau]$$

while each of the sets $X[\tau]$, $X_*[\tau]$, $X^*[\tau]$ lies within

$$Y(\tau) = \{\, x : G(\tau) x \in Q(\tau) \,\}, \quad \tau \geq k_0 + 1,$$

The sets $X[\tau]$, $X_*[\tau]$, $X^*[\tau]$ may therefore be obtained by solving sequences of problems

$$x(k+1) \in F(k\,, x(k)) \tag{20.6}$$

$$G(k+1)\, x(k+1) \in Q(k)\,, \qquad k \geq k_o \tag{20.7}$$

for $X[s]$, (20.1), (20.2) for $X^*[s]$ and (20.3) - (20.5) for $X_*[s]$

In order to solve the "guaranteed" filtering problem with $Q(k) = y(k) - \tilde{Q}(k)$ one may follow *the first scheme* of § 19, considering the multistage system

$$Z(k+1) = (I_n - M(k+1) G(k+1)) F^0(k, S(k)) + M(k+1)(y(k+1) - \tilde{Q}(k+1)) \tag{20.8}$$

$$S(k) = \{\cap Z(k) \mid M(k)\}, \quad k > k_0, \; S(k_0) = X^0, \tag{20.9}$$

where $M(k+1) \in \mathbb{R}^{n \times m}$.

From Theorem 19.1 one may now deduce the following result

Theorem 20.1 The solving relations for $X[s]$, $X_*[s]$, $X^*[s]$ are as follows

$$X[s] = S(s) \quad \text{for} \quad F^0(k, S(k)) = F(k, S(k)) \tag{20.10}$$

$$X^*[s] = S(s) \quad \text{for} \quad F^0(k, S(k)) = \operatorname{co} F(k, S(k)) \tag{20.11}$$

$$X_*[s] = \operatorname{co} S(s) \quad \text{for} \quad F^0(k, S(k)) = F(k, \operatorname{co} S(k))\,. \tag{20.12}$$

It is obvious that $X[\tau]$ is the *exact solution* for the *guaranteed filtering* problem while $X_*[\tau]$, $X^*[\tau]$ are *upper convex majorants* for $X[\tau]$. It is clear that by interchanging and combining relations (20.11), (20.12) from stage to stage it is possible to construct a broad variety of other convex majorants for $X[\tau]$. However for the linear case they will all coincide with $X[\tau]$.

Lemma 20.1 *Assume $F^0(k,S) = A(k)S + P(k)$ with $P(k)$, X^0 being convex and compact. Then $X[k] = X^*[k] = X_*[k]$ for any $k \geq k_0$.*

Consider the nonlinear system

$$\begin{aligned} Z(k+1) &= (I_n - M(k+1)G(k+1))F^0(k,Z(k)) \\ &\quad + M(k+1)(y(k+1) - \tilde{Q}(k+1)) , \qquad\qquad Z(k_0) = X^0 , \end{aligned}$$

having denoted its solution as

$$\begin{aligned} &Z(k;M_k(\cdot)) \quad \text{for} \quad F^0(k,Z) = F(k,Z) \\ &Z_*(k,M_k(\cdot)) \quad \text{for} \quad F^0(k,Z) = F(k,\mathrm{co}\, Z) \\ &Z^*(k,M_k(\cdot)) \quad \text{for} \quad F^0(k,Z) = \mathrm{co}\, F(k,Z) \end{aligned}$$

Then theorem 20.1 yields the following conclusion

Theorem 20.2 *Whatever is the sequence $M_s(\cdot)$, the following solving inclusions are true*

$$\begin{aligned} &X[s] \subseteq Z(s,M_s(\cdot)) \\ &X_*[s] \subseteq Z_*(s,M_s(\cdot)) \\ &X^*[s] \subseteq Z^*(s,M_s(\cdot)) , \quad s > k_0 , \end{aligned} \tag{20.13}$$

with $Z(s,M_s(\cdot)) \subseteq Z_*(s,M_s(\cdot)) \subseteq Z^*(s,M_s(\cdot))$.

Hence we also have

$$X[s] \subseteq \bigcap \{ Z(s,M_s(\cdot)) \mid M_s(\cdot)) \} \tag{20.14}$$

$$X_*[s] \subseteq \bigcap \{ Z_*(s,M_s(\cdot)) \mid M_s(\cdot)) \} \tag{20.15}$$

$$X^*[s] \subseteq \bigcap \{ Z^*(s,M_s(\cdot)) \mid M_s(\cdot)) \} \tag{20.16}$$

over all $M_s(s)$.

However a question arises which is whether (20.14)-(20.16) could turn into exact equalities.

Lemma 20.2 Assume the system (16.1), to be linear: $F(k,x) = A(k)x + P(k)$ *with sets* $P(k)$, $Q(k)$ *convex and compact. Then*

$$X[s] = X^*[s] = \bigcap \{ Z_s(\cdot, M_s(\cdot)) \} \tag{20.17}$$

where $Z_s(\cdot\; M_s(\cdot))$ *is the solution tube for the equation*

$$\begin{aligned} Z(k+1) = (I_n - M(k+1)G(k+1))\,(A(k)\,Z(k) + P(k)) + \\ + M(k+1)(y(k+1) - \tilde{Q}(k+1)),\ Z(k_o) = X^o \end{aligned} \tag{20.18}$$

Hence in this case the intersections over $M(k)$ could be taken *either at each stage* as in (20.10), (20.11) *or at the final stage* as in (20.17).

Let us now follow *the second scheme* of § 19, considering the equation

$$x(k+1) \in \tilde{F}_{Y(k)}(k,x(k)),\quad x^0 = x(k_0),\quad x^0 \in X^0\,, \tag{20.19}$$

and denoting the set of its solutions that start at $x^0 \in X^0$ as $X^0(k,k_0,x^0)$ as

$$\bigcup \{ x^0(k,k_0,x^0) \mid x^0 \in X^0 \} = X^0(k,k_0,X^0) = X^0[k]\,.$$

According to Lemma 19.1 we may substitute (20.19) by equation

$$x(k+1) \in \bigcap_L (\tilde{F}\,(k,x(k)) - LG(k)x(k) + LQ(k))\ ,\quad x^0 \in X^0,$$

The calculation of $X^0[k]$ should hence follow the procedure of (19.10)

$$\tilde{X}[k+1] = \bigcup_{x \in \tilde{X}(k)} \bigcap_L (\tilde{F}(k,x) - LG(k)x + LQ(k)),\qquad X(k_0) = X^0\,. \tag{20.20}$$

Denote the "whole" solution tube for this solution ($k_0 \le k \le s$) as $\tilde{X}^s_{k_0}[\cdot]$. Then the following assertion will be true.

Theorem 20.3 Assume $\tilde{X}^s_{k_0}[k]$ *to be the cross-section of the tube* $\tilde{X}^s_{k_0}[\cdot]$ *at instant* k *and* $X^o = X^o \bigcap Y(k_o)$. *Then*

$$\begin{aligned} X[s] &= \tilde{X}^{s+1}_{k_0}[s] \ \text{ if } \ \tilde{F}(k,x) = F(k,x)\,, \\ X^* &= \tilde{X}^{s+1}_{k_0}[s] \ \text{ if } \ \tilde{F}(k,x) = \operatorname{co} F(k,x) \end{aligned}$$

Here $\tilde{X}^s_{k_0}[s] \supseteq \tilde{X}^{s+1}_{k_0}[s]$ and the set $\tilde{X}^s_{k_0}[s]$ may not lie totally within $Y(s)$, while always $\tilde{X}^{s+1}_{k_o}\,[s] \subseteq Y(s)$.

Solving equation (20.19) is equivalent to finding all the solutions for the inclusion

$$x(k+1) \in \bigcap_L (\tilde{F}(k,x(k)) + L(y(k) - G(k)\, x(k) - \tilde{Q}(k))\ , \ x(k_0) \in X^0 \tag{20.21}$$

Equation (20.21) may now be "decoupled"into a system of "simpler" inclusions

$$x(k+1) \in \tilde{F}(k,x(k)) + L(k)\ (y(k) - G(k)x(k)) - L(k)\ \tilde{Q}(k)\ , \quad x(k_0) \in X^0 \tag{20.22}$$

for each of which the solution set for $k_0 \leq k \leq s$ will be denoted as

$$\tilde{X}^s_{k_0}(\cdot, k_0, X^0, L(\cdot)) = \tilde{X}^s_{k_0}[\cdot, L(\cdot)]$$

Theorem 20.4 *The set $X^s_{k_0}[\cdot]$ of solutions to the inclusion*

$$x_{k+1} \in \tilde{F}(k,x(k))\ , \quad x(k_0) \in X^0$$
$$y(k) \in G(k)x(k) + \tilde{Q}(k), \quad k_0 \leq k \leq s$$

is the part of the solution tube

$$X^{s+1}_{k_0}[\cdot] = \bigcap_L \tilde{X}^{s+1}_{k_0}[\cdot, L]\ , \ [k_0, \ldots, s+1]$$

which is restricted to stages $[k_0\ , \ s]$. Here the intersection may be taken only over all constant matrices $L(k) \equiv L$.

This scheme also allows to calculate the cross sections $X^s_{k_0}[s]$. Obviously

$$X^s_{k_0} \subseteq \bigcap_{L[\cdot]} \tilde{X}^{s+1}_{k_0}[s, L[\cdot]] \tag{20.23}$$

over all sequences $L[\cdot] = \{\, L(k_0),\ L(k_0+1), \ldots, L(s+1)\,\}$. Moreover the following proposition is true, and may be compared with [5, 9-11].

Theorem 20.5 *Assume $\tilde{F}(k,x)$ to be linear-convex: $\tilde{F}(k,x) = A(k)x + P(k)$, with $P(k)$, $Q(k)$ convex and compact. Then (20.23) turns to be an equality.*

The next estimation problems are those of "prediction" and "refinement".

21. THE "GUARANTEED PREDICTION" PROBLEM

The solution to the guaranteed prediction problem is to specify set $X(s \mid t\ , k_o\ , X^o)$ for $s \geq t$ It may be deduced from the previous relations due to (17.2) since

$$X(s \mid t , k_o , X^o) = X(s \mid t , X(t , k_o , X^o))$$

Similarly we may introduce set

$$X^*(s \mid t , k_o , X^o) = X^*(s \mid t , X^*(s \mid t , X^*(t , k_o , X^o))$$

where $X^*(s \mid t , x)$ is the *attainability domain* for the inclusion

$$x(k+1) \in \text{co } F(k , x(k))$$

with $t \leq k \leq s , \qquad x(t) = x$

The description of $X(s \mid t , k_o X^o)$, $X^*(s \mid t , k_o X^o)$ may be given through a modification of theorems 20.1 - 20.5, by the following assertion

Theorem 21.1 The solving relations for the prediction problem are

$$X(s \mid t , k_o, X^o) = X[s]$$

$$X^*(s \mid t , k_o, X^o) = X^*[s]$$

where $X[s]$, $X^*[s]$ are determined through (20.10), (20.12), (20.8), under the condition

$$S(k) = \{\bigcap Z(k) \mid M(k) \in \mathbf{R}^{n \times m}\}$$
$$S(k) = Z(k) \textit{ for } k > t$$

For the linear convex case an alternative presentation is true. Denote $L_t^s(\cdot) = \{L(k_o),\ldots,L(s)\}$ to be a sequence of $(n \times m)$ - matrices $L(i)$, $k_o \leq i \leq s$, such that $L(i) \equiv 0$ for $t < i \leq s$.

Theorem 21.2 Assume $F(k , x) = A(k)x + P$ with P , X^o convex and compact. Then

$$X(s \mid t , k_o , X^o) = \{\bigcap \tilde{X}_{k_o}^s [s , L_s^t(\cdot)] \mid L_s^t(\cdot)\} \tag{21.1}$$

The solution to the prediction problem may therefore be decoupled into the calculation of the attainability domains $\tilde{X}_{k_o}^s [s , L_t^s(\cdot)]$ for the variety of systems

$$x(k+1) \in (A(k) - L(k)\, G(k))\, x(k) + L(k)\, y(k) + L(k)\, \tilde{Q}(k) + P(k) \tag{21.2}$$
$$L(k) \equiv 0 \text{ for } k > t$$

each of which starts its evolution from X^o.

The forthcoming "refinement" problem is a deterministic version of the interpolation problem of stochastic filtering theory.

22. THE "GUARANTEED" REFINEMENT PROBLEM

Assume the sequence $y[k\ ,\ t]$ to be fixed. Let us discuss the means of constructing sets $X(s \mid t\ ,\ k\ ,\ \mathbf{F})$, with $s \in [k\ ,\ t]$. From relation (17.2) one may deduce the assertion

Lemma 22.1 The following equality is true

$$X(s \mid t\ ,\ k\ ,\ \mathbf{F}) = X(s \mid s\ ,\ t\ ,\ X(t\ ,\ k\ ,\ \mathbf{F})) \tag{22.1}$$

Here the symbol $X(s \mid s\ ,\ t\ ,\ \mathbf{F})$, taken for $s \leq t$, stands for the set of states $x(s)$ that serve as starting points for all the solutions $x(k\ ,\ s\ ,\ x(s))$ that satisfy the relations

$$x(k+1) \in F(k\ ,\ x(k))\ ,\ x(t) \in \mathbf{F}$$
$$x(k) \in Y(k)\ , \qquad s \leq k \leq t$$

Corollary 22.1 Formula (22.1) may be substituted for

$$X(s \mid t\ ,\ k\ ,\ \mathbf{F}) = X(s\ ,\ k\ ,\ \mathbf{F}) \bigcap X(s \mid s\ ,\ t\ ,\ \mathbf{K}) \tag{22.2}$$

where $\mathbf{K}$ *is any subset of* $\mathbf{R}^n$ *that includes* $X(t\ ,\ k\ ,\ \mathbf{F})$.

Thus the set $X(s \mid t\ ,\ k\ ,\ \mathbf{F})$ is described through the solutions of two problems the first of which is to define $X(s\ ,\ k\ ,\ \mathbf{F})$ (along the techniques of the above) and the second is to define $X(s \mid s\ ,\ t\ ,\ \mathbf{K})$. The solution of the second problem will be further specified for $\mathbf{F} \in \text{comp}\, \mathbf{R}^n$ and for a closed convex Y.

The underlying elementary operation is to describe $\boldsymbol{X}$ - the set of all the vectors $x \in \mathbf{R}^n$ that satisfy the system

$$z \in F(x)\ , \qquad z \in Y$$
$$(\boldsymbol{X} = \{x : F(x) \bigcap Y \neq \emptyset\})$$

Using suggestions similar to those applied in Lemma 19.1 we come to

Lemma 22.2 The set $\boldsymbol{X}$ *may be described as*

$$\boldsymbol{X} = \bigcup \{ \bigcap \{Ex - MF(x) + MY \mid M \in \mathbb{M}^{n \times n}\} \mid x \in \mathbf{R}^n\}$$

From here it follows:

Theorem 22.1 The set $X(s \mid s\ ,\ t\ ,\ \mathbf{R})$ *may be described as the solution of the multistage system (in backward "time")*

$$X[k] = Y(k) \bigcap X[k] \tag{22.3}$$

where

$$X[k] = \bigcup\{ \bigcap \{Ex - MF(x) + MX[k+1] \mid M \in \mathbf{M}^{n\times n} \} \mid x \in \mathbf{R}^n\} ,$$
$$s \le k \le t \, , \, X[t] = Y[t] \, .$$

Finally we will specify the solution for the linear case

$$x(k+1) \in A(k)\, x(k) + P(k) \, , \, Y(k) = \{x : y(k) \in G(k)x + Q(k)\}$$

Assume

$$X = \{x : z \in Ax - P \, , x \in Y \, , z \in Z\} \ , \ Y = \{x : Gx \in Q - y\} \tag{22.4}$$

where $A \in \mathbf{M}^{n\times n}$, $G \in \mathbf{M}^{m\times n}$, P , Q , Z are convex and compact.

Lemma 22.3 The set X may be defined as

$$\rho(l \mid X) = \inf\{\rho(\lambda \mid P) + \rho(\lambda \mid Z) + \rho(p \mid Q - y)\}$$

over all the vectors $\lambda \in \mathbf{R}^n$, $p \in \mathbf{R}^m$ that satisfy the equality $l = A'\,\lambda + G'p$.

The latter relation yields:

Lemma 22.4 The set X may be defined as

$$X \subseteq L'(Z+P) + M'(Q-y) = H(L\,,M) \tag{22.5}$$

whatever are the matrices $L \in \mathbf{M}^{n\times n}$ and $M \in \mathbf{M}^{m\times n}$ that satisfy the equality $L'\,A + M'G = E_n$. Moreover the following equalities are true

$$X = \{\bigcap H\,(L\,,M) \mid L\,,M\} \tag{22.6}$$
$$\rho(l \mid X) = \inf\,\{\rho(l \mid H(L\,,M)) \mid L,M\}$$

over all $L \in \mathbf{M}^{n\times n}$, $M \in \mathbf{M}^{m\times n}$

Corollary 22.2 Suppose $\mid A \mid \neq 0$. Then conditions (22.5), (22.6) may be substituted for

$$X \subseteq (E_n - M'G)\, A^{-1}\,(Z+P) + M'(Q-y) = H(M) \, ,$$
$$X = \bigcap\{H(M) \mid M\} \, , \rho(l \mid X) = \inf\,\{\rho(l \mid H\,(M)) \mid M\}$$

where

$$M \in \mathbf{M}^{m\times n} \, .$$

The latter relations may be used for recurrent procedures. These are either

$$X[k] = \bigcap\{H_k(L, M) \mid L A(k) + MG(k) = E_n\}, \tag{22.7}$$
$$H_k(L, M) = L'(X[k+1] + P(k)) + M'(y(k) - Q(k), X[t] = Y[t],$$
$$s \leq k \leq t$$

with

$$X(s \mid s, t, Y[t]) = X[s] \tag{22.8}$$

or

$$\boldsymbol{X}[k] \subseteq H_k(L(k), M(k)), \boldsymbol{X}[t] = Y[t] \tag{22.9}$$
$$s \leq k \leq t$$

with

$$X(s \mid s, t, Y(t)) = \bigcap \{\boldsymbol{X}[s] \mid L_s(\cdot), M_s(\cdot)\} \tag{22.10}$$

where

$$L_s(\cdot) = (L(s), \ldots, L(t)); M_s(\cdot) = (M(s), \ldots, M(t))$$

Theorem 22.2 The set $X(s \mid s, t, Y)$ may be derived due to either equations (22.7) or (22.9), (22.10).

23. CONCLUSIONS

This paper gives an introduction to the theory of *guaranteed identification and state estimation* under uncertainty with unknown but bounded observation "noise" and input disturbances. The whole problem is considered within a deterministic setting so that the results are given in the form of set-valued estimates the description of whose evolution is the objective of the solution schemes.

The mathematical techniques applied here are mainly those of convex analysis, set theory and related topics [21, 27]. A respective continuous version of the given problems would thus further lead us to the techniques of differential inclusions and viability theory of nonlinear analysis, [28, 29].

An important issue is that the purely deterministic solutions to the guaranteed filtering problems may be well approximated by solutions to related problems of stochastic filtering as shown in § § 14, 15. (This idea also applies to the identification problems of § § 7, 8.) Thus the well-developed computational techniques of stochastic estimation theory may be modified through some procedures of parallel computations to solve the problems of the above. Basically this gives a robust procedure for solving the specific class of *inverse problems* discussed in this paper (see also [30, 31]).

One may raise the question of what is more adequate in the analysis of systems - a stochastic or a "set-membership", deterministic description of uncertainty? The author's opinion is that this question is not correct - the specific informational assumptions for a given problem may require either of these approaches and techniques or perhaps a combination and interaction of both, [32–34]. It is the specific modelling problem that should dominate the tools.

The author wishes to thank F. Stettinger for the computer simulations.

1. REFERENCES

[1] Cramer, H.: *Mathematical Methods of Statistics*. Princeton, 1946.

[2] Cox, D.R. and D.V. Hinkley: *Theoretical Statistics*. Chapman and Hall, London, 1974.

[3] Kalman, R.E.: A new approach to linear filtering and prediction problems, *Trans. ASME*. Ser. D, Vol. 82, pp. 35-45, 1960.

[4] Eykhoff, P.: *System Identification: Parameter and State Estimation*, J. Wiley & Sons, New York, 1974.

[5] Krasovski, N.N.: *The Control of a Dynamic System*. Nauka, Moscow, 1986.

[6] Lee, C.S. and G. Leitmann: On optimal long-term management of some ecological systems subject to uncertain disturbances. *Int. J. Systems Sci.,* Vol. 14, No. 8, pp. 979-994, 1983.

[7] Chernousko, F.L.: Ellipsoidal Bounds for Sets of Attainability and Uncertainty in Control. *Optimal Control, Appl. & Methods*, Vol. 3, pp. 87-202, 1982.

[8] Ackerman, J. and R. Muench: Robustness analysis in a plant parameter plane. 10th World IFAC Congress (preprints) Vol. 8, pp. 230-235, 1987.

[9] Byrnes, C. and A. Isidori: Heuristics in Nonlinear Control and Modelling. Adaptive Control. Eds. C.I. Byrnes, A.B. Kurzhanski. Springer Lecture Notes in Control and Information Sciences, Vol. 105, pp. 48-70, 1988.

[10] Usoro, P.B., F.C. Schweppe, D.N. Wormley and L.A. Gould: Ellipsoidal Set-Theoretic Control Synthesis. *Trans. ASME. Ser. G, Journal of Dynamic Systems Measurement & Control*, Vol. 104, No. 4, pp. 331-336, 1982.

[11] Krasovskii, N.N.: On the theory of controllability and observability of linear dynamic systems. *Prikl. Math. Mech.*, Vol. 28, No. 1, pp. 1-14, 1964. (in Russian)

[12] Witsenhausen, H.S.: Sets of possible states of linear systems given perturbed observations. *IEEE Trans. Automat. Control*, Vol. AC-3, pp. 556-558, 1968.

[13] Kurzanskii, A.B.: On the duality of the problems of control and observation, *Prikl. Math. Mech.*, Vol. 34, No. 3, pp. 429-439, 1970. (in Russian)

[14] Bertsekas, D. and I.B. Rhodes: Recursive state estimation for a setmembership description of uncertainty. *IEEE Trans. Automat. Control*, Vol. AC-16, No. 2, pp. 117-128, 1971.

[15] Schweppe, F.C.: *Uncertain Dynamic Systems.* Prentice Hall, 1973.

[16] Kurzanskii, A.B.: *Control and Observation Under Conditions of Uncertainty.* Nauka, Moscow, 1977. (in Russian)

[17] Pschenichnyi, B.N. and V.G. Pokotilo: The Minmax Approach to the Estimation of the Parameters of Linear Regression. *Izv. Akad. Nauk SSR, Tech. Cybernetics*, No. 2, pp. 94-106, 1983 (translated as "Engineering Cybernetics").

[18] Fogel, E.: System Identification via membership set constraints with energy constrained noise. *IEEE Trans. Automat. Control*, Vol. AC-24, No. 5, pp. 752-758, 1979.

[19] Kurzhanski. A.B. and A. Yu. Hapalov: On the state estimation problem for distributed systems. Analysis & Optimization of Systems Proc. of the VIIth Int. INRIA Conf. Springer Lecture Notes in Control & Information Sciences, Vol. 83, pp. 102-113, 1986.

[20] Kurzhanski, A.B. and Filippova: On the description of the set of viable trajectories of a differential inclusion. *Soviet Math. Doklady* (Engl. Transl.), Vol. 34, No. 1, 1987 (Vol. 283, No. 1, pp. 37-41, 1986).

[21] Rockafellar, R.T.: *Convex Analysis*, Princeton University Press, Princeton. 1970.

[22] Shiriayev, A.N.: *Probability.* Nauka, Moscow, 1980 (Engl. Transl.).

[23] Kurzhanski, A.B.: On evolution equations in estimation problems for systems with uncertainty. WP-82-49, 1982 IIASA, Laxenburg, Austria.

[24] Ustyuzhanin, A.M.: On the problem of matrix parameter identification. *Problems of Control & Information Theory*, Vol. 15, No. 4, pp. 265-273, 1986.

[25] Albert, A.: *Regression and the Moor-Penrose Pseudo-inverse*, Academic Press, New York, 1972.

[26] Koscheev, A.S. and A.B. Kurzanskii: Adaptive estimation of multistage systems under uncertainty, *Izv. Akad. Nauk SSR, Tech. Cybernetics*, No. 2, pp. 72-93, 1983 (translated as "Engineering Cybernetics").

[27] Kuratowski, K.: *Topology*, Vol. 1, No. 2, Academic Press, New York, 1968.

[28] Aubin, J.-P. and A. Cellina: *Differential Inclusions.* Springer-Verlag, 1984.

[29] Aubin, J.-P. and J. Ekeland: *Applied Nonlinear Analysis.* Wiley-Interscience, 1984.

[30] Tikhonov, A.N. and V.A. Arsenin: *Methods of Solving Ill-Posed Problems.* Nauka, Moscow, 1979.

[31] Huber, P.J.: *Robust Statistics.* Wiley, 1981.

[32] Katz, I.J. and A.B. Kurzanskii: Minimax multistage filtering in statistically uncertain situations, *Automation and Remote Control*, Vol. 11, pp. 79-87, 1978. (in Russian)

[33] Polyak, B.T. and Ya. Z. Tsypkin: Robust Identificaion. *Automatica*, Vol. 16, No. 1, pp. 53-63, 1980.

[34] Anan'ev, B.I. and A.B. Kurzhanski: The nonlinear filtering problem for a multistage system with statistical uncertainty. Second IFAC Symposium on Stochastic Control. Vilnius, USSR, Pt. I, pp. 205-210, 1986.

STATISTICAL ASPECTS OF MODEL SELECTION

RITEI SHIBATA

Abstract

Various aspects of statistical model selection are discussed from the view point of a statistician. Our concern here is about selection procedures based on the Kullback Leibler information number. Derivation of AIC (Akaike's Information Criterion) is given. As a result a natural extension of AIC, called TIC (Takeuchi's Information Criterion) follows. It is shown that the TIC is asymptotically equivalent to Cross Validation in a general context, although AIC is asymptotically equivalent only for the case of independent identically distributed observations. Next, the maximum penalized likelihood estimate is considered in place of the maximum likelihood estimate as an estimation of parameters after a model is selected. Then the weight of penalty is also the one to be selected. We will show that, starting from the same Kullback-Leibler information number, a useful criterion RIC (Regularization Information Criterion) is derived to select both the model and the weight of penalty. This criterion is in fact an extension of TIC as well as of AIC. Comparison of various criteria, including consistency and efficiency is summarized in Section 5. Applications of such criteria to time series models are given in the last section.

Keywords

Statistical modelling, model selection, information criterion, cross validation.

1. INTRODUCTION

In any science *modeling* is a way of approximation of reality. As far as it yields a *good* approximation, a simpler model is better than complex one both for understanding the phenomena and for various applications, for example, forecasting, control, making decision and so on. The principle is the same for selecting a statistical model. One specific point is that we, statisticians, assume that the number of observation is limited and only partial information is available through *data* which possibly involve random fluctuations. Random fluctuation means here various measurement errors as well as fluctuations of the system which generates data. By introducing randomness into a model, the model becomes much more flexible than a deterministic model and resistant to unexpected fluctuations of the system. Another advantage is that we may leave the error of approximation as a part of random fluctuations which are introduced beforehand into a model as long as the former is compatible order of magnitude to the latter. This often results in a simplification of a model. A desirable procedure of statistical model selection is, therefore, to reject a model which is far from the reality and pick up a model in which the error of approximation and the error due to random fluctuations are well balanced. There may be cases where we have to satisfy with a model, not the best one but the best possible one in a given family of models when only very poor information is available for the underlying phenomena.

Complexity of a model is restricted both by the size of observation and by the signal to noise ratio. Needless to say, complete specification might be possible if an infinite number of observations were available for a quite simple system. Otherwise a practical procedure is, starting from a simple model, to increase the complexity until a trade off between the error of approximation and the error due to random fluctuations. To do this systematically, a convenient way is to introduce a criterion to compare models. In this chapter, we discuss various criteria, some of which are based on an information measure. Although in system sciences, time series models, AR, MA or ARMA are quite common, to clarify the point, we first restrict our attention into models for independent observations. Extensions to time series models or state space models are rather technical. Some of them are explained in the last section.

2. INFORMATION CRITERIA

Let $\mathbf{Y}_n' = (y_1,...,y_n)$ be n independent observations but not necessarily identically distributed, whose joint density is denoted by $g(\mathbf{Y}_n)$. Hereafter $'$ denotes transpose of a vector or of a matrix, and E denotes the expectation with respect to the vector of random variables, $\mathbf{Y}_n$. We mean by statistical model a parametric family of densities $\mathbf{F} \doteq \{f(\mathbf{Y}_n;\theta),\theta\in\Theta\}$. The part, usually called *model*, for example, linear or non-linear relation between input and output, is described by parametrization of densities through θ in $\mathbf{F}$. A regression equation $y = x'\beta + \varepsilon$ with explanatory variable x and Gaussian error ε with mean 0 and variance σ^2 is formulated as the model

$$\mathbf{F} = \left\{ \prod_{i=1}^{n} \frac{1}{\sigma} \phi \left[\frac{y_i - x_i'\beta}{\sigma} \right], \ \theta = (\beta, \sigma)' \in \mathbf{R}^m \times (0, \infty) \right\},$$

where ϕ is the standard normal density. A natural way of evaluating goodness of a model $\mathbf{F}$ is to introduce a kind of distance of the estimated density $f(\cdot;\hat{\theta})$, an approximation to the true $g(\cdot)$ based on $\mathbf{Y}_n$, from the $g(\cdot)$. For a while, to simplify the problem, $\hat{\theta} = \hat{\theta}(\mathbf{Y}_n)$ is taken to be the maximum likelihood estimate of θ under the model $\mathbf{F}$, based on $\mathbf{Y}_n$. As a distance, a natural choice is the Kullback-Leibler information number:

$$K_n(g(\cdot), f(\cdot;\hat{\theta})) = \int g(\mathbf{x}_n) \log \frac{g(\mathbf{x}_n)}{f(\mathbf{x}_n;\hat{\theta})} d\mathbf{x}_n .$$

Note that this is a pseudo-distance since the triangular inequality does not hold true. This varies with the observation $\mathbf{Y}_n$ through $\hat{\theta}(\mathbf{Y}_n)$. As a measure of closeness between two densities $g(\cdot)$ and $f(\cdot;\hat{\theta})$, the measure has been widely accepted. It is known that the measure is nonnegative, zero if two densities coincide, and additive for independent samples $\mathbf{x}_n = (x_1, \ldots, x_n)$. More importantly, as is shown below, this has a close connection with the *maximum likelihood principle* or the *minimum entropy principle* which is a basic in statistical inference. If

$$\left| \int g(\mathbf{x}_n) \log g(\mathbf{x}_n) d\mathbf{x}_n \right| < \infty,$$

then the expectation of the Kullback-Leibler information number $K_n(g(\cdot), f(\cdot;\hat{\theta}))$ can be rewritten as

$$\int g(\mathbf{x}_n) \log g(\mathbf{x}_n) d\mathbf{x}_n - \mathrm{E} \int g(\mathbf{x}_n) \log f(\mathbf{x}_n;\hat{\theta}) d\mathbf{x}_n . \tag{2.1}$$

A problem in using (2.1) as a criterion of comparing models is that the second term of (2.1) depends on unknown $g(\cdot)$. We demonstrate that a useful approximation is obtained by expanding it for a large number of observations under the following assumptions A1 to A4.

A1. The parameter space Θ is a Euclidean p-dimensional space $\mathbf{R}^p$ or an open subspace of it. Both the Gradient vector

$$\mathrm{g}_n(\theta)' = \left(\frac{\partial}{\partial \theta_l} l(\theta), l = 1, \ldots, p\right)$$

and the Hessian matrix

$$H_n(\theta) = \left(\frac{\partial^2}{\partial \theta_l \partial \theta_m} l(\theta), 1 \le l, m \le p\right)$$

of the log-likelihood function $l(\theta) = \log f(\mathbf{Y}_n;\theta)$, are well defined with probability 1, and both continuous with respect to θ.

A2. $E|g_n(\theta)|<\infty$ and $E|H_n(\theta)|<\infty$, where $|\cdot|$ denotes the absolute value of each component of a vector or of a matrix.

A3. There exists a unique θ^* in Θ, which is the solution of $E g_n(\theta^*)=0$. For any $\varepsilon>0$,

$$\sup_{\|\theta-\theta^*\|>\varepsilon} l(\theta) - l(\theta^*)$$

diverges to $-\infty$ a.s..

A4. For any $\varepsilon>0$, there exists $\delta>0$ such that

$$\sup_{\|\theta-\theta^*\|<\delta} | E(\hat{\theta}-\theta^*)' J_n(\theta)(\hat{\theta}-\theta^*) - \mathrm{tr}(I_n(\theta^*) J_n(\theta^*)^{-1}) | < \varepsilon$$

for large enough n. Here

$$I_n(\theta^*) = E\, g_n(\theta^*) g_n(\theta^*)' \quad \text{and} \quad J_n(\theta) = -E\, H_n(\theta)$$

are assumed to be positive definite matrices and continuous with respect to θ.

The assumption A3 assures that $\hat{\theta}-\theta^*$ converges to 0 a.s. as n tends to infinity. That is, $\hat{\theta}$ is a consistent estimate of θ^*. The assumption A3 together with A2 implies that $K_n(g(\cdot), f(\cdot;\theta))$ is minimized at θ^*. This means that $f(\cdot;\theta^*)$ is the best approximation to $g(\cdot)$. However such a θ^* completely relies on unknown $g(\cdot)$. The situation can be further understood by looking at the decomposition,

$$K_n(g(\cdot), f(\cdot;\hat{\theta})) = K_n(g(\cdot), f(\cdot;\theta^*)) + \int g(\mathbf{x}_n) \log \frac{f(\mathbf{x}_n;\theta^*)}{f(\mathbf{x}_n;\hat{\theta})} d\mathbf{x}_n .$$

The first term on the right hand side is the least error of approximation by the model **F**, and the second term is the error due to the estimation of parameter θ^* by $\hat{\theta}$.

We note that all assumptions above are commonly used regularity conditions. By expanding $\log f(\mathbf{x}_n;\hat{\theta})$ around θ^*, we have

$$\log f(\mathbf{x}_n;\hat{\theta}) = \log f(\mathbf{x}_n;\theta^*) + (\hat{\theta}-\theta^*)' \frac{\partial}{\partial\theta} \log f(\mathbf{x}_n;\theta^*)$$

$$+ \frac{1}{2} (\hat{\theta}-\theta^*)' \frac{\partial^2}{\partial\theta\partial\theta'} \log f(\mathbf{x}_n;\theta^{**})\, (\hat{\theta}-\theta^*),$$

where θ^{**} is a value between $\hat{\theta}$ and θ^*. We should note that the Gradient vector $\frac{\partial}{\partial\theta} \log f(\mathbf{x}_n;\theta)$ and the Hessian matrix $\frac{\partial^2}{\partial\theta\partial\theta'} \log f(\mathbf{x}_n;\theta)$ are not of the log likelihood $\log f(\mathbf{Y}_n;\theta)$ of the observations $\mathbf{Y}_n$, but of the log likelihood $\log f(\mathbf{x}_n;\theta)$ of *test sample* $\mathbf{x}_n$. Since

$$\int g(\mathbf{x}_n) \frac{\partial}{\partial\theta} \log f(\mathbf{x}_n;\theta^*) d\mathbf{x}_n = 0,$$

the assumption A3 justifies the expansion;

$$\int g(\mathbf{x}_n) \log f(\mathbf{x}_n;\hat{\theta}) d\mathbf{x}_n = \int g(\mathbf{x}_n) \log f(\mathbf{x}_n;\theta^*) d\mathbf{x}_n$$

$$+\frac{1}{2}(\hat{\theta}-\theta^*)'\left\{\int g(\mathbf{x}_n)\frac{\partial^2}{\partial\theta\partial\theta'}\log f(\mathbf{x}_n;\theta^{**})\right\}(\hat{\theta}-\theta^*). \tag{2.2}$$

From the assumption A4, the expectation of (2.2) is

$$\mathrm{E}\int g(\mathbf{x}_n)\log f(\mathbf{x}_n;\hat{\theta})d\mathbf{x}_n = \int g(\mathbf{x}_n)\log f(\mathbf{x}_n;\theta^*)d\mathbf{x}_n - \frac{1}{2}\mathrm{tr}(I_n(\theta^*)J_n(\theta^*)^{-1}) + o(1)$$

$$= \mathrm{E}\,l(\theta^*) - \frac{1}{2}\mathrm{tr}(I_n(\theta^*)J_n(\theta^*)^{-1}) + o(1).$$

Furthermore, by expanding $l(\theta^*)$ around $\hat{\theta}$, from the fact that $g_n(\hat{\theta})=0$ we have

$$l(\theta^*) = l(\hat{\theta}) + \frac{1}{2}(\theta^*-\hat{\theta})'H_n(\theta^{**})(\theta^*-\hat{\theta}), \tag{2.3}$$

and then

$$\mathrm{E}\int g(\mathbf{x}_n)\log f(\mathbf{x}_n;\hat{\theta})d\mathbf{x}_n = \mathrm{E}\,l(\hat{\theta}) - \mathrm{tr}(I_n(\theta^*)J_n(\theta^*)^{-1}) + o(1).$$

Thus the expected Kullback-Leibler information number (2.1), is written as

$$\mathrm{E}\,K_n(g(\cdot),f(\cdot;\hat{\theta})) = \int g(\mathbf{x}_n)\log g(\mathbf{x}_n)d\mathbf{x}_n + \mathrm{E}(-l(\hat{\theta})) + \mathrm{tr}(I_n(\theta^*)J_n(\theta^*)^{-1})) + o(1). \tag{2.4}$$

The first term on the right hand side of (2.4) is independent of any model, and we may omit it. Therefore a practical procedure for selecting a model is to compare values of

$$-l(\hat{\theta}) + \overline{t_n(\theta^*)}, \tag{2.5}$$

for various models **F**, where $\overline{t_n(\theta^*)}$ is an estimate of $t_n(\theta^*) = \mathrm{tr}(I_n(\theta^*)J_n(\theta^*)^{-1})$ which is the sum of the second term on the right hand side of (2.2), the penalty for the increasing model size and the bias correction appeared in (2.3).

There are various ways of estimating $t_n(\theta^*)$, and different criteria may follow. If $g(\cdot)$ is equal to one of densities in **F**, say $f(\cdot;\theta_0)$, then $\theta^*=\theta_0$, $I_n(\theta_0)=J_n(\theta_0)$ and $t_n(\theta^*)=p$. Therefore, for the case when $g(\cdot)$ is expected equal to or very close to one of the densities in **F**, the criterion known as Akaike's Information Criterion [2],

$$\mathrm{AIC} = -2l(\hat{\theta}) + 2p$$

follows from (2.5) since $t_n(\theta^*)=p$. Multiplication by 2 is only a convention.

A more general procedure of estimating $t_n(\theta^*)$, suggested by Takeuchi [36] is the following. An example may illustrate his idea.

Example 2.1

Let us consider a simple location and scale family

$$\mathbf{F} = \left\{\prod_{i=1}^{n} f(y_i;\theta),\ \theta\in\Theta\right\},$$

Here, $\theta'=(\mu,\sigma)$, $\Theta = (-\infty,\infty)\times(0,\infty)$, and $f(y_i;\theta)=\frac{1}{\sigma}\phi\left[\frac{y_i-\mu}{\sigma}\right]$ with the standard normal density ϕ. In other words, this is exactly the same as the observational equation $y_i=\mu+\varepsilon_i$ with normal error $\varepsilon_i \sim N(\mu,\sigma^2)$. Note that the assumptions above are only for specifying a model but not for restricting the observation generating mechanism $g(\cdot)$. We only assume that the y_i's are independent observations with the same first and second moments. Since $\mu^* = \sum Ey_i/n$ and $\sigma^{*2} = \sum E(y_i-\mu^*)^2/n$, we have

$$\frac{1}{n} I_n(\theta^*) = \begin{bmatrix} 1/\sigma^{*2} & \mu(3)/\sigma^{*5} \\ \mu(3)/\sigma^{*5} & \mu(4)/\sigma^{*6}-1/\sigma^{*2} \end{bmatrix}$$

and

$$\frac{1}{n} J_n(\theta^*) = \begin{bmatrix} 1/\sigma^{*2} & 0 \\ 0 & 2/\sigma^{*2} \end{bmatrix},$$

where $\mu(l)=\sum E(y_i-\mu^*)^l/n$ for $l>1$. Then,

$$t_n(\theta^*) = 1 + \frac{1}{2}(\mu(4)/\sigma^{*4} - 1).$$

By replacing $\mu(4)$ by the 4th sample central moment $\hat{\mu}(4)=\sum(y_i-\bar{y})^4/n$ and σ^{*2} by the maximum likelihood estimate $\hat{\sigma}^2=\sum(y_i-\bar{y})^2/n$ respectively, we have an estimate of $t_n(\theta^*)$,

$$\overline{t_n(\theta^*)} = 1 + \frac{1}{2}(\hat{\mu}(4)/\hat{\sigma}^4 - 1).$$

A statistic which follows from (2.5) is then

$$\begin{aligned} \mathrm{TIC}_0(\mathbf{F}) &= -2l(\hat{\theta}) + 2 + (\hat{\mu}(4)/\hat{\sigma}^4 - 1) \\ &= n + n\log(2\pi\hat{\sigma}^2) + 2 + (\hat{\mu}(4)/\hat{\sigma}^4 - 1). \end{aligned}$$

Multiplication by 2 is again a convention as is in AIC. The difference between TIC_0 and

$$\mathrm{AIC} = -2l(\hat{\theta}) + 4$$

is clear. The discrepancy of the shape of $g(\cdot)$ from the normal density is counted in TIC_0.

By applying the same technique we can derive TIC_0 for the problem of selecting a sample transformation ψ. Consider models;

$$\mathbf{F}_\psi = \left\{ \prod_{i=1}^{n} \left| \frac{\psi'(y_i)}{\sigma} \right| \phi\left(\frac{\psi(y_i)-\mu}{\sigma} \right) \right\},$$

where ψ' is the derivative of ψ. Then

$$\mathrm{TIC}_0(\mathbf{F}_\psi) = -2l(\hat{\theta}) + 2 + (\hat{\mu}(4)/\hat{\sigma}^4 - 1)$$

follows, where $\hat{\mu}(4)=\sum(\psi(y_i)-\hat{\mu})^4/n$ and $\hat{\sigma}^2=\sum(\psi(y_i)-\hat{\mu})^2/n$ with $\hat{\mu}=\sum\psi(y_i)/n$. Comparing

$\mathrm{TIC}_0(\mathbf{F}_\psi)$, we may select a transformation ψ.

However, such a procedure of deriving an estimate of $t_n(\theta^*)$ is not widely applicable. It is laborious to find an estimate of $t_n(\theta^*)$ model by model. And there is no definite answer, what kind of assumption is appropriate for the y_i's. Before proceeding to a generalization of TIC_0, let us consider another example.

Example 2.2

A Gaussian regression model $y_i=x_i'\beta+\varepsilon_i$ with m-dimensional regression parameter β is denoted by

$$\mathbf{F}_m = \left\{ \prod_{i=1}^{n} \frac{1}{\sigma}\phi\left[\frac{y_i-x_i'\beta}{\sigma}\right],\ \theta=(\beta',\sigma)'\in\mathbf{R}^m\times(0,\infty) \right\}.$$

We first only assume independence of y_i's. Then

$$I_n(\theta^*) = \begin{bmatrix} \sum(\mu_i(2)-\mu_i(1)^2)x_ix_i'/\sigma^{*2} & \sum(\mu_i(3)-\mu_i(1)\mu_i(2))x_i'/\sigma^{*3} \\ \sum(\mu_i(3)-\mu_i(1)\mu_i(2))x_i/\sigma^{*3} & \sum(\mu_i(4)-\mu_i(2)^2)/\sigma^{*6} \end{bmatrix}$$

and

$$J_n(\theta^*) = \begin{bmatrix} X'X/\sigma^{*2} & 0 \\ 0 & 2n/\sigma^{*2} \end{bmatrix}.$$

Here $\mu_i(l)=\mathrm{E}(e_i)^l$ for $l>1$ with $e_i=y_i-x_i'\beta^*$, $i=1,\ldots,n$ and $X=(x_1,\ldots,x_n)'$ is the design matrix. Denoting the hat matrix by $H=(h_{ij})=X(X'X)^{-1}X'$ we have

$$t_n(\theta^*) = \sum(\mu_i(2)-\mu_i(1)^2)h_{ii}/\sigma^{*2} + \frac{1}{2}\{\frac{1}{n}\sum(\mu_i(4)-\mu_i(2)^2)/\sigma^{*4}\}.$$

If we assume that the first and the second moments of e_i's are the same, then

$$t_n(\theta^*) = m + \frac{1}{2}(\frac{1}{n}\sum \mu_i(4)/\sigma^{*4}-1) \tag{2.6}$$

and so we have

$$\mathrm{TIC}_0(\mathbf{F}_m) = -2l(\hat{\theta}) + 2m + \frac{1}{2}(\frac{1}{n}\sum \hat{e}_i^4/\hat{\sigma}^4-1),$$

where $\hat{e}_i=y_i-x_i\hat{\beta}$, and $\hat{\beta}$ and $\hat{\sigma}$ are the maximum likelihood estimates.
One possible way to avoid such assumptions on $g(\cdot)$ that the second moments are all the same, is to make use of the following inequality.

$$t_n(\theta^*) \le \sum\mu_i(2)h_{ii}/\sigma^{*2} + \frac{1}{2}(\frac{1}{n}\sum\mu_i(4)/\sigma^{*4}-1). \tag{2.7}$$

Here the equality holds true if and only if $\mu_i(1)=\mathrm{E}(e_i)=0$ and $\mu_i(2)=\mathrm{E}(e_i^2)=\sigma^{*2}$ for all i, and the value becomes the same as in (2.6). The right hand side of (2.7) can be estimated by

$$\hat{t}_n = \sum \hat{e}_i^2 h_{ii}/\hat{\sigma}^2 + \frac{1}{2}(\frac{1}{n}\sum \hat{e}_i^4/\hat{\sigma}^4 - 1).$$

This estimate is possibly biased. However it is toward a safer direction. More penalty is put on models which are far from the best fitting. The resulting criterion is

$$\mathrm{TIC}(\mathbf{F}_m) = -2l(\hat{\theta}) + 2\hat{t}_n .$$

This example leads to a general definition of TIC. We only assume that $\mathbf{Y}_n$ is a vector of independent observations. Since we are modeling independent observations, it is natural to assume that the joint likelihood can be decomposed into

$$l(\theta) = \sum l_i(\theta),$$

where $l_i(\theta)$=log $f_i(y_i;\theta)$. Estimate $I_n(\theta^*)$ by

$$\hat{I} = \sum_i \frac{\partial}{\partial\theta} l_i(\hat{\theta}) \frac{\partial}{\partial\theta'} l_i(\hat{\theta})$$

and $J_n(\theta^*)$ by

$$\hat{J} = -H_n(\hat{\theta}) = -\sum_i \frac{\partial^2}{\partial\theta\partial\theta'} l_i(\hat{\theta}).$$

Then we have a general definition of TIC;

$$\mathrm{TIC} = -2l(\hat{\theta}) + 2\mathrm{tr}(\hat{I}\hat{J}^{-1}).$$

As noted in the previous example, since

$$\sum_i \mathrm{E}\left\{\frac{\partial}{\partial\theta} l_i(\theta^*) \frac{\partial}{\partial\theta'} l_i(\theta^*)\right\} = I_n(\theta^*) + \sum_i \mathrm{E}\left\{\frac{\partial}{\partial\theta} l_i(\theta^*)\right\} \mathrm{E}\left\{\frac{\partial}{\partial\theta'} l_i(\theta^*)\right\}, \tag{2.8}$$

$\mathrm{tr}(\hat{I}\hat{J}^{-1})$ tends to over-estimate $t_n(\theta^*)$ by the last term on the right hand side of (2.8). We can not expect any stable behavior of the maximum likelihood estimate $\hat{\theta}$, as long as such an over estimation is significant. The observations contribute unequally to the Gradient of the log likelihood function at θ^*, which is the solution of

$$\sum_i \mathrm{E} \frac{\partial}{\partial\theta} l_i(\theta^*) = 0.$$

Thus such a bias does not affect our objectives to select a model which well balances the approximation error and the error due to random fluctuations. It is worth noting that $\mathrm{tr}(\hat{I}\hat{J}^{-1})$ is the well known Lagrange-multiplier test statistics [15]. TIC consists of two parts, -2 log (*maximum likelihood*) plus twice of the test statistic.

3. EQUIVALENCE BETWEEN CROSS VALIDATION AND INFORMATION CRITERIA

Cross validation is one of naive methods of checking goodness of fit of a model. The observations obtained are divided into two parts. One of them is used for estimation and the other is used for goodness test of fit. Detailed analyses and discussions can be found in Stone[32].

In this section we will show that the cross validation is asymptotically equivalent to TIC. We restrict our attention into a *simple* cross validation. By $\hat{\theta}(-i)$, we denote the maximum likelihood estimate of θ based on $\mathbf{Y}_n$ without using the ith observation y_i. The cross validation is then defined as

$$\mathrm{CV} = -2 \sum_i l_i(\hat{\theta}(-i)).$$

It is shown by Stone[33] that CV is asymptotically equivalent to AIC, when $y_1, \ldots, y_n$ are independent and identically distributed and $g(\cdot)$ is a member of the underlying model **F**. It does not hold true otherwise. However we can show an equivalence of CV to TIC. Necessary assumptions are the following A5 to A7 besides A1 to A3.

A5. For any $\varepsilon > 0$,

$$\max_i \sup_{\|\theta-\theta^*\|>\varepsilon} (l_{-i}(\theta) - l_{-i}(\theta^*))$$

diverges to $-\infty$ a.s. as n tends to infinity, where $l_{-i}(\theta) = l(\theta) - l_i(\theta)$. This implies that $\hat{\theta}(-i)$'s, the solutions of

$$\frac{\partial}{\partial\theta} l_{-i}(\hat{\theta}(-i)) = 0, \quad i=1,\ldots,n,$$

uniformly converge to θ^* as n tends to infinity.

A6. For any $\varepsilon > 0$, there exists $\delta > 0$ such that

$$\sup_{\|\theta-\theta^*\|<\delta} \| I - H_n(\theta) H_n(\theta^*)^{-1} \| < \varepsilon$$

for large enough n, where $\|\cdot\|$ is the Euclidean norm of a vector or the operator norm of a matrix.

A7. For any $\varepsilon > 0$, there exists $\delta > 0$ such that

$$\max_i \sup_{\|\theta-\theta^*\|<\delta} \left\| \left\{ \frac{\partial^2}{\partial\theta\partial\theta'} l_i(\theta) \right\} H_n(\theta^*)^{-1} \right\| < \varepsilon$$

for large enough n.

From the definition of $\hat{\theta}(-i)$ we have

$$\begin{aligned}
\frac{\partial}{\partial\theta} l_i(\hat{\theta}(-i)) &= \frac{\partial}{\partial\theta} l(\hat{\theta}(-i)) \\
&= \frac{\partial}{\partial\theta} l(\hat{\theta}) + \left\{ \frac{\partial^2}{\partial\theta\partial\theta'} l(\hat{\theta}) \right\} (\hat{\theta}(-i) - \hat{\theta}) \; (1 + o_p(1)). \\
&= -\hat{J} \cdot (\hat{\theta}(-i) - \hat{\theta})(1 + o_p(1))
\end{aligned}$$

and

$$l_i(\hat\theta(-i)) = l_i(\hat\theta) + (\hat\theta(-i) - \hat\theta)'\frac{\partial}{\partial\theta} l_i(\hat\theta)$$

$$+ (\hat\theta(-i) - \hat\theta)'\frac{\partial^2}{\partial\theta\partial\theta'} l_i(\theta^{**})(\hat\theta(-i) - \hat\theta)$$

$$= l_i(\hat\theta) - \left\{\frac{\partial}{\partial\theta'} l_i(\hat\theta(-i))\, \hat J^{-1}\, \frac{\partial}{\partial\theta} l_i(\hat\theta)\right\}(1 + o_p(1))$$

$$= l_i(\hat\theta) - \left\{\frac{\partial}{\partial\theta'} l_i(\hat\theta)\, \hat J^{-1}\, \frac{\partial}{\partial\theta} l_i(\hat\theta)\right\}(1 + o_p(1)).$$

Therefore

$$\mathrm{CV} = -2 \sum_i l_i(\hat\theta(-i))$$

$$= -2l(\hat\theta) + 2 \sum_i \left\{\frac{\partial}{\partial\theta'} l_i(\hat\theta)\, \hat J^{-1}\, \frac{\partial}{\partial\theta} l_i(\hat\theta)\right\}(1 + o_p(1))$$

$$= -2l(\hat\theta) + 2\,\mathrm{tr}(\hat I \hat J^{-1})\,(1 + o_p(1))$$

is equivalent to TIC.

Example 3.1

Consider the same regression model as in Example 2.2. To simplify our discussion, we regard σ as a nuisance parameter and estimate it by $\hat\sigma$. From the well known equality [32], we have

$$\mathrm{CV} = n\ \log(2\pi\hat\sigma^2) + \sum_i (y_i - x_i'\hat\beta(-i))^2/\hat\sigma^2$$

$$= n\ \log(2\pi\hat\sigma^2) + \sum_i \{\, \hat e_i/(1-h_{ii})\,\}^2/\hat\sigma^2.$$

To assure the consistency of $\hat\theta$, we may assume that $\max_i(h_{ii})$ converges to 0 as n tends to infinity, which is equivalent to the assumption A6. We then have

$$\mathrm{CV} = n\ \log(2\pi\hat\sigma^2) + \sum_i \hat e_i^2(1 + 2h_{ii})/\hat\sigma^2 + o_p(1)$$

$$= n\ \log(2\pi\hat\sigma^2) + n + \sum_i \hat e_i^2 h_{ii}/\hat\sigma^2 + o_p(1),$$

which is asymptotically equivalent to TIC when σ is regarded as a nuisance parameter. The term $(\frac{1}{n}\sum \hat e_i^4/\hat\sigma^4 - 1)$ will appear in CV, if $\hat\sigma(-i)$ is used in place of $\hat\sigma$.

GCV proposed by G. Wahba [40, 18] is a variant of cross validation. It is known that GCV is asymptotically equivalent to AIC at least in the context of regression. Actually

$$\begin{aligned} \mathrm{GCV} &= \sum_i \hat{e}_i^2/(1 - m/n)^2 \\ &= \sum_i \hat{e}_i^2(1 + 2m/n) + O_p(1/n) \\ &= n\ \{\hat{\sigma}^2(1 + 2m/n) + O_p(1/n^2)\}, \end{aligned}$$

and

$$\begin{aligned} \mathrm{AIC} &= n + n\ \log 2\pi + 2 + n\ \log\{\exp(2m/n)\hat{\sigma}^2\} \\ &= n + n\ \log 2\pi + 2 + n\ \log\{((1+2m/n)\hat{\sigma}^2) + O_p(1/n^2)\}. \end{aligned}$$

Therefore GCV may behave differently from TIC.

Although the equivalence shown above is only for the case of large enough n, this allows us more freedom to choose one of two equivalent criteria, CV or TIC. An advantage of the use of TIC is that the calculation is simpler than that of CV. A simple reduction is possible for CV in the case of regression, but it is generally not true. We have to search for n maximums $l_i(\hat{\theta}(-i))$ $i=1,\ldots,n$ to obtain CV. On the other hand, only one time maximization of the likelihood is necessary to obtain TIC. Another advantage of TIC is that meaning of the value is clear as an estimate of the Kullback-Leibler information number. Note that CV and TIC cover wider area than the AIC does.

4. FURTHER EXTENSION OF INFORMATION CRITERIA

Estimation of parameters in previous sections is always based on the maximum likelihood principle. In statistical literature, it is common to use such an estimate because of the proof of its efficiency or asymptotic efficiency. However the optimality is only valid for the class of unbiased estimators of $\hat{\theta}$. Since we are measuring the closeness of estimated density $f(\cdot;\hat{\theta})$ to $g(\cdot)$ by the Kullback-Leibler information number, there is no definite reason why we restrict our attention into such unbiased estimators. In this section, we trace the derivation of AIC or TIC for the case when a more general estimate, the maximum penalized likelihood estimate, is used for estimating θ.

The penalized likelihood is defined as

$$L_\lambda(\mathbf{Y}_n;\theta) = \log f(\mathbf{Y}_n;\theta) + \lambda k(\theta),$$

or

$$L_\lambda(\mathbf{Y}_n;\theta) = \sum_i L_\lambda(y_i;\theta) = \sum_i \{\log f(y_i;\theta) + \lambda k_i(\theta)\}$$

where $k(\theta)\le 0$ is an arbitrary penalty function which may depend on n and is twice differentiable. The weight $\lambda\ge 0$ controls the amount of penalty.
The maximum penalized likelihood estimate $\hat{\theta}(\lambda)$ is the solution of

$$\frac{\partial}{\partial\theta}L_\lambda(\mathbf{Y}_n;\theta)=0.$$

We assume that $\hat\theta(\lambda)$ converges to $\theta^*(\lambda)$ which is the unique solution of

$$\mathrm{E}\,\frac{\partial}{\partial\theta}L_\lambda(\mathbf{Y}_n;\theta)=0.$$

By similar expansions as in Section 2, we can show

$$\mathrm{E}\int L_\lambda(\mathbf{x}_n;\hat\theta(\lambda))g(\mathbf{x}_n)d\mathbf{x}_n=\mathrm{E}\,L_\lambda(\mathbf{Y}_n;\hat\theta(\lambda))-\mathrm{E}\,(\hat\theta(\lambda)-\theta^*(\lambda))'J_n(\lambda)(\hat\theta(\lambda)-\theta^*(\lambda))+o(1),\quad (4.1)$$

where

$$J_n(\lambda)=-\mathrm{E}\,\frac{\partial^2}{\partial\theta\partial\theta'}L_\lambda(\mathbf{Y}_n;\theta^*(\lambda)).$$

Subtracting $\lambda k(\hat\theta(\lambda))$ from the both sides of (4.1), we have

$$\mathrm{E}\int g(\mathbf{x}_n)\log f(\mathbf{x}_n;\hat\theta(\lambda))d\mathbf{x}_n=\mathrm{E}\{\,l(\hat\theta(\lambda))-(\hat\theta(\lambda)-\theta^*(\lambda))'J_n(\lambda)(\hat\theta(\lambda)-\theta^*(\lambda))\,\}+o(1).$$

Since the expansion

$$0=\frac{\partial}{\partial\theta}L_\lambda(\mathbf{Y}_n;\theta^*(\lambda))+\frac{\partial^2}{\partial\theta\partial\theta'}L_\lambda(\mathbf{Y}_n;\theta^*(\lambda))(\hat\theta(\lambda)-\theta^*(\lambda))$$

asymptotically holds true, we can rewrite the expectation of the Kullback-Leibler information number as

$$\mathrm{E}\int\log\frac{g(\mathbf{x}_n)}{f(\mathbf{x}_n;\hat\theta(\lambda))}\,g(\mathbf{x}_n)d\mathbf{x}_n=\int g(\mathbf{x}_n)\log g(\mathbf{x}_n)d\mathbf{x}_n-\mathrm{E}\,l(\hat\theta(\lambda))+\mathrm{tr}(I_n(\lambda)J_n(\lambda)^{-1})+o(1),$$

where

$$I_n(\lambda)=\mathrm{E}\left\{\frac{\partial}{\partial\theta}L_\lambda(\mathbf{Y}_n;\theta^*(\lambda))\,\frac{\partial}{\partial\theta'}L_\lambda(\mathbf{Y}_n;\theta^*(\lambda))\right\}.$$

Then the TIC is extended as a regularization information criterion,

$$\mathrm{RIC}=-2l(\hat\theta(\lambda))+2\mathrm{tr}(\hat I(\lambda)\hat J(\lambda)^{-1}),$$

where

$$\hat I(\lambda)=\sum_i\{\frac{\partial}{\partial\theta}L_\lambda(y_i;\hat\theta(\lambda))\,\frac{\partial}{\partial\theta'}L_\lambda(y_i;\hat\theta(\lambda))\}$$

and

$$\hat J(\lambda)=-\frac{\partial^2}{\partial\theta\partial\theta'}L_\lambda(\mathbf{Y}_n;\hat\theta(\lambda)).$$

When $\lambda=0$, RIC is reduced to TIC. Then RIC is in fact an extension of TIC. By RIC we can choose λ as well as to select a model. One practical procedure is to choose λ for each model so as to minimize RIC and compare the minimized value of RIC for each model. We may overcome instability of the estimate when the model happen to be overfitted.

Example 4.1

Consider the same regression model as in Example 2.2. To simplify the problem, we regard σ as a nuisance parameter. As a penalty function we adopt

$$k(\theta) = -\ \|X\beta\|^2/2\sigma^2.$$

The maximum penalized likelihood estimate of β is then a shrinkage estimate, $\hat\beta(\lambda)=\hat\beta(0)/(1+\lambda)$, where $\hat\beta(0)$ denotes the maximum likelihood estimate of β. Since

$$\hat I(\lambda) = \sum \hat e_i^2\, x_i x_i'/\sigma^4$$

and

$$\hat J(\lambda) = (1+\lambda)\, X'X/\sigma^2,$$

we have

$$\mathrm{RIC}(\mathbf{F}_m,\lambda) = n\ \log 2\pi\sigma^2 + \sum (y_i - x_i'\hat\beta(\lambda))^2/\sigma^2 + \frac{2}{1+\lambda}\sum \hat e_i^2 h_{ii}/\sigma^2$$

$$= n\ \log 2\pi\sigma^2 + \{\ \sum \hat e_i^2 + (\frac{\lambda}{1+\lambda})^2 \sum \hat y_i^2 + \frac{2}{1+\lambda}\ \sum \hat e_i^2 h_{ii}\ \}/\sigma^2,$$

where $\hat y_i = y_i - \hat e_i$. Here

$$\frac{\partial}{\partial\theta}\ \mathrm{RIC}(\mathbf{F}_m,\lambda) = \frac{1}{(1+\lambda)^3\sigma^2}\{\ \lambda(\sum \hat y_i^2 - \sum \hat e_i^2 h_{ii}) - \sum \hat e_i^2 h_{ii}\ \}. \qquad (5.2)$$

The $\hat\lambda$ which minimizes RIC is then

$$\hat\lambda = \frac{\sum \hat e_i^2 h_{ii}}{\sum \hat y_i^2 - \sum \hat e_i^2 h_{ii}} \qquad \text{if} \quad \sum \hat y_i^2 > \sum \hat e_i^2 h_{ii},$$

$$= \infty \qquad\qquad \text{otherwise.}$$

The resulting estimate of β is

$$\hat\beta(\hat\lambda) = (1 - \sum \hat e_i^2 h_{ii} / \sum \hat y_i^2)^+\ \hat\beta(0),$$

where $(\alpha)^+ = \max(\alpha,0)$. It is interesting to note that a non-negative shrinkage factor automatically follows from minimizing RIC. As a special case, for the model with a single location parameter μ as in Example 2.1,

$$\hat\mu(\hat\lambda) = (1 - \hat\sigma^2/n\bar y^2)^+\ \bar y,$$

which is a natural shrinkage estimate.
The minimum value of the RIC for each model is

$$\mathrm{RIC}(\mathbf{F}_m,\hat\lambda) = n\ \log 2\pi\sigma^2 + \frac{1}{\sigma^2}\left\{ \sum \hat e_i^2 + 2\left[1 - \frac{1}{2}\ \frac{\sum \hat e_i^2 h_{ii}}{\sum \hat y_i^2}\right](\sum \hat e_i^2 h_{ii}) \right\} \quad \text{if } \hat\lambda < \infty,$$

$$= n \log 2\pi\sigma^2 + \sum y_i^2/\sigma^2 \quad \text{otherwise.}$$

Thus

$$\mathrm{RIC}(\mathbf{F}_m,\infty) \le \mathrm{RIC}(\mathbf{F}_m,\hat{\lambda}) \le \mathrm{RIC}(\mathbf{F}_m,0).$$

Therefore RIC($\mathbf{F}_m$,λ) decreases as λ increases from 0 and attains the minimum at $\hat{\lambda}$. Particularly when $\hat{\lambda}=\infty$, the complete shrinkage estimate $\hat{\beta}(\infty)=0$ follows. By using such an estimate we can always decrease the value of RIC except for the case when all $\hat{e}_i$'s are 0. We then compare such minimized value for different models $\mathbf{F}_m$, and choose one of them.

More generally if the penalty function is of the form of $k(\theta) = -\|A\beta\|^2/2\sigma^2$, then

$$\mathrm{RIC}(\mathbf{F}_m,\lambda) = n \log 2\pi\sigma^2 + \{ \|y - X\hat{\beta}(\lambda)\|^2 + 2\sum h_{ii}(\lambda)\hat{e}_i^2 \}/\sigma^2,$$

where

$$H(\lambda) = (h_{ij}(\lambda)) = X(X'X + \lambda A'A)^{-1}X'$$

and

$$\hat{\beta}(\lambda) = (X'X + \lambda A'A)^{-1}X'y.$$

As a result, in this regression context, RIC is closely related to a criterion $\hat{T}(h)$ which is mentioned in Titterington [38]. A more closely related criterion is C_L proposed by Mallows [19]. As far as in the context of regression, RIC is almost equivalent to C_L and AIC is equivalent to the C_p proposed by the same author.

The effect of introducing maximum penalized estimator and selecting both model and the λ can be seen in Fig.1. Hundred random samples are generated from

$$y = 1 + 0.8x - 1.8x^2 + x^3 + \varepsilon$$

for $0 \le x \le 1$. Here ε is a random number normally distributed with mean 0 and standard deviation 0.04. The selected order of the polynomial is 5 by TIC or AIC. By RIC, the order 4 and λ=0.003 are selected. The order 4 is still overfitting but it is compensated by choosing λ as 0.003.

It is also possible to extend RIC for the case of more than one penalty function. Still much works have to be done for this criterion. We leave these for future investigations.

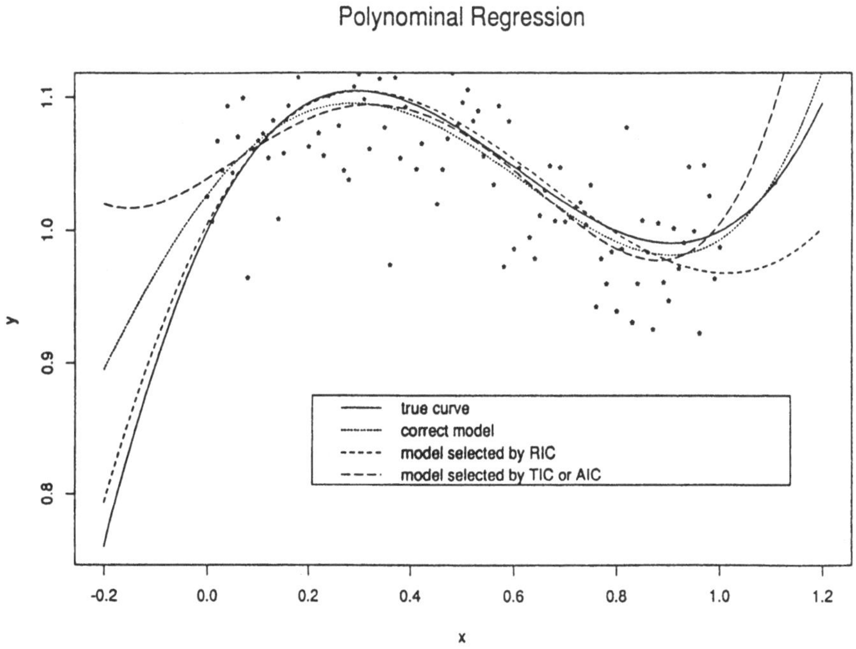

Fig.1 Comparison of three information criteria

5. COMPARISON OF CRITERIA

5.1. Consistency

A lot of papers are devoted to the consistency of various model selection procedures. Particular interest is in the *inconsistency* of the minimum AIC procedure. However, we may raise a question whether it is always meaningful to only discuss the consistency of model selection. In other words, is the correctness of the selected model is always required first? Recall that a model is only an approximation to the reality. It is a tradition of statistics to discuss correctness of the estimated parameter by assuming that the data are generated from a fixed model. Model selection is however somewhat different from typical estimation problem. We are dealing with different models and looking for a model which gives us a good approximation. Therefore, we should note that the following discussion of consistency is meaningful only when the true system is known to be quite simple and one of the underlying model can describe the system without error. Furthermore, as is seen later, a tricky point is that consistency of the selection is not compatible with goodness of the resulting estimate of parameters.

To discuss the consistency the following generalization of AIC [5, 4, 9 pp.366-367] is convenient.

$$\mathrm{AIC}_\alpha = -2l(\hat{\theta}) + \alpha\, p,$$

where α is a pre-determined value which controls the amount of penalty for an increasing size of the model and may depend on the size n of observations. The result by Hannan and Quinn [10] suggests that under suitable regularity conditions a necessary and sufficient condition for the strong consistency is , putting $\alpha = \alpha_n$,

$$\liminf_n \alpha_n/(2\log\log n) > 1$$

and

$$\limsup_n \alpha_n/n = 0.$$

That for the weak consistency is

$$\liminf_n \alpha_n = \infty$$

and

$$\limsup_n \alpha_n/n = 0.$$

The result above is not yet generally proved, but intuitively clear if we note that $2\{l(\hat{\theta})-l(\theta_0)\}$ is χ^2 distributed with degree of freedom p if $g(\cdot)$ is equal to $f(\cdot;\theta_0)$, a density in the underlying model, otherwise χ^2 distributed with degree of freedom of the order of n. The condition for strong consistency comes from the law of iterated logarithm. An implication of the result above is that the AIC, TIC or CV introduced in the previous section are not consistent. For the asymptotic distribution, see Shibata [25], Bhansali & Downham[5], and Woodroofe[42]. They obtained the asymptotic distribution of the selected model by applying theorems of random walk. Some consistent criterion procedures have been proposed, BIC by Schwarz[24] and HQ by Hannan and Quinn[10], which are AIC_α with $\alpha = \log n$ and $\alpha = c \log\log n$ for $c > 2$, respectively. It is interesting to note the result by Takada[35], that any procedure so as to minimize AIC_α is admissible under the 0-1 loss. In other word, if our main concern is the correctness of the selection, there is no dominant selection procedure in such class of selection procedures.

5.2. Optimality

If we are interested in goodness of a model selection procedure, a natural way is to check the Kullback-Leibler distance of $f(\cdot;\hat{\theta})$ from $g(\cdot)$ where $\hat{\theta}$ is an estimate of the parameter θ under the selected model. Although not exactly the same, an optimality of the AIC has been shown in terms of such a distance. The key point for proving an optimality property of AIC is that the trade off between the bias and the variance remains significant even when n is large enough. If we restrict our attention to the estimation of regression parameters, such trade off mechanism is rigorously formulated. The result by Shibata [26] shows that if the regression variables are selected so as to minimize one of AIC_α then the selection is asymptotically optimal if and only if $\alpha=2$, that is the case of AIC. Necessary assumptions for the proof are that the shape of $g(\cdot)$ is the same as that of **F**, and the mean vector of observations is parametrized by infinitely many regression parameters. Otherwise, AIC is not necessarily optimal. But TIC is instead expected to be optimal under a loss function like the Kullback-Leibler information number as well as under the squared loss, even when the shape of $g(\cdot)$ does not coincide with that in **F**. This result is not yet completely proved.

For admissibility under the squared loss with an additional penalty p, Stone [34] proved local asymptotic admissibility, and Kempthorne [17] proved the admissibility under the squared loss. Such results are corresponding to the result by Takada [35] in the case of 0-1 loss function.

All of the results above are in the sense of asymptotics. If the size n is fixed, theoretical comparison is difficult and the only available results are by simulations. Recent paper by Hurvich [16] will help the understanding of the behavior in small samples, for example, consistency does not necessarily imply the goodness of selection. One practical procedure might be obtained by choosing α according to the size n (see [31]). For more detailed discussion on incompatibility between consistency and efficiency, see [30], and for comparisons with testing procedures see [29].

6. SELECTION OF TIME SERIES MODELS

There have been a lot of articles on the problem of selecting a time series model. In this section, we will review some more criteria of selection of time series models and related works, in connection with the general problem of model selection. The reader can consult some other review papers on this topic [28, 8].

6.1. Autoregressive models

Autoregressive process with order p, AR(p), is a weakly stationary process, which satisfies the equation,

$$A_p(B)z_t = \varepsilon_t, \quad t=\cdots,-1,0,1,\cdots$$

where $A_p(z)=1+a_1z+a_2z^2+\cdots+a_pz^p$ is the associate polynomial, B is the backward shift operator and $\{\varepsilon_t\}$ is a sequence of innovations with mean 0 and variance σ^2. To completely specify a model, the shape of the distribution of ε_t have to be specified. It is typically assumed Gaussian, but not restricted to it. A different shape of the density yields different kind of estimates.

By denoting the joint density of consecutive observations $\mathbf{z}_n=(z_1,z_2,\ldots,z_n)'$ by $f(\mathbf{z}_n;\theta)$, we can explicitly specify a model by

$$\mathbf{F}_p = \{ f(\mathbf{z}_n;\theta);\ \theta=(\sigma,a_1,\ldots,a_p,0,\ldots,0)'\in(0,\infty)\times\mathbf{R}^P \}.$$

We then have a nested family of AR models $\{\mathbf{F}_p;\ 0\le p\le P\}$ for given P. Denote the maximum likelihood estimate of θ under each model $\mathbf{F}_p$ by

$$\hat{\theta}(p) = (\hat{\sigma}(p),\hat{a}_1(p),\ldots,\hat{a}_p(p),0,\ldots,0)'.$$

To obtain an estimate, the exact maximum likelihood procedure is desirable. The approximation error may affect the behavior. Hereafter, we assume that the shape of densities in the underlying model is normal. For AR models, we can replace it by the conditional maximum likelihood estimate, given $z_1,\ldots,z_P$, or the estimate which minimizes

$$f(z_{P+1},\ldots,z_n;\theta \mid z_1,\ldots,z_P).$$

Then $\hat{a}(p)=(\hat{a}_1(p),\ldots,\hat{a}_p(p),0,\ldots,0)'$ is the solution of the Yule-Walker equation,

$$R\,\hat{a}(p) = -r$$

and $\hat{\sigma}^2(p) = \frac{1}{N}\sum_{P+1}^{n}\hat{\varepsilon}_t(p)^2$ with $N=n-P$, where

$$R = \left[\frac{1}{N}\sum_{P+1}^{n} z_{t-l}z_{t-m};\ 1\le l,m\le P \right],$$

is the sample autocovariance matrix,

$$r = \left[\frac{1}{N}\sum_{P+1}^{n} z_t z_{t-m};\ 1\le m\le P \right]$$

is the vector of sample autocovariances, and

$$\hat{\varepsilon}_t(p) = z_t+\hat{a}_1(p)z_{t-1}+\cdots+\hat{a}_p(p)z_{t-p} \quad t=P+1,\ldots,n,$$

are residuals.

Similarly as in the case of multiple regression, AIC_α for the model $\mathbf{F}_p$ is

$$\mathrm{AIC}_\alpha = N + N\log 2\pi\hat{\sigma}^2(p) + \alpha(p+1).$$

Note that the matrix R and the vector r are defined the same for any order $1\le p\le P$ since the normalization is with $N=n-P$ but not with $n-p$. This is a crucial point when a quasi-maximum likelihood estimate is used in place of the exact one. For example, if $\hat{\sigma}^2(p) = \frac{1}{n-p} \sum_{p+1}^{n} \hat{\varepsilon}_t(p)^2$ is used, then AIC$_\alpha$ behaves differently.

On the other hand, TIC becomes

$$\mathrm{TIC} = N + N \log 2\pi\hat{\sigma}^2(p) + \left\{\frac{1}{N} \sum \hat{\varepsilon}_t(p)^4/\hat{\sigma}^4(p) - 1\right\}$$

$$+ 2 \sum_t \left\{ \hat{\varepsilon}_t(p)^2 \frac{1}{N} \sum_{l,m} (z_{t-l}\, R^{lm}\, z_{t-m}) \right\} / \hat{\sigma}^2(p),$$

where $\{ R^{lm};\ 1 \le l, m \le p \}$ is the inverse of the p by p principal submatrix of R. Although little is known about TIC, it is clear that TIC is close to AIC if the true $g(\cdot)$ is close to one of densities in $\mathbf{F}_p$, since

$$\frac{1}{N} \sum_{l,m} (z_{t-l}\, R^{lm}\, z_{t-m})$$

is corresponding to the diagonal element h_{ii} of the hat matrix in the case of multiple regression and has the expectation p/N.

To evaluate the behavior of the selection, we need some assumptions on the true density $g(\mathbf{z}_n)$. As was mentioned before, it is meaningless to consider consistency of the selection unless $g(\mathbf{z}_n)$ is expected to be equal to one of densities in $\mathbf{F}_p$, that is, the true model AR(p_0) exists with an order $p_0 \le P$. Under this assumption, the asymptotic distribution of the selected order $\hat{p}$ which minimizes AIC has been obtained by Shibata [25]. The distribution is nondegenerate and concentrated on $p \ge p_0$, so that the minimum AIC procedure is inconsistent and tends to select a higher order than p_0. This also holds true for AIC$_\alpha$ (Bhansali and Downham [5] with any fixed α. This has been extended to multiple AR models by Sakai[23], to ARMA models by Hannan [10, 11, 14], to ARIMA models by Yajima [43], and to AR models with a time dependent variance by Tjøstheim and Paulsen [39]. It is known that the minimum AIC$_\alpha$ procedure is consistent if $\alpha = \alpha_n$ is increased with n at the rate that $\liminf_n (\alpha_n / 2 \log \log n) > 1$ (see [10]).

However, for the case when the true $g(\cdot)$ is not expected to be in $\mathbf{F}_p$ for any $p \le P$, our main concern may be about the goodness of the resulting inference rather than the correctness of the selection. In this case, one natural assumption on $g(\cdot)$ is that z_n comes from an autoregressive process with infinite order. That is, z_n is generated by the process,

$$A_\infty(B) z_t = \varepsilon_t, \tag{6.1}$$

where $\{\varepsilon_t\}$ is a sequence of innovations with variance σ_∞^2, $A_\infty(B) = 1 + a_1 B + a_2 B^2 + \cdots$ is a nondegenerate infinite order transfer function with $\sum |a_l| < \infty$, and $A_\infty(z) \ne 0$ for $|z| \le 1$. Then we can show an optimality property of the minimum AIC procedure $\hat{p}$.

Theorem 6.1

Assume that $\{\varepsilon_t\}$ is a sequence of innovations which are independent and normally distributed and z_n is generated by the process (6.1). If P is taken to be P_n which diverges to infinity with the order of $o(n^{1/2})$, then

$$\lim_{n\to\infty} \mathrm{P}\left[\frac{\|\hat{a}(\tilde{p}) - a\|_c^2}{\min\limits_{p} \mathrm{E}\, \|\hat{a}(p) - a\|_c^2} \geq 1-\varepsilon \right] = 1.$$

for any selection $\tilde{p}$ from $1 \leq p \leq P$. Here, $\|x\|_c^2 = \sum x_l\, c_{lm}\, x_m$ is the norm with the autocovariances $c_{lm} = \mathrm{E}(z_{t+l} z_{t+m})$, and $a' = (a_1, a_2, \cdots)$ is the infinite dimensional vector of the coefficients of the transfer function $A_\infty(B)$. Thus $\|\hat{a}(p) - a\|_c^2 + \sigma_\infty^2$ is the one step ahead prediction error of the estimated predictor $(1 - \hat{A}_p(B))z_{t+1}^*$ of z_{t+1}^*, a realization of a process $\{z_t^*\}$ which is independent of $\{z_t\}$ but has the same covariance structure as that of $\{z_t\}$. The lower bound is attained in probability for large enough n by the selection $\hat{p}$ which minimizes AIC_α with $\alpha=2$. Any other choice of α does not yield any selection which always attains the bound.

Keys for the proof of the theorem are the following facts. The prediction error is decomposed into two parts, the squared bias and the variance;

$$\begin{aligned} \|\hat{a}(p) - a\|_c^2 &= \|a(p) - a\|_c^2 + \|\hat{a}(p) - a(p)\|_c^2, \\ &= \sigma^2(p) - \sigma_\infty^2 + \|\hat{a}(p) - a(p)\|_c^2, \end{aligned} \tag{6.2}$$

where $a(p)$ is the projection of the infinite dimensional vector a on the p-dimensional subspace $\{a = (a_1, a_2, \ldots, a_p, 0, \cdots)\}$ with respect to the norm $\|\cdot\|_c$, and $\sigma^2(p) = \mathrm{E}\,(A_p(B)z_t)^2$ is the residual variance for the transfer function $A_p(B)$ with the coefficients $a(p)$. Note that

$$V = N\, \|\hat{a}(p) - a(p)\|_c^2 / \sigma_\infty^2. \tag{6.3}$$

is asymptotically χ^2 distributed with degree of freedom p. The normalized prediction error,

$$N\, \|\hat{a}(p) - a\|_c^2/\sigma_\infty^2 = N\, (\sigma^2(p)/\sigma_\infty^2 - 1) + V \tag{6.4}$$

is close to

$$N\, (\hat{\sigma}^2(p)/\sigma_\infty^2 - 1) + p.$$

For large p, $\hat{\sigma}^2(p)/\sigma_\infty^2$ is close to 1 and then the estimate above is approximately equal to $N \log \hat{\sigma}^2(p) - N \log \sigma_\infty^2 + p$. This means that AIC is estimating (6.4) as well as estimating the Kullback-Leibler information number for large p. It is enough to consider the case when p is large. The $\hat{p}$ diverges to infinity and the bias term will be dominant for a fixed p. A remaining problem in the proof of the theorem is how well p behaves as an estimate of V in (6.3) for large p. The estimation error is relatively small and negligible, because V/p converges to 1 in probability as p tends to infinity simultaneously with n.

In the theorem, the assumption of normality of $\{\varepsilon_t\}$ which generate z_n is not essential. In fact, this theorem was extended by Taniguchi [37] to the case of ARMA model selection without the normality assumption on $\{\varepsilon_t\}$. In our case, instead of the normality it is enough to assume that

$$\sum_j j^{\beta} |a_j| < \infty \text{ for some } \beta > 1,$$

and $\{\varepsilon_t\}$ is an independent identically distributed sequence with finite moments up to the 16th. The shape of densities in each model $\mathbf{F}_p$ is assumed to be normal.

One other possible extension of the theorem is to the case of subset selection, that is, to select a model from the family of models

$$\mathbf{F}_j = \{f(z_n;\theta);\ \theta = (\sigma, 0, .., a_{j_1}, \ldots, 0, a_{j_p}, 0, \ldots, 0)' \in (0,\infty) \times \mathbf{R}^p\}$$

which is specified by a set of indices $j=(j_1, \ldots, j_p)$. For the case of multiple regression, Shibata [26] has already proved that the theorem still holds true. However as far as I know there is no rigorous proof for autoregressive models.

We can also show an optimal property of the minimum AIC in terms of the integrated relative squared error of autoregressive spectral estimate. A fundamental relation between two autoregressive spectral densities, $g(\lambda) = \sigma^2/|A(e^{2\pi i\lambda})|^2$ and $h(\lambda) = s^2/|B(e^{2\pi i\lambda})|^2$, is

$$\int \left\{ \frac{h(\lambda) - g(\lambda)}{g(\lambda)} \right\}^2 d\lambda - 2\frac{\|a-b\|_h^2}{s^2} = (1 - \frac{s^2}{\sigma^2})^2 + 2(2\frac{s^2}{\sigma^2} - \frac{\sigma^2}{s^2} - 1)\frac{\|a-b\|_h^2}{\sigma^2}$$
$$+ \frac{s^4}{\sigma^4} \int \left[\frac{|A-B|^2}{|B|^2} + 2\frac{\Delta|A|^2}{|B|^2} \right] \frac{|A-B|^2}{|B|^2} d\lambda, \tag{6.5}$$

where $\Delta|A|^2 = B(\overline{A-B}) + \overline{B}(A-B)$, a and b are vectors of coefficients of A and B, and $\|x\|_h^2 = \sum (x_l\ h_{lm}\ x_m)$ is the norm with

$$h_{lm} = \int e^{2\pi i(l-m)\lambda} h(\lambda) d\lambda$$

(see [1,27]). In (6.5) the order of transfer functions A and B can be infinite. Since

$$\frac{|A-B|^2}{|B|^2} \le \frac{|a-b|^2 H}{s^2}, \quad \left| \frac{\Delta|A|^2}{|B|^2} \right| \le \frac{2|a-b|\ H^{1/2}}{s}$$

and

$$\int \frac{|A-B|^2}{|B|^2} d\lambda = \frac{\|a-b\|_h^2}{s^2},$$

where $|x| = \sum |x_l|$ is the absolute norm of the vector x and $H = \max_\lambda h(\lambda)$. The last term on the right hand side of (6.5) is bounded by

$$\|a-b\|_h^2 \, (|a-b|^2 H + 4|a-b| H^{1/2} s)/\sigma^4$$

in absolute value.

Consider the autoregressive spectral estimate

$$\hat{f}_p(\lambda) = \hat{\sigma}^2(p) / |\hat{A}_p(e^{2\pi i \lambda})|^2$$

and the true spectral density

$$f_\infty(\lambda) = \sigma_\infty^2 / |A_\infty(e^{2\pi i \lambda})|^2.$$

Putting $h(\lambda)=\hat{f}_p(\lambda)$ and $g(\lambda)=f_\infty(\lambda)$ in (6.5), we can show optimality of $\hat{p}$ from Theorem 6.1, since $\hat{\sigma}^2(p)$ and $\hat{A}_p(e^{2\pi i \lambda})$ converge to σ_∞^2 and $A_\infty(e^{2\pi i \lambda})$ respectively, as p increases to infinity simultaneously with n.

Theorem 6.2 (shibata[27])

$$\lim_{n \to \infty} P \left[\frac{\int \left[(\hat{f}_{\tilde{p}}(\lambda) - f_\infty(\lambda))/f_\infty(\lambda) \right]^2 d\lambda}{2 \min_p E \, \|\hat{a}(p) - a\|_c^2 / \sigma_\infty^2} \geq 1-\varepsilon \right] = 1.$$

for any selection $\tilde{p}$. The bound is attained by $\hat{p}$ in probability for large enough n.

The criterion autoregressive transfer function, CAT [20] is derived from the principle to select the order p so as to minimize the integrated relative squared error as above,

$$\frac{1}{2} \int \left[(\hat{f}_p(\lambda) - f_\infty(\lambda))/f_\infty(\lambda) \right]^2 d\lambda,$$

which is approximately equal to $\|\hat{a}(p) - a\|_{\hat{f}_p}^2 / \hat{\sigma}^2(p)$ from (6.5). Noting the consistency of $\hat{f}_p(\lambda)$ and the decomposition (6.2), we have an estimate,

$$\mathrm{CAT}_0 = 1 - \hat{\sigma}_\infty^2/\hat{\sigma}^2(p) + (p/N)\, \hat{\sigma}_\infty^2/\hat{\sigma}^2(p),$$

provided that an estimate $\hat{\sigma}_\infty^2$ of σ_∞^2 is available. By replacing $(p/N)\hat{\sigma}_\infty^2/\hat{\sigma}^2(p)$ by p/N, we have the criterion

$$\mathrm{CAT} = 1 - \hat{\sigma}_\infty^2/\hat{\sigma}^2(p) + p/N.$$

It is clear that CAT is equivalent to AIC for large p, so that the theorems 6.1 and 6.2 also hold true for the minimum CAT procedure. In fact,

$$\begin{aligned}
\mathrm{CAT} &= 1 - \hat{\sigma}_\infty^2/\hat{\sigma}^2(p) + p/N \\
&= \log \hat{\sigma}^2(p) - \log \hat{\sigma}_\infty^2 + p/N + O_p((1-\hat{\sigma}_\infty^2/\hat{\sigma}^2(p))^2) \\
&= \log \hat{\sigma}^2(p) + 2p/N + O((p/N)^2) + O_p((1-\hat{\sigma}_\infty^2/\hat{\sigma}^2(p))^2) - \log \hat{\sigma}_\infty^2.
\end{aligned}$$

As is easily seen from the derivation, CAT_0 and CAT are more closely connected with the integrated relative squared error than AIC. As an estimate of $\hat{\sigma}_\infty^2$, Parzen suggested the use of

$$\hat{\sigma}_{\infty}^2 = \exp\left[\frac{2}{m}\sum_{j=1}^{m}\log I\left(\frac{j}{m}\right) + \gamma\right],$$

where m is integral part of $n/2$, γ is Euler's constant and

$$I(\lambda) = \frac{1}{n}\left|\sum_t z_t e^{2\pi i t\lambda}\right|^2$$

is the periodogram. An alternative is to use $\hat{\sigma}^2(P)$ which does not depend on each model and goes to σ_{∞}^2 as P tends to infinity.

Later Parzen [21] proposed a modified CAT,

$$\mathrm{CAT}^* = \frac{1}{N}\sum_{j=1}^{p}\frac{1}{\hat{\sigma}^2(j)} - \frac{1}{\hat{\sigma}^2(p)}.$$

This does not require any estimate of σ_{∞}^2 like $\hat{\sigma}_{\infty}^2$. Note that

$$\hat{\sigma}_{\infty}^2\mathrm{CAT}^* + 1 = 1 - \frac{\hat{\sigma}_{\infty}^2}{\hat{\sigma}^2(p)} + \frac{1}{N}\sum_{j=1}^{p}\frac{\hat{\sigma}_{\infty}^2}{\hat{\sigma}^2(p)}.$$

The behavior of the order which minimizes a criterion is determined only by the differences of values of the criterion. Therefore the behavior of the minimum CAT^* is almost equal to that of the minimum CAT for large p, or for $p \geq p_0$ when the true order p_0 is assumed. Theorems 6.1 and 6.2 will also hold true for CAT^*.

6.2. Autoregressive moving average models

Autoregressive moving average process with order p and q, ARMA(p,q), is a weakly stationary process, which satisfies the equation,

$$A_p(B)z_t = B_q(B)\varepsilon_t, \qquad t=\ldots,-1,0,1,\ldots$$

where $A_p(z)$ and $\{\varepsilon_t\}$ are the same as in AR models, and $B_p(z) = 1 + b_1 z + b_2 z^2 + \cdots + b_q z^q$ is the associated polynomial for the moving average part. Similarly as in AR models, we can construct a family of models $\{\mathbf{F}_{p,q};\ 1 \leq p \leq P, 1 \leq q \leq Q\}$, in which $\mathbf{F}_{p,q}$ signifies the ARMA(p,q) model.

In each model $\mathbf{F}_{p,q}$, densities are parametrized by $\theta' = (\sigma, a_1, \ldots, a_p, b_1, \ldots, b_q)$. Denote the covariance matrix of $\mathbf{z}_n$ by $Q(\theta)^{-1}$ or shortly Q^{-1}. Assuming that the shape of the densities in the model is normal, we have AIC for $\mathbf{F}_{p,q}$,

$$\mathrm{AIC} = -2l(\hat{\theta}) + 2(p+q+1)$$

with

$$-2l(\hat{\theta}) = n\log 2\pi\hat{\sigma}^2 - \log|\hat{Q}| + \mathbf{z}_n'\hat{Q}\,\mathbf{z}_n/\hat{\sigma}^2,$$

where $\hat{Q}$ and $\hat{\sigma}^2$ are the maximum likelihood estimates of Q and σ^2 respectively under the model, and $|\hat{Q}|$ is the determinant of $\hat{Q}$. There are various methods for obtaining the maximum likelihood estimate [22]. Some of them are:

a) Exact maximum likelihood [3].

b) Conditional likelihood. $z_t, t \leq 0$ are put zero or extrapolated by backward forecasting. Maximization is only for the quadratic term and the remaining terms are disregarded in the log likelihood function [6].

c) Whittle's approximation of the log likelihood function [41].

$$\frac{n}{2} \int \left\{ \log 2\pi f(\lambda) + \frac{I(\lambda)}{f(\lambda)} \right\} d\lambda$$

d) Three-stage approximation [13].

i) Fit an AR(P),

ii) obtain initial estimates of parameters by least squares based on the innovations $\{\hat{\varepsilon}_t\}$ obtained by using the AR coefficients estimated in i),

iii) apply a correction to the initial estimates.

We should be careful to apply an approximation like b), c) and d). Special attention should be given to the estimates $\hat{\sigma}^2$ and $|\hat{Q}|$, which should be equal to the exact ones up to the order of $O(1/n)$, except the constant which does not depend on p and q. Otherwise, AIC will behave differently.

Simple expressions of TIC_0 and TIC have not been obtained. Findley [7] evaluated the bias of AIC as an estimate of the Kullback-Leibler information number for the case when the true model is an infinite order moving average process. His result suggests a simple expression of TIC_0.

A specific problem arises in ARMA model selection, i.e., *identifiability*. If an ARMA(p,q) is fitted to ARMA(p_0,q_0) with $p_0<p$ and $q_0<q$, then the transfer functions $A_p(B)$ and $B_q(B)$ have common roots, which are not identifiable. Then, the maximum likelihood estimates of parameters behave differently. In fact, Hannan [12, 14] proved that the exact maximum likelihood estimates $\hat{a}_1$ and $\hat{b}_1$ converge to ±1 if ARMA(1,1) is fitted to ARMA(0,0), and Shibata [30] proved that they are asymptotically Cauchy distributed if three-stage approximation procedure d) is employed. Therefore, as far as the true model is expected to be or close to a finite order ARMA model, inconsistency of the selection is troublesome. For example, the minimum AIC is inconsistent even when $p_0<P$ and $q_0<Q$. A modification of AIC may solve this problem [30]. The use of a consistent selection procedure like the minimum BIC or HQ may solve this problem, too. But it increases the error of the resulting parameter estimate. Whereas, under the assumption that z_n is generated from an infinite order moving average process which is not a degenerate finite order ARMA, such problem never arises and an optimality property holds true similarly as in AR models [37].

How to select a moving average model whose associate polynomial has roots on the unit circle is also an interesting problem which has to be investigated in the future.

7. REFERENCES

[1] Akaike, H., A fundamental relation between predictor identification and power spectrum estimation, *Ann. Inst. Statist. Math.*, Vol. 22, pp. 219-223, 1970.

[2] Akaike, H., Information theory and an extension of the maximum likelihood principle, pp. 267-281 in *2nd Int. Symposium on Information Theory*, Eds. B. N. Petrov and F. Csáki, Akadémia Kiado, Budapest, 1973.

[3] Ansley, C. F., An algorithm for the exact likelihood of a mixed autoregressive-moving average process, *Biometrika*, Vol. 66, pp. 59-65, 1979.

[4] Atkinson, A. C., A note on the generalized information criterion for choice of a model, *Biometrika*, Vol. 67, pp. 413-418, 1980.

[5] Bhansali, R. J. and D. Y. Downham, Some properties of the order of an autoregressive model selected by a generalization of Akaike's EPF criterion, *Biometrika*, Vol. 64, pp. 547-551, 1977.

[6] Box, G. E. P. and G. M. Jenkins, *Time Series Analysis: Forecasting and Control*, Holden-Day, 1976.

[7] Findley, D. F., On the unbiasedness property of AIC for exact or approximating linear stochastoic time series models, *J. Time Series Analysis*, Vol. 6, pp. 229-252, 1985.

[8] Gooijer, J. G. de, B. Abraham, A. Gould and L. Robinson, Method for determining the order of an autoregressive-moving average process: A survey, *Int. Statist. Rev.*, Vol. 53, pp. 301-329, 1985.

[9] Hampel, F. R., E. M. Ronchetti, P. J. Rousseeuw and W. A. Stahel, *Robust Statistics: the Approach Based on Influence Functions*, John Wiley, 1986.

[10] Hannan, E. J. and B. G. Quinn, The determination of the order of an autoregression, *J. Roy. Statist. Soc.*, Vol. B 41, pp. 190-195, 1979.

[11] Hannan, E. J., The estimation of the order of an ARMA process, *Ann. Statist.*, Vol. 8, pp. 1071-1081, 1980.

[12] Hannan, E. J., Testing for autocorrelation and Akaike's criterion, pp. 403-412 in *Essays in Statistical Science (Papers in honour of P.A.P. Moran)*, Eds. J. M. Gani and E. J. Hannan, Applied Probability Trust, Sheffield, 1982.

[13] Hannan, E. J. and J. Rissanen, Recursive estimation of mixed autoregressive-moving average order, *Biometrika*, Vol. 69, pp. 81-94, 1982.

[14] Hannan, E. J., Fitting multivariate ARMA models, pp. 307-316 in *Statistics and Probability (Essays in Honor of C. R. Rao)*, Eds. G. Kallianpur, P. R. Krishnaiah and J. K. Ghosh, North-Holland Publishing Company, Amsterdam, 1982.

[15] Hosking, J. R. M., Lagrange-multiplier tests of time-series models, *J. R. Statist. Soc.*, Vol. B42, pp. 170-181, 1980.

[16] Hurvich, C. M., Data-Driven choice of a spectraum estimate: Extending the applicability of cross-validation methods, *J. Amer. Statist. Soc.*, Vol. 80, pp. 933-940, 1985.

[17] Kempthorne, P. J., Admissible variable-selection procedures when fitting regression models by least squares for prediction, *Biometrika*, Vol. 71, pp. 593-597, 1984.

[18] Li, K. C., From Stein's unbiased estimates to the method of generalized cross validation, *Ann. Statist.*, Vol. 13, pp. 1352-1377, 1985.

[19] Mallows, C. L., Some comments on C_p, *Technometrics*, Vol. 15, pp. 661-675, 1973.

[20] Parzen, E., Some recent advances in time series modeling, *IEEE*, Vol. AC-19, pp. 723-730, 1974.

[21] Parzen, E., Multiple time series: determining the order of approximating autoregressive schemes, pp. 283-295 in *Multivariate Analysis-IV*, North-Holland, 1977.

[22] Priestly, M. B., *Spectral Analysis and Time Series*, Academic Press, 1981.

[23] Sakai, H., Asymptotic distribution of the order selected by AIC in multivariate autoregressive model fitting, *Int. J. Control*, Vol. 33, pp. 175-180, 1981.

[24] Schwarz, G., Estimating the dimension of a model, *Ann. Statist.*, Vol. 6, pp. 461-464, 1978.

[25] Shibata, R., Selection of the order of an autoreegressive model by Akaike's information criterion, *Biometrika*, Vol. 63, pp. 117-126, 1976.

[26] Shibata, R., An optimal selection of regression variables, *Biometrika*, Vol. 68, pp. 45-54, Correction 69, p.492, 1981.

[27] Shibata, R., An optimal autoregressive spectral estimate, *Ann. Statist. 9*, pp. 300-306, 1981.

[28] Shibata, R., Various model selection techniques in time sereis analysis, pp. 179-187 in *Handbook of Statistics*, Eds. E. J. Hannan and P. R. Krishnaiah, Elsevier, 1985.

[29] Shibata, R., Selection of regression variables, pp. 709-714 in *Encyclopedia of Statistical Sciences*, John Wiley & Sons, 1986.

[30] Shibata, R., Consistency of model selection and parameter estimateion, pp. 127-141 in *Essays in Time Series and Allied Processes*, Eds. J. M. Gani and M. B. Priestley, Applied Probability Trust, Sheffield, 1986.

[31] Shibata, R., Selection of the number of regression variables; a minimax choice of generalized FPE, *Ann. Inst. Statist. Math.*, Vol. 38 A, pp. 459-474, 1986.

[32] Stone, M., Cross-validatory choice and assessment of statistical predictions, *J. Roy. Statist. Soc.*, Vol. 36, pp. 111-133, 1974.

[33] Stone, M., An asymptotic equivalence of choice of model by cross-validation and Akaike's criterion, *J. Roy. Statist. Soc.*, Vol. B 39, pp. 44-47, 1977.

[34] Stone, C. J., Local asymptotic admissibility of a generalization of Akaike's model selection rule, *Ann. Inst. Statist. Math.*, Vol. 34, pp. 123-133, 1982.

[35] Takada, Y., Admissibility of some variable selection rules in linear regression model, *J. Japan Statist. Soc.*, Vol. 12, pp. 45-49, 1982.

[36] Takeuchi, K., Distribution of information statistics and a criterion of model fitting, *Suri-Kagaku (Mathematical Sciences)*, Vol. 153, pp. 12-18, (in Japanese), 1976.

[37] Taniguchi, M., On selection of the order of the spectral density model for a stationary process, *Ann. Inst. Statist. Math.*, Vol. 32 A, pp. 401-419, 1980.

[38] Titterington, D. M., Common structure of smoothing techniques in statistics, *Int. Statist. Rev.*, Vol. 53, pp. 141-170, 1985.

[39] Tjφstheim, D. and J. Paulsen, Least squares estimates and order determination procedures for autoregressive processes with a time dependent variance, *J. Time Series Analysis*, Vol. 6, pp. 117-133, 1985.

[40] Wahba, G., A comparison of GCV and GML for choosing the smoothing parameter in the generalized spline smoothing problem, *Ann. Statist.*, Vol. 13, pp. 1378-1402, 1985.

[41] Walker, A. M., Asymptotic properties of least squares estimates of parameters of the spectrum of a stationary non-deterministic time series, *Austral. Math. Soc.*, Vol. 4, pp. 363-384, 1964.

[42] Woodroofe, M., On model selection and the arc sine laws, *Ann. Statist.*, Vol. 10, pp. 1182-1194, 1982.

[43] Yajima, Y., Estimation of the degree of differencing of an ARIMA process, *Ann. Inst. Statist. Math.*, Vol. 37, pp. 389-408, 1985.

INDEX

ADDRESSES OF AUTHORS

M. Deistler: Institute of Econometrics
University of Vienna
Argentinierstrasse 8
A-1040 Vienna
AUSTRIA

K. Glover: Cambridge University Engineering Department
Control & Management Systems Division
Trumpington Street
Cambridge CB2 1RX
ENGLAND

C. Heij: Econometrics Institute
Erasmus University Rotterdam
P.O. Box 1738
3000 DR Rotterdam
THE NETHERLANDS

A.B. Kurzhanski: International Institute of Applied Systems
A-2361 Laxenburg
AUSTRIA

R. Shibata: Department of Mathematics
Keio University
3-14-1 Hiyoshi Kohuko
Yokohama
223 JAPAN

J.C. Willems: Department of Mathematics
Groningen University
P.O. Box 800
9700 AV Groningen
THE NETHERLANDS